国外优秀数学著作
原版丛书

应用与计算拓扑学进展（英文）

Advances in Applied and Computational Topology

［美］阿弗拉·佐莫罗迪安（Afra Zomorodian） 主编

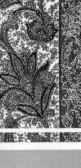

哈尔滨工业大学出版社
HARBIN INSTITUTE OF TECHNOLOGY PRESS

黑版贸登字 08-2022-064 号

This work was originally published by the American Mathematical Society under the title Advances in Applied and Computational Topology, © 2012 by the American Mathematical Society. The present edition was produced by Harbin Institute of Technology Press under authority of the American Mathematical Society and is published under license.

Special Edition for People's Republic of China distribution only. This edition has been authorized by the American Mathematical Society for sale in People's Republic of China only and is not for export therefrom.

图书在版编目(CIP)数据

应用与计算拓扑学进展=Advances in Applied and Computational Topology:英文/(美)阿弗拉·佐莫罗迪安(Afra Zomorodian)主编. —哈尔滨:哈尔滨工业大学出版社,2025.2. —ISBN 978-7-5767-1776-1

Ⅰ.O189

中国国家版本馆 CIP 数据核字第 2025MH1818 号

YINGYONG YU JISUAN TUOPUXUE JINZHAN

策划编辑　刘培杰　杜莹雪
责任编辑　张嘉芮　李兰静
版权编辑　李　丹
封面设计　孙茵艾
出版发行　哈尔滨工业大学出版社
社　　址　哈尔滨市南岗区复华四道街10号　邮编150006
传　　真　0451-86414749
网　　址　http://hitpress.hit.edu.cn
印　　刷　哈尔滨起源印务有限公司
开　　本　787 mm×1 092 mm　1/16　印张16.25　字数322千字
版　　次　2025年2月第1版　2025年2月第1次印刷
书　　号　ISBN 978-7-5767-1776-1
定　　价　98.00元

(如因印装质量问题影响阅读,我社负责调换)

To Dr. Benjamin Mann

Contents

Preface	v
Topological Data Analysis AFRA ZOMORODIAN	1
Topological Dynamics: Rigorous Numerics via Cubical Homology MARIAN MROZEK	41
Euler Calculus with Applications to Signals and Sensing JUSTIN CURRY, ROBERT GHRIST, and MICHAEL ROBINSON	75
On the Topology of Discrete Planning with Uncertainty MICHAEL ERDMANN	147
Combinatorial Optimization of Cycles and Bases JEFF ERICKSON	195
Index	229
编辑手记	233

Preface

This volume is the proceedings of the *AMS Short Course on Computational Topology*, organized for the *Joint Mathematics Meetings* in New Orleans on January 4 – 5, 2011. *Computational topology* emerged in response to topological impediments within geometric problems, such as extraneous holes and tunnels in surfaces reconstructed by the computer graphics and computational geometry communities. Topological problems arise naturally, however, in many areas of science. In robotics, we need to capture the *connectivity* of the configuration space of a robot for planning. In sensor networks, we wish to deduce *global* information from *local* sensing. In dynamical systems, we want to understand *qualitative* properties of a system via computation. In data analysis, we look for robust features of an underlying space given a finite set of noisy samples.

Like most emerging areas, computational topology is claimed by several communities, and naturally, each community defines the area in its own image. With this course, I wanted to broaden the definition of this field to include any area that resolves a topological question using computational techniques. To this end, I invited speakers from a broad spectrum of specializations, including algebraic topology, dynamical systems, applied topology, robotics, and computational geometry. I also wanted the course to cover the rich development of computational topology from theory, to algorithm design and analysis, implementation of fast software, and applications.

The structure of the book mirrors that of the course. We dedicated the first day to topological data analysis. One of the speakers, Gunnar Carlsson, did not contribute a chapter, as he is currently developing a book on the subject. The day culminated with a software session: Henry Adams provided a tutorial on *JPlex*, a Java software package for topological data analysis; Marian Mrozek demonstrated *RedHom*, a C++ library for computing cubical homology. The second day of the short course concentrated on applications of topology to sensor networks, robotics, and geometry. Robert Ghrist contributed his chapter in collaboration with two colleagues.

The short course concluded with a panel session, during which the speakers and the attendees discussed the state and future of computational topology. Ten years ago, publishing in this area was difficult as the required mathematics was unknown to computer scientists, while the value of the applications was unappreciated by mathematicians. By now, a number of conferences and journals have recognized computational topology as a subarea. As it matures through ad-hoc workshops and programs, computational topology will require dedicated conferences and journals so that researchers have centralized forums for disseminating their research.

I am grateful to Dan Rockmore for soliciting the short course proposal and for his many helpful comments. I thank the speakers for their excellent presentations as well as developing their contributed chapters during the last year. The book went through a two-stage peer-review process in which the speakers and anonymous reviewers participated. I thank all reviewers for their on-time and thorough critiques. Finally, I thank Sergei Gelfand and Christine Thivierge from the American Mathematical Society for shepherding this project.

During his tenure at DARPA, Dr. Benjamin Mann was an energetic champion of researchers in applied and computational topology. We dedicate this volume to him in deep appreciation of his persistent support.

Afra Zomorodian
February 2012
New York, NY

Topological Data Analysis

Afra Zomorodian

ABSTRACT. Scientific data is often in the form of a finite set of noisy points, sampled from an unknown space, and embedded in a high-dimensional space. Topological data analysis focuses on recovering the topology of the sampled space. In this chapter, we look at methods for constructing combinatorial representations of point sets, as well as theories and algorithms for effective computation of robust topological invariants. Throughout, we maintain a computational view by applying our techniques to a dataset representing the conformation space of a small molecule.

1. Introduction

Topological data analysis is a subarea of computational topology that develops topological techniques for robust analysis of scientific data. To clarify our task, we begin this chapter by examining the three words that constitute the title. We then lay out a two-step pipeline around which the rest of the chapter is organized. We focus on intuition in this section, formalizing the concepts in the remainder of the chapter.

1.1. Topology. Geometry studies shapes. For instance, we think of the closed curve in Figure 1(a) as having the same shape as the curve in Figure 1(b), even though the two curves are not identical pointwise. If we translate the first curve by about an inch, and rotate it by 30 degrees, we get the second curve. Even though we have transformed the curve, we believe its shape has not changed. In this sense, geometry classifies objects according to properties that do not change under certain permissible transformations. Felix Klein introduced this expansive definition of geometry in his famous *Erlangen Program* in 1872 [**45**]. Restricting to the group of *rigid* transformations yields *Euclidean geometry*. Through its rigidity, this geometry has a fine granularity when viewed as a classification system. If we enlarge the group of approved transformations, we may obtain other classifications that are coarser and may capture more *qualitative* information about shapes.

2010 *Mathematics Subject Classification.* Primary 55N35, 55U05, 55-04; Secondary 55U10, 68T10, 62H99.

Key words and phrases. Čech, alpha, Vietoris-Rips, witness, simplicial complex, cubical complex, persistent homology, multidimensional persistence, zigzag, tidy sets.

This chapter was completed at Dartmouth College, where the author's research was partially supported by ONR N 00014-08-1-0908 and NSF CAREER CCF-0845716.

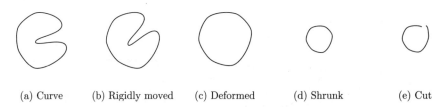

(a) Curve (b) Rigidly moved (c) Deformed (d) Shrunk (e) Cut

FIGURE 1. The closed curve (a) is transformed under rigid motion (b), deformed using homeomorphisms (c) including scaling (d), and finally cut (e). The final transformation changes the curve's connectivity as the closed loop becomes a path.

Topology allows the larger group of *homeomorphisms* that deform an object by stretching or shrinking, as we do for the curve in Figures 1(c) and (d) Under any homeomorphism, the curve remains a *Jordan curve* that divides the plane into two regions. It is only by cutting the curve that we change its *topology* from a closed loop to a path. Neither *cutting*, nor its inverse, *gluing*, are permissible in topology as they change the way an object is connected. Topology, then, classifies a shape according to its *connectivity*, such as its number of pieces, loops, or presence of boundary. The main object of study in topology is a *topological space*: the most general form of a space that still retains a notion of connectivity.

1.2. Data. Data is processed mainly on digital computers and communicated via packet switching. As such, data is stored in a finite representation, such as the IEEE standard for floating-point arithmetic, resulting in *discretization error*. Moreover, acquisition devices are usually imperfect, adding *noise* to data. Therefore, we think of *data* as a finite set of discrete noisy samples, such as two-dimensional images from digital cameras, terrains from satellite observations [**61**], sampled three-dimensional surfaces from laser scanners [**67**], voxelized MRI scans of the human body [**59**], or snapshots from simulated protein folding trajectories [**32**]. Abstract spaces may also be modeled with discrete samples, as the following example demonstrates.

EXAMPLE 1.1 (conformation space). To understand molecular motion, we need to characterize the molecule's possible shapes. For instance, consider the molecule *cyclooctane* with formula C_8H_{16}. Structurally, cyclooctane has a ring of eight carbon atoms, each bonded to a pair of hydrogen atoms, as shown in its chemical diagram in Figure 2(a) [**4**]. A *conformation* of a molecule is a potential shape it may assume. We visualize a conformation of cyclooctane using two models in Figures 2(b) and (c). To specify a conformation, we need to map every atom in the molecule to a point in $\mathbb{R}^3$. Since cyclooctane has 24 atoms, each conformation, such as the one in Figure 2, may be viewed as a single point in $\mathbb{R}^{72}$. The set of all physically realizable conformations of a molecule is its *conformation space*. We may model this space with a set of finite samples. Figure 3 shows three-dimensional projections of a dataset of 6,400 samples from the cyclooctane conformation space [**4**].

1.3. Analysis. Suppose that we are given a set of data points S embedded in some d-dimensional space $\mathbb{Y}$. We assume that this data is sampled from some

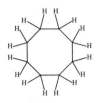  

(a) Chemical Diagram (b) One conformation (Stick) (c) Hard Sphere Model

FIGURE 2. Cyclooctane: Chemical diagram (a) and two visualizations of a conformation (b) and (c).

unknown k-dimensional subspace $\mathbb{X} \subseteq \mathbb{Y}$, where $k \leq d$. Both the geometry and the topology of $\mathbb{X}$ are lost during sampling. Our goal in *analysis* is recovering information about $\mathbb{X}$ from the given dataset S. Properties of the embedding space $\mathbb{Y}$ are *extrinsic*, while properties of the unknown space $\mathbb{X}$ are *intrinsic*. For example, S has extrinsic dimension d, but intrinsic dimension k. In analysis, we try to recover intrinsic information, given only extrinsic information.

EXAMPLE 1.2 (spiral). Consider the spiral in Figure 4. The data points (a) are embedded in $\mathbb{R}^2$, so, the extrinsic dimension is 2. We may use the Euclidean metric of the embedding space to compute distances between points (b). The points are sampled, however, from a spiral (c). Since the spiral is a one-dimensional curve, its intrinsic dimension is 1. Note that the *geodesic* distance (d) may be very different from the *embedding* distance in (b).

Every analysis method makes fundamental assumptions about the unknown space $\mathbb{X}$. *Principal Component Analysis (PCA)* assumes that $\mathbb{X}$ is a linear subspace, a flat hyperplane with no curvature [40, 73]. ISOMAP assumes that $\mathbb{X}$ is intrinsically flat, but is isometrically embedded, like the spiral in Figure 4(c). The method also assumes that $\mathbb{X}$ is a single convex patch with the topology of a disc [24]. The method of *Hessian eigenmaps*, a refinement of *locally linear embeddings (LLE)* [65], also assumes isometric embedding, but relaxes the restriction on topology [24]. A large class of methods from computer graphics and computational geometry focus on

FIGURE 3. Three-dimensional projections of 6,400 samples of the conformation space of cyclooctane [4].

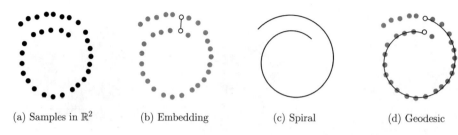

(a) Samples in $\mathbb{R}^2$ (b) Embedding (c) Spiral (d) Geodesic

FIGURE 4. Spiral. Two-dimensional samples (a) and extrinsic embedding distance (b). Original spiral (c) and intrinsic geodesic distance (d).

surface reconstruction from samples. These methods assume that $\mathbb{X}$ is a closed surface without self-intersection. Additionally, they often assume that $\mathbb{X}$ is smooth and that the sampling is sufficiently dense, respecting the unknown *local feature size* of the original surface [23].

The methods above are all instances of *manifold learning*, where the key assumption is that $\mathbb{X}$ is a *manifold*, that is, it is locally Euclidean [60]. But most real-world point sets are sampled from spaces that violate nearly all the above assumptions.

EXAMPLE 1.3 (reconstruction). It is already clear from Figure 3 that the conformation space of cyclooctane in Example 1.1 has non-manifold structure, visible as potential self-intersections in the three-dimensional embeddings. Indeed, the reconstructed conformation space is a two-dimensional surface with non-manifold structure [50]. We embed this surface in $\mathbb{R}^3$ using ISOMAP in Figure 5. Topologically, the conformation space is the Klein bottle glued to the two-dimensional sphere along two rings [49].

Note that the conformation space of cyclooctane violates every assumption made by prior analysis techniques. Reconstructing a surface with non-manifold

FIGURE 5. The reconstructed conformation space of cyclooctane is a two-dimensional surface with self-intersections [50]. Topologically, it is the Klein bottle glued to the sphere along two rings [49].

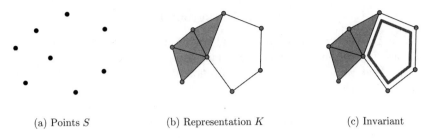

(a) Points S (b) Representation K (c) Invariant

FIGURE 6. Analysis Pipeline. The input is a set of points S (a). Section 3 describes the first step, the geometric process of going from (a) to a representation K (b). Sections 4 and 5 describe the second step, the combinatorial process of going from (b) to a topological invariant, such as the cycle (c).

structure is a challenging problem in computational geometry [**50**]. Once we have the surface, we may also recover the topology of the conformation space. But if we are only interested in topology, surface reconstruction is excessive as topology is a much coarser classification system.

Having examined the three words in *topological data analysis*, we may now define our task in this chapter.

DEFINITION 1.4 (topological data analysis). Given a finite dataset $S \subseteq \mathbb{Y}$ of noisy points sampled from an unknown space $\mathbb{X}$, *topological data analysis* recovers the topology of $\mathbb{X}$, assuming both $\mathbb{X}$ and $\mathbb{Y}$ are topological spaces.

The assumption here is much weaker than those of geometric analysis techniques: We do not assume manifold structure, smoothness, lack of curvature, or the existence of a metric. Correspondingly, our goal is modest and coarse.

1.4. Pipeline. Traditional topological analysis uses the two-step pipeline summarized in Figure 6. Given a finite set of points (a):

(1) We first approximate the unknown space $\mathbb{X}$ in a combinatorial structure K, as shown in Figure 6(b). We devote Section 3 to such structures and the methods for constructing them.
(2) We then compute topological invariants of K, such as the cycle in Figure 6(c). We devote Section 4 to classic topological invariants and Section 5 to modern multiscale invariants.

Topological invariants of K provide approximations to properties of $\mathbb{X}$ as finitely represented by S. While the pipeline is effective, it is not computationally feasible for large point sets embedded in high dimensions. For this reason, we describe methods for combining the two steps in Section 6.

Topological data analysis is an applied field, concerned with theory that facilitates analysis of real-world datasets. Therefore, we return at the end of each section to our motivating dataset, the cyclooctane conformation space, applying techniques toward its analysis and providing empirical results. Throughout this chapter, all computation is on a 64-bit GNU/Linux machine with a 2.4 GHz dual-core Xeon processor and 2 GB RAM. Our software is not threaded and uses only one core.

2. Background

We begin by formalizing topological spaces and describing two topological classifications. We then introduce simplicial complexes, the primary combinatorial structure that we will use for representation We end this section by specifying a general scheme that is the basis for a few of the methods for constructing complexes in the next section. Throughout, our aim is not to be comprehensive, but pedagogical. Recent surveys on topological analysis include Ghrist [35] and Carlsson [5]. For a broader introduction to computational topology, see [78].

2.1. Topology. Intuitively, a topological space is a set of points, each of whom knows its neighbors. A *topology* on a set X is a subset $T \subseteq 2^X$ such that:
  (1) If $S_1, S_2 \in T$, then $S_1 \cap S_2 \in T$.
  (2) If $\{S_J \mid j \in J\} \subseteq T$, then $\cup_{j \in J} S_j \in T$.
  (3) $\emptyset, X \in T$.

The pair $\mathbb{X} = (X, T)$ is a *topological space*. A set $S \in T$ is an *open set* and its complement in X is *closed*. We often abuse notation by using $p \in \mathbb{X}$ for $p \in X$ when the topology is clear from context. A subset $A \subseteq X$ with *induced topology* $T_A = \{S \cap A \mid S \in T\}$ is a *subspace* $\mathbb{A}$ of $\mathbb{X}$. A familiar example of a topological space is the *d-dimensional Euclidean space* $\mathbb{R}^d$, where we use the Euclidean metric to measure distances and define open sets. We may also turn any subset of a Euclidean space into a topological space by using the induced topology.

A function $f : \mathbb{X} \to \mathbb{Y}$ is *continuous* if for every open set A in $\mathbb{Y}$, $f^{-1}(A)$ is open in $\mathbb{X}$. A *homeomorphism* $f \colon \mathbb{X} \to \mathbb{Y}$ is a bijection such that both f and f^{-1} are continuous. Given a homeomorphism $f \colon \mathbb{X} \to \mathbb{Y}$, we say that $\mathbb{X}$ is *homeomorphic* to $\mathbb{Y}$. As homeomorphism is an equivalence relation on topological spaces, we also say that $\mathbb{X}$ and $\mathbb{Y}$ have the same *topological type*, denoted $\mathbb{X} \approx \mathbb{Y}$. The topological type is the finest level of classification available in topology.

A *homotopy* is a family of maps $f_t : \mathbb{X} \to \mathbb{Y}$, $t \in [0, 1]$, such that the associated map $F : \mathbb{X} \times [0, 1] \to \mathbb{Y}$ given by $F(x, t) = f_t(x)$ is continuous. Here, $\mathbb{X} \times [0, 1]$ is a topological space whose open sets are products of the open sets of $\mathbb{X}$ and the open sets of $[0, 1]$, viewed as a subspace of $\mathbb{R}$ [38]. Then, $f_0, f_1 : \mathbb{X} \to \mathbb{Y}$ are *homotopic* via the homotopy f_t, denoted $f_0 \simeq f_1$. A map $f : \mathbb{X} \to \mathbb{Y}$ is a *homotopy equivalence* if there exists a map $g : \mathbb{Y} \to \mathbb{X}$, such that $f \circ g \simeq 1_\mathbb{Y}$ and $g \circ f \simeq 1_\mathbb{X}$, where $1_\mathbb{X}, 1_\mathbb{Y}$ are the identity maps on the respective spaces. Given a homotopy equivalence $f : \mathbb{X} \to \mathbb{Y}$, we say that $\mathbb{X}$ and $\mathbb{Y}$ are *homotopy equivalent* and have the same *homotopy type*, denoted $\mathbb{X} \simeq \mathbb{Y}$, as homotopy equivalence is an equivalence relation on topological spaces. A space with the homotopy type of a point is *contractible*. Homotopy type is a coarser classification than topological type. For example, a disc is contractible, but not homeomorphic, to a point.

2.2. Simplicial Complex. Simplicial complexes are popular in topological data analysis due to their structural simplicity. Intuitively, a simplicial complex is similar to a hypergraph, where we represent a relationship between $(n + 1)$ nodes with a n-dimensional simplex. Formally, a *simplicial complex* is a set K of finite sets closed under the subset relation: If $\sigma \in K$ and $\tau \subseteq \sigma$, then $\tau \in K$. Here, σ is a *simplex* (plural *simplices*) and τ is a *face* of σ, its *coface*. The (-1)-*simplex* $\emptyset$ is a face of any simplex. A simplex is *maximal* if it has no proper coface in K. If $\sigma \in K$ has cardinality $|\sigma| = n + 1$, we call σ a n-*simplex* of *dimension* n, denoted

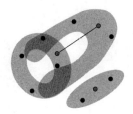

FIGURE 7. A set of 8 black points, an open cover of 3 sets, and the nerve of the cover: a simplicial complex with 3 vertices and 1 edge.

$\dim(\sigma) = n$. Generalizing, if the maximum dimension of a simplex in K is d, we call K a d-dimensional complex, $\dim(K) = d$.

Our notion of dimensionality for simplices stems from our ability to realize a n-simplex geometrically as a n-dimensional subspace of $\mathbb{R}^d$, $d \geq n$, namely, the convex hull of $(n+1)$ affinely-independent points [**38**]. In this view, an n-simplex is called a *vertex*, an *edge*, a *triangle*, or a *tetrahedron* for $0 \leq n \leq 3$, respectively. A key property of a realized simplex is that it is contractible. A simplicial complex K may be embedded in Euclidean space as the union of its geometrically realized simplices such that they only intersect along shared faces. This union is the *underlying space* $|K| = \cup_{\sigma \in K} \sigma$ of K, a topological space. Topological invariants, such as homotopy type, do not depend on a particular geometric realization of a complex.

EXAMPLE 2.1. Figure 6(b) displays a geometric realization of a simplicial complex with 8 vertices, 11 edges, and 3 triangles. The triangles and four of the edges defining the hole are maximal.

A *subcomplex* is a subset $L \subseteq K$ that is also a simplicial complex. An important subcomplex is the *n-skeleton* consisting of simplices in K of dimension less than or equal to n. The 1-skeleton of a simplicial complex is a graph.

2.3. Cover and Nerve. For the rest of this chapter, we assume we are given a finite set of data points S, sampled from some unknown space $\mathbb{X}$, and embedded in some topological space $\mathbb{Y}$, as described in Section 1.3. A key idea in topological analysis is to approximate $\mathbb{X}$ locally using pieces of the embedding space $\mathbb{Y}$. An *open cover* of S is
$$U = \{\mathbb{U}_i\}_{i \in I}, \quad \mathbb{U}_i \subseteq \mathbb{Y},$$
where I is an indexing set, $S \subseteq \cup_i \mathbb{U}_i$, and $\mathbb{U}_i$ are open. The *nerve* N of U is

(1) $\emptyset \in N$, and
(2) If $\cap_{j \in J} \mathbb{U}_j \neq \emptyset$ for $J \subseteq I$, then $J \in N$.

Clearly, the nerve is a simplicial complex.

EXAMPLE 2.2. Figure 7 displays a set of 8 black points, an open cover of 3 sets, and the nerve of this cover: a simplicial complex with 3 vertices and 1 edge.

The union of the sets in an open cover is our approximation of the unknown $\mathbb{X}$. Its nerve serves as a finite combinatorial representation to be used in computation. If the sets in the cover do not hide interesting topology, either within themselves or in their intersection patterns, all topology is exposed within the nerve. Formally, a cover U is *good* if all $\mathbb{U}_i$ are contractible and so are all their nonempty finite

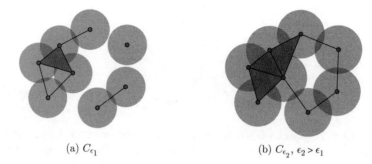

FIGURE 8. The Čech complex C_ϵ is the nerve of a cover of ϵ-balls. We show the complex at two scales $0 < \epsilon_1 < \epsilon_2$.

intersections. Clearly, the cover in Figure 7 is not good as the leftmost set is an annulus, and its intersection with the middle set has two pieces. By Leray's *Nerve Lemma*, the nerve of a good cover is homotopy equivalent to the cover, that is, the union of the sets in the cover [3, 64]. This lemma is the basis of a few of the methods for representing point sets in the next section.

3. Combinatorial Representations

In this section, we focus on the first step of the analysis pipeline in Figure 6. Our input is a finite point set $S \subseteq \mathbb{Y}$. We construct combinatorial representations K that approximate the space $\mathbb{X}$ from which S was sampled. In the remainder of this section, we assume that $\mathbb{Y}$ is a metric space with metric $d\colon \mathbb{Y} \times \mathbb{Y} \to \mathbb{R}$. A number of the algebraic methods may be extended to non-metric spaces easily, while the geometric methods, such as the alpha complex, require the Euclidean metric. As promised, we end this section by constructing a representation for the cyclooctane dataset.

3.1. Čech Complex. Let $B_\epsilon(x)$ be the *open ball* of radius ϵ centered at x. That is, for $\epsilon \in \mathbb{R}$ and $x \in \mathbb{Y}$,

$$B_\epsilon(x) = \{y \in \mathbb{Y} \mid d(x,y) < \epsilon\}.$$

Given $S \subseteq \mathbb{Y}$ and $\epsilon \in \mathbb{R}$, we center an ϵ-ball at each point to get a cover:

$$U_\epsilon = \{B_\epsilon(x) \mid x \in S\}.$$

The *Čech complex* C_ϵ is the nerve of this cover [38]. Since balls are convex and convex sets are contractible, the cover is good and its nerve captures the topology of the cover.

EXAMPLE 3.1. Figures 8 show covers U_{ϵ_1} and U_{ϵ_2} at two scales $0 < \epsilon_1 < \epsilon_2$. The nerve of each cover is drawn above it. Note that each nerve is homotopy equivalent to its cover: The cover and nerve in (a) both have 3 components and 1 hole, while the cover and nerve in (b) both have 1 component and 1 hole.

We may compute a Čech complex at each scale ϵ. Clearly, $C_0 = \emptyset$ and C_∞ is an $(|S|-1)$-simplex. That is, the Čech complex may have a much higher dimension than the embedding space $\mathbb{Y}$. Since an n-simplex has 2^{n+1} faces, the complex may become massive at higher scales.

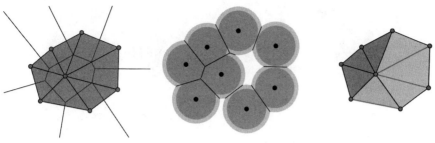

(a) Voronoi & Delaunay   (b) Restricted Voronoi regions   (c) Alpha subcomplex A_{ϵ_2}

FIGURE 9. A dataset, its Voronoi diagram and its nerve, the Delaunay complex (a). The restricted Voronoi regions (b) form a cover whose nerve is the alpha complex, (c) shown as a subcomplex of the lighter Delaunay complex.

The Čech complex is not computed in practice due to its computational complexity. The uniform ball radii imply an assumption of uniform sampling on the input, which is not valid in real-world datasets. We could use non-uniform radii to form a cover, and this idea has been explored in other methods, such as the alpha complex described in the next section.

3.2. Alpha Complex. To reduce the size of the complex, we limit its dimension by using the geometry of the embedding space. Given $S \subseteq \mathbb{Y}$, the *Voronoi region* $R(x)$ of a point $x \in S$ is the set of points in $\mathbb{Y}$ closest to it:
$$R(x) = \{y \in \mathbb{Y} \mid d(x,y) \leq d(x',y), \forall x' \in S, x' \neq x\}.$$
The *Voronoi diagram* is the set of all Voronoi regions for points in S. This diagram may be viewed as a closed cover for $\mathbb{Y}$. The *Delaunay complex* is the nerve of the Voronoi diagram. The Voronoi cover and its nerve are fundamental geometric objects and have been extensively studied within computational geometry [20].

EXAMPLE 3.2. Figure 9(a) displays the Voronoi diagram for our example point set, and overlays its nerve, the Delaunay complex.

We now use the Voronoi diagram to restrict the interactions of the ϵ-ball cover from the previous section. For each point $x \in S$, we intersect its ϵ-ball and Voronoi region to get a *restricted Voronoi region*. The set of all restricted regions forms a new cover:
$$U_\epsilon = \{B_\epsilon(x) \cap R(x) \mid x \in S\}.$$
The *alpha complex* A_ϵ is the nerve of this cover [27, 28]. By construction, $A_0 = \emptyset$, A_∞ is the Delaunay complex, and A_ϵ is a subcomplex of the Delaunay complex, for any ϵ. Moreover, the alpha and Čech complexes are homotopy equivalent. Unlike the Čech complex, however, the maximum dimension of the alpha complex is limited to the embedding dimension, provided S is in *general position*, a theoretical assumption that may be enforced computationally [76].

EXAMPLE 3.3. Figure 9(b) overlays the restricted Voronoi regions for our example point set at the two scales used in Figure 8. Figure 9(c) shows the alpha complex A_{ϵ_2} as a subcomplex of the Delaunay complex. At this scale, the complex is the same as the Čech complex C_{ϵ_2} in Figure 8(b).

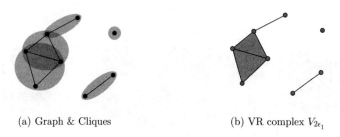

(a) Graph & Cliques (b) VR complex $V_{2\epsilon_1}$

FIGURE 10. The highlighted maximal cliques of the 2ϵ-neighborhood graph (a) become the maximal simplices of the the VR complex $V_{2\epsilon_1}$ (b).

We construct alpha complexes by first building the Delaunay complex. For each simplex of the Delaunay complex, we compute the minimum scale at which the simplex enters the Alpha complex. Then, we sort the simplices by their minimum scale to get a partial order of simplices. We may now form the alpha complex at any scale ϵ using this ordering. Since the Delaunay complex is finite, the alpha complex may change only at a finite number of critical scales as we increase the scale ϵ from 0 to infinity.

Using uniform radii implies an implicit assumption of uniform sampling. This assumption may be removed by generalizing the alpha complex. We may assume, for instance, that the point set is weighted, where the weight of a point is related to the local feature size. We may then use the *power metric* to define a *power diagram* cover and its nerve, the *regular triangulation* [1, 29]. Alternatively, we may define non-uniform radii using the local density of the point set itself to get the *conformal alpha complex* [12].

Efficient algorithms and software exist for computing Delaunay complexes, and in turn, alpha complexes in 2 and 3 dimensions [14], so the complex is well-suited for topological analysis in low dimensions. The construction of the Delaunay complex is difficult in higher dimensions, although progress is being made [2].

3.3. Vietoris-Rips Complex.

The Vietoris-Rips complex is popular in topological analysis due to the ease of its construction even in higher dimensions. Unlike the previous complexes, it is based on a graph, instead of a cover. Given $S \subseteq \mathbb{Y}$ and $\epsilon \in \mathbb{R}$, let $G_\epsilon = (S, E_\epsilon)$ be the ϵ-*neighborhood graph* on S, where

$$E_\epsilon = \{\{u, v\} \mid \mathrm{d}(u, v) \leq \epsilon,\ u \neq v \in S\}.$$

A *clique* in a graph is the subset of vertices that induces a complete subgraph [18]. A clique is *maximal* if it cannot be made any larger. The *clique complex*, also called the *flag complex*, has the maximal cliques of a graph as its maximal simplices [46]. The *Vietoris-Rips complex* V_ϵ is the clique complex of the ϵ-neighborhood graph [37, 74]. We refer to the complex as the *VR complex* for brevity.

EXAMPLE 3.4. Figure 10 shows the construction of a VR complex for our point set. We begin with a $2\epsilon_1$-neighborhood graph (a) so that the VR complex is comparable to Čech and alpha complexes at scale ϵ_1. The graph has 5 maximal cliques, highlighted by gray ovals. Each maximal clique becomes a maximal simplex in the VR complex $V_{2\epsilon_1}$ (b). Note that the VR complex is different than the Čech complex C_{ϵ_1} in Figure 8(a).

As the example illustrates, the VR complex is not always homotopy equivalent to the Čech complex, so we may view it as an approximation. Clearly, $C_\epsilon \subseteq V_{2\epsilon}$. Moreover, the Čech and VR complexes can be shown to be related homologically [36]. We will describe this classification level of topology in Section 4.

Like its Čech counterpart, the VR complex may be as large as a $(|S| - 1)$-dimensional simplex. This extremity occurs whenever the neighborhood graph is complete. In practice, we usually only require and construct a n-skeleton for some $n \leq |S|$. We also compute the VR complex $V_{\hat{\epsilon}}$ at some maximum scale $\hat{\epsilon} \in \mathbb{R}$. For each simplex $\sigma \in V_{\hat{\epsilon}}$, we compute the minimum ϵ at which the simplex enters the VR complex, with the vertices entering at $\epsilon = 0$ and the edges at their length. We then sort the simplices according to this value, extracting the VR complex for any $0 < \epsilon \leq \hat{\epsilon}$ as a prefix of this ordering. As for the alpha complex, since the VR complex is finite, there is only a finite number of critical scales at which the complex changes.

For analysis of small point sets in low dimensions, the VR complex is usually computed using ad-hoc methods. For an in-depth study of its construction for large point sets in higher dimensions, see [79]. Public software for building VR complexes is available [56, 66]. Currently, the VR complex is one of the few practical methods for topological analysis in high dimensions.

3.4. Witness Complex. Since the VR complex may be massive, we try to approximate it with smaller number of vertices. We motivate the complex in this section by reinterpreting the Delaunay complex from Section 3.2.

EXAMPLE 3.5. Consider the spiral point set in Figure 11(a). The triangle is in the Delaunay complex because the Voronoi regions of its three vertices intersect in the white Voronoi vertex. The white vertex is equidistant from the three vertices of the triangle.

Given $S \subseteq \mathbb{Y}$, a *strong witness* $w \in \mathbb{Y}$ is equidistant from the points in $\sigma \subseteq S$, *witnessing* the creation of a Delaunay simplex σ, such as the triangle in the example. We are motivated to search for strong witnesses within $\mathbb{Y}$ to construct the Delaunay complex, but this approach is not feasible as the set of strong witnesses has measure zero. So, we relax the definition of a witness: A *weak witness* $w \in \mathbb{Y}$ is closer to points in $\sigma \subseteq S$ than $S - \sigma$. The set of weak witnesses for a n-simplex form its region in the order-$(n+1)$ Voronoi diagram of S and has positive measure [20]. Moreover, if a simplex and all its faces have weak witnesses, the simplex also has a strong witness [21].

We may build the Delaunay complex on a sample set $S \subseteq \mathbb{Y}$, allowing witnesses to be anywhere in $\mathbb{Y}$. The resulting Delaunay complex captures the topology of $\mathbb{Y}$. But we are interested in the unknown space $\mathbb{X}$ with our only knowledge being the set of samples $S \subseteq \mathbb{Y}$. Therefore, we mimic the process above and replace $\mathbb{Y}$ with S. Let $L \subseteq S$ be the set of *landmarks* and the remaining points, $W = S - L$ be the set of potential witnesses. For $\epsilon \in \mathbb{R}$, the ϵ-*witness graph* is the graph $G = (L, E_\epsilon)$, where $\{l_1, l_2\} \in E_\epsilon$ if there exists a weak witness $w \in W$ that is closer to l_i than any other landmark, and $d(w, l_i) \leq \epsilon$ for $i = 1, 2$. The *(weak) witness complex* W_ϵ is the clique complex of this ϵ-witness graph.

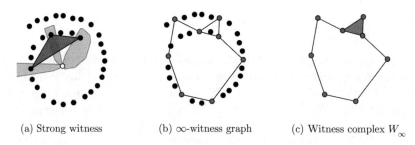

(a) Strong witness   (b) ∞-witness graph   (c) Witness complex W_∞

FIGURE 11. The triangle in (a) is Delaunay as the Voronoi regions of its vertices intersect in the white vertex. The witness graph (b) is built on the gray landmark points, with the remaining black points acting as witnesses. The witness complex (c) is the clique complex of this graph.

EXAMPLE 3.6. Figure 11(b) shows the witness graph for the spiral point set with $\epsilon = \infty$. The graph is built on the gray landmarks, with the remaining black points acting as potential witnesses for the edges. The witness complex W_∞ in Figure 11(c) is the clique complex of this graph.

It is clear that the witness complex depends on the chosen landmarks. The choice and size of the landmark set remains an art rather than a science. For analysis, it is best to bootstrap by choosing multiple sets of landmark points and seeing if the result is replicable.

The weak witness complex is the simplest of a family of witness complexes [**21**]. The software package *JPlex* [**66**] can compute several types of witness complexes. The witness graph is easily constructable by computing the $|L| \times |S|$ distance matrix using *k-nearest neighbors* [**57**]. Since it is a clique complex, the weak witness complex may be expanded from this graph using the algorithms designed for the Vietoris-Rips complex [**79**]. Currently, the witness complex is one of the few practical methods for topological analysis of large datasets.

3.5. Cubical Complex. A cubical complex is another type of combinatorial structure used in topological analysis. Informally, a cubical complex is a *cell complex*, where the cells are now *cubes* of different dimensions, rather than simplices [**38**]. A n-cube is called a *vertex*, an *edge*, a *square*, or a *cube* for $0 \leq n \leq 3$, respectively. Like a geometrically realized simplicial complex, a pair of cubes in a cubical complex only intersect along shared faces, which are lower-dimensional cubes.

Given $S \subseteq \mathbb{Y}$, where $\mathbb{Y}$ is a d-dimensional Euclidean space, we may easily construct a cubical complex at scale ϵ by covering $\mathbb{Y}$ with a grid of d-dimensional cubes with side ϵ. The *cubical complex* Q_ϵ is simply the rasterization of S on this grid: If an ϵ-cube c contains any point $s \in S$, then $c \in Q_\epsilon$. The cubical complex is dependent on the orientation of the grid. Also, all cubes are maximal and of the same dimension, so the complex is *pure*.

Alternatively, we may view a cubical complex as a cover, taking the nerve to get a simplicial complex. However, there is no need to do this, as all algorithms that require a simplicial complex as input extend easily to other cell complexes, such as the cubical complex.

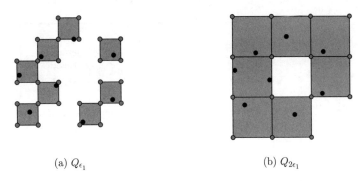

(a) Q_{ϵ_1} (b) $Q_{2\epsilon_1}$

FIGURE 12. Cubical complexes Q_ϵ on top of grids for our point set at scales ϵ_1 and $2\epsilon_1$. Compare with Čech, alpha, and VR complexes in Figures 8, 9(c), and 10(b), respectively.

EXAMPLE 3.7. Figure 12 shows cubical complexes Q_{ϵ_1} and $Q_{2\epsilon_1}$ extracted from conforming grids. While the cubical complexes are not comparable to simplicial complexes combinatorially, they may capture similar topological features. The cubical complex Q_{ϵ_1} has the same homotopy type as the VR complex $V_{2\epsilon_1}$ in Figure 10(b), and the cubical complex $Q_{2\epsilon_1}$ has the same homotopy type as the Čech complex C_{ϵ_2} in Figure 8(b), as well as the alpha complex A_{ϵ_2} in Figure 9(c).

Cubical complexes arise naturally in analysis of two- and three-dimensional rasterized images, such as in the discrete simulation of *dynamical systems* [**42**]. Thresholding a grayscale image, we get an black and white image, where we may interpret the set of black pixels or voxels as a cubical complex. Since these complexes are based on grids, each cube may only be connected to neighboring cubes. This regularity in connectivity allows for tailored algorithms and heuristics to compute topological invariants of cubical complexes [**15, 63**]. We recommend the chapter by Marian Mrozek in this volume as an introduction to current techniques in dynamical systems and cubical homology.

3.6. Analysis. We have now provided multiple structures and methods for the first step of the analysis pipeline depicted in Figure 6. We end this section by completing this step for our motivating dataset of the cyclooctane conformation space from Example 1.1. Recall that the cyclooctane has 8 carbon and 16 hydrogen atoms. The locations of the carbons determine the locations of the hydrogens through energy minimization, so we limit our parameterization to the coordinates of the carbons [**4**]. Therefore, the *cyclooctane dataset* $\mathcal{S}$ is a set of 6,400 points embedded in $\mathbb{Y} = \mathbb{R}^{24}$.

We use the VR complex from Section 3.3 as $\mathcal{S}$ is not too large and is embedded in a high-dimensional Euclidean space [**79**]. To get an idea of scale in $\mathcal{S}$, we compute the maximum interpoint distance between closest pairs of points. This distance is 0.18, so we set the maximum scale to be $\hat{\epsilon} = 0.4$. We first build the neighborhood graph at this scale in 0.29 seconds. With 6,400 vertices and 76,657 edges, the graph is sparse with only 0.4% of the possible edges. We construct the 4-skeleton of the VR complex in 10.83 seconds using the INCREMENTAL-VR algorithm [**79**]. The resulting complex has 3,034,973 simplices, but only 66,179 critical ϵ values at which the complex grows. As described in Section 3.3, we may now extract the

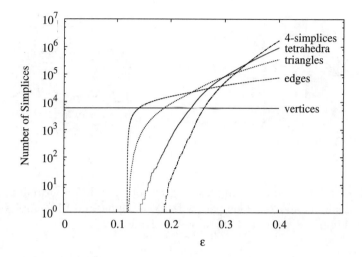

FIGURE 13. Number of n-simplices in the 4-skeleton of the VR complex V_ϵ for the cyclooctane dataset, with $0 \leq n \leq 4$ and $0 \leq \epsilon \leq 0.4$.

complex V_ϵ for any $0 < \epsilon \leq 0.4$. Figure 13 plots the number of n-simplices in this ϵ range for $0 \leq n \leq 4$. The size of the complex grows exponentially with dimension, as expected.

4. Topological Invariants

We now assume that we have a combinatorial representation K, such as a simplicial or cubical complex. In this section, we start on the second step of the analysis pipeline depicted in Figure 6: Computing topological invariants. We begin by formally defining an invariant. We then introduce two classic topological invariants: the Euler characteristic and homology. We conclude the section by returning to the cyclooctane dataset, analyzing the VR complex just built in the previous section.

4.1. Definition. Recall from Section 2.1 that homeomorphisms provide the finest level of classification under topology. The *Homeomorphism Problem* asks whether topological spaces $\mathbb{X}$ and $\mathbb{Y}$ are homeomorphic. This problem is undecidable even when restricted to manifolds of dimension greater than three [48]. Since we remain interested in analyzing real-world datasets, we need effective algorithms, so we must lower our expectations and look for partial solutions. These partial solutions come in the form of topological invariants.

Formally, a *topological invariant* is a map f that assigns the same object to homeomorphic spaces, that is:

$$\mathbb{X} \approx \mathbb{Y} \implies f(\mathbb{X}) = f(\mathbb{Y})$$

Note that an invariant is only useful through its contrapositive,

$$f(\mathbb{X}) \neq f(\mathbb{Y}) \implies \mathbb{X} \not\approx \mathbb{Y},$$

TABLE 1. Cell complexes K that are homotopy equivalent to the 1-sphere $\mathbb{S}^1$ have Euler characteristic $\chi(K) = 0$.

Figure	K	type	# n-cells 0	1	2	$\chi(K)$
8(b)	C_{ϵ_2}	Čech	8	11	3	0
11(c)	W_∞	witness	8	9	1	0
12(b)	$Q_{2\epsilon_1}$	cubical	15	22	7	0

as the converse of an implication is not true. Therefore, the *trivial* invariant that assigns the same object to all spaces is useless. On the other hand, the *complete* invariant that assigns different objects to non-homeomorphic spaces solves the homeomorphism problem. Most invariants are *incomplete*, falling in the spectrum between these two extremes. An incomplete topological invariant is a classification that is coarser than, but respects, the topological type. In general, the more powerful an invariant, the harder it is to compute it. Naturally, we look for invariants that assign finitely representable objects, as we intend to store them on computers.

We have already seen one topological invariant: homotopy equivalence. Since it is an invariant, we have $\mathbb{X} \approx \mathbb{Y} \implies \mathbb{X} \simeq \mathbb{Y}$. Unfortunately, the general problem of homotopy equivalence is also intractable [48], so we must look for less powerful invariants.

4.2. Euler Characteristic. Our first invariant assigns a single integer to a topological space. Let K be any cell complex, such as a simplicial or a cubical complex. The *Euler characteristic* $\chi(K)$ is

$$(4.1) \qquad \chi(K) = \sum_{\sigma \in K} (-1)^{\dim \sigma} = \sum_{n=0}^{\dim K} (-1)^n c_n,$$

where c_n is the number of n-dimensional cells in K. The Euler characteristic is an integer invariant for $|K|$ up to homotopy type, so we get the same integer for different complexes whose underlying spaces are homotopy equivalent. That is, for cell complexes $K_1, K_2, |K_1| \simeq |K_2| \implies \chi(K_1) = \chi(K_2)$.

EXAMPLE 4.1 ($\mathbb{S}^1$). The *1-dimensional sphere (1-sphere)* $\mathbb{S}^1$ is a space that is homeomorphic to a circle, such as any closed curve in Figure 1. The 1-sphere has Euler characteristic $\chi(\mathbb{S}^1) = 0$. Table 1 lists three complexes from the previous section that are homotopy equivalent to the 1-sphere. It also verifies that their Euler characteristic is zero using Equation (4.1).

4.3. Simplicial Homology. Instead of an integer, the *homology* invariant assigns a group to a topological space. Homology is quite popular in topological data analysis as it is effectively computable. We define homology for simplicial complexes, but the theory
    extends to arbitrary topological spaces, and the algorithms extend to arbitrary cell complexes, such as cubical complexes [38].

Let K be a simplicial complex, and suppose we fix an order on its set of vertices. An *orientation* of a n-simplex $\sigma = \{v_0, v_1, \ldots, v_n\} \in K$, is an equivalence class of orderings on its vertices, where $(v_0, v_1, \ldots, v_n) \sim (v_{\tau(0)}, v_{\tau(1)}, \ldots, v_{\tau(n)})$ if the parity of the permutation τ is even. An *oriented simplex* is a simplex with an orientation, denoted as sequence $[\sigma]$. For notation brevity, we list oriented simplices as strings

FIGURE 14. A 2-dimensional simplicial complex for Example 4.2.

rather than sequences, so $[v_0, v_1, v_2] \equiv v_0 v_1 v_2$. The *nth chain group* $C_n(K)$ *of* K is the free Abelian group on K's set of oriented n-simplices. We will abuse notation by dropping K in the notation when the complex is clear from context. An element $c \in C_n$ is an n-chain, $c = \sum_i c_i[\sigma_i]$, with n-simplices $\sigma_i \in K$ and *coefficients* $c_i \in \mathbb{Z}$. Given such a chain c, the *boundary homomorphism* $\partial_n \colon C_n \to C_{n-1}$ is a homomorphism defined linearly by its action on any oriented simplex in c:

$$\partial_n[v_0, \ldots, v_n] = \sum_i (-1)^i [v_0, \ldots, \hat{v}_i, \ldots, v_n],$$

where $\hat{v}_i$ indicates that v_i is deleted from the vertex sequence. We also define $\partial_0 \equiv 0$. A fundamental property of the boundary operator is that $\partial_n \circ \partial_{n+1} \equiv 0$ for all $n \geq 0$. The boundary operator connects the chain groups into a *chain complex* C_*:

$$\cdots \to C_{n+1} \xrightarrow{\partial_{n+1}} C_n \xrightarrow{\partial_n} C_{n-1} \to \cdots.$$

Given any chain complex, the *nth homology group* H_n is:

(4.2) $$H_n = \ker \partial_n \,/\, \mathrm{im}\, \partial_{n+1},$$

where ker and im are the *kernel* and *image* of a linear operator, respectively. An n-chain z is an n-*cycle* if $z \in \ker \partial_n$; it is also an n-*boundary* if $z \in \mathrm{im}\, \partial_{n+1}$. Since $\partial_n \circ \partial_{n+1} \equiv 0$, all boundaries are cycles and $\mathrm{im}\, \partial_{n+1}$ forms a subgroup of $\ker \partial_n$. Two cycles in the same homology class are *homologous*. Homology is a invariant for $|K|$ up to homotopy type. That is, for simplicial complexes K_1, K_2, $|K_1| \simeq |K_2| \implies H_n(K_1) = H_n(K_2)$ for all $n \geq 0$.

EXAMPLE 4.2. Consider the simplicial complex with labeled vertices in Figure 14. We place the alphabetic ordering on the vertices. The triangle $\{a, b, f\}$ has two orientations: $[a, b, f] = -[b, a, f]$, or $abf = -baf$ using our string notation. The 1-chain $ab + bc$ has boundary

$$\partial_1(ab + bc) = \partial_1(ab) + \partial_1(bc) = (b - a) + (c - b) = c - a,$$

so the 0-chain $c - a$ is a 0-boundary as it is in $\mathrm{im}\, \partial_1$. The boundary of the 1-chain $h = bc + ce + ef - bf$ is 0, so $h \in \ker \partial_1$ is a 1-cycle. Since h does not bound a 2-chain, h belongs to a non-trivial homology class. The 1-cycle $h' = bc + cd + de + ef - bf$ is homologous to h as their difference is the boundary $cd + de - ce$. Both cycles describe the single hole in the complex.

In order to understand the structure of algebraic invariants, such as homology, we use the following three-step approach:

(1) Correspondence,
(2) Classification, and
(3) Parameterization.

In the first step, we identify the algebraic structure. In the second step, we obtain a complete classification of the structure, up to isomorphism. In the third step, we parameterize the classification. We follow this approach for homology of a simplicial complex [26, 38]:

(1) Correspondence: The nth homology H_n of a simplicial complex is a group, or equivalently, a $\mathbb{Z}$-module, where $\mathbb{Z}$ is the *ring of coefficients*. We may, instead, construct modules over other rings R. Since the complex K is finite, H_n becomes a finitely generated R-module.

(2) Classification: Suppose R is a principle ideal domain (PID), such as $\mathbb{Z}$. Any finitely generated R-module decomposes uniquely into the form:

$$\bigoplus_{i=1}^{\beta_n} R \oplus \bigoplus_{j=1}^{m} R/t_j R,$$

for integers $\beta_n \geq 0$ and nonzero nonunit elements $t_j \in R$, such that $t_j | t_{j+1}$.

(3) Parameterization: The left direct sum is the *free* submodule and is characterized by its *Betti number* $\beta_n = \operatorname{rank} H_n$. The right direct sum is the *torsion* submodule and is characterized by its *torsion coefficients* t_j. The set of $m+1$ elements $\{\beta_n\} \cup \{t_j\}_j$ is the parameterization. Over a field k of coefficients, H_n simplifies to a k-vector space with *dimension* $\beta_n = \dim H_n$, so the parameterization is simply the integer β_n.

There is a one-to-one correspondence between the parameterization and finitely generated R-modules, so this parameterization is a complete invariant up to isomorphism. We have a full characterization of homology, provided we compute over PIDs.

The invariance of the Euler characteristic is derived from the invariance of homology. For a topological space $\mathbb{X}$, the *Euler-Poincaré formula* states that

(4.3) $$\chi(\mathbb{X}) = \sum_n (-1)^n \beta_n.$$

Compare the formula with the previous definition in Equation (4.1) in Section 4.2. This formula emphasizes that χ can be defined purely in terms of homology and depends only on the homotopy type of $\mathbb{X}$. That is, $\chi(\mathbb{X})$ is independent of the choice of cell complex representing $\mathbb{X}$.

For torsion-free spaces in three-dimensions, the Betti numbers have intuitive meaning as a consequence of the *Alexander Duality*: β_0 counts the number of connected *components*; β_1 is the rank of any basis for the *tunnels*; β_2 counts the number of enclosed spaces or *voids*.

EXAMPLE 4.3. Table 2 lists the Betti numbers for some of the topological spaces and cell complexes that we have seen so far. For instance, the reconstructed surface has one component, one tunnel, and two voids. The table also lists the Euler characteristics for the spaces, this time computed by Equation (4.3). We see that homology is a more refined invariant than the Euler characteristic. For example, the surface and C_{ϵ_1} have the same χ, but different β_0 and β_2. We can distinguish the two spaces with homology, but not with the Euler characteristic.

Having characterized homology, we next turn to its computation. Since the boundary operator $\partial_n \colon C_n \to C_{n-1}$ is linear, it has a matrix M_n in terms of a choice of bases for C_n and C_{n-1}. We may use oriented n-simplices as a basis for C_n

TABLE 2. Topological spaces and cell complexes, and their Betti numbers β_n and Euler characteristics χ.

Figure	Space	β_0	β_1	β_2	χ
1(a)	curve	1	1	0	0
4(c)	spiral	1	0	0	1
5	surface	1	1	2	2
8(a)	C_{ϵ_1}	3	1	0	2
8(b)	C_{ϵ_2}	1	1	0	0
10(b)	$V_{2\epsilon_1}$	3	0	0	3
11(c)	W_∞	1	1	0	0
12(a)	Q_{ϵ_1}	3	0	0	3
12(b)	$Q_{2\epsilon_1}$	1	1	0	0
14	complex	1	1	0	0

in each dimension. Computing the kernel and image in Equation (4.2) is equivalent to computing the *null space* of the matrix for ∂_n, and the *range space* of the matrix for ∂_{n+1}, respectively.

Over PIDs, the *reduction* algorithm reduces each matrix to the *Smith normal form*, from which the parameterization may be read [26]. Over $\mathbb{Z}$, neither the size of the matrix entries nor the number of operations in $\mathbb{Z}$ is polynomially bounded for reduction. There are sophisticated polynomial algorithms based on modular arithmetic [69], although reduction is still preferred in practice [25].

Over fields, C_n is a vector space in each dimension and we compute its dimension with *Gaussian elimination* matches that of matrix multiplication [72]. In practice, topological analysis nearly always uses the field of two elements $\mathbb{Z}_2 = \mathbb{Z}/2\mathbb{Z}$ for coefficients, which simplifies computation even further. Each simplex is its own inverse so there is no need for orientation. The matrices have 0 or 1 entries, so the columns may be stored sparsely as lists of simplices with coefficient 1. We use only *elementary column operations* in Gaussian elimination to reduce the matrix to *column echelon form* and read off the dimension.

EXAMPLE 4.4. Over $\mathbb{Z}_2$, the matrix for ∂_1 for the complex in Figure 14 is:

$$\left[\begin{array}{c|cccccccc} & ab & bc & cd & de & ef & af & bf & ce \\ \hline a & 1 & 0 & 0 & 0 & 0 & 1 & 0 & 0 \\ b & 1 & 1 & 0 & 0 & 0 & 0 & 1 & 0 \\ c & 0 & 1 & 1 & 0 & 0 & 0 & 0 & 1 \\ d & 0 & 0 & 1 & 1 & 0 & 0 & 0 & 0 \\ e & 0 & 0 & 0 & 1 & 1 & 0 & 0 & 1 \\ f & 0 & 0 & 0 & 0 & 1 & 1 & 1 & 0 \end{array}\right],$$

where we augment the matrix to show the bases for C_1 and C_0. Applying Gaussian elimination, we reduce the matrix to column echelon form:

$$\left[\begin{array}{c|cccccccc} & ab & bc & cd & de & ef & z_1 & z_2 & z_3 \\ \hline a & 1 & 0 & 0 & 0 & 0 & 0 & 0 & 0 \\ b & 1 & 1 & 0 & 0 & 0 & 0 & 0 & 0 \\ c & 0 & 1 & 1 & 0 & 0 & 0 & 0 & 0 \\ d & 0 & 0 & 1 & 1 & 0 & 0 & 0 & 0 \\ e & 0 & 0 & 0 & 1 & 1 & 0 & 0 & 0 \\ f & 0 & 0 & 0 & 0 & 1 & 0 & 0 & 0 \end{array}\right],$$

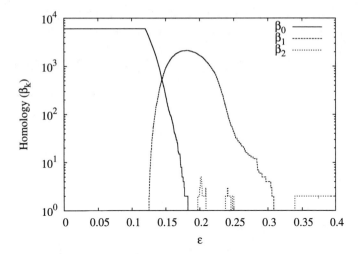

FIGURE 15. The Betti numbers β_n of the VR complex V_ϵ for the cyclooctane dataset S, with $0 \leq n \leq 2$ and $0 \leq \epsilon \leq 0.4$.

where the basis elements

$$z_1 = af + ab + bc + cd + de + ef,$$
$$z_2 = bf + ab + bf,$$
$$z_3 = ce + cd + de,$$

are the generators of ker ∂_1, so $\dim(\ker \partial_1) = 3$. Note that the three generators form a vector space of cycles in the 1-skeleton of the complex in Figure 14. Repeating the process for ∂_2, we get $\dim(\operatorname{im} \partial_2) = 2$. Therefore,

$$\dim H_1 = \dim(\ker \partial_1) - \dim(\operatorname{im} \partial_2) = 3 - 2 = 1,$$

and homology has captured the central hole in the complex.

4.4. Single-Scale Analysis. We have now looked at two topological invariants for the second step of the analysis pipeline. We end this section by completing this step for the cyclooctane dataset S using the 4-dimensional VR complex built in Section 3.6.

We compute homology over $\mathbb{Z}_2$ coefficients in 13.35 seconds using the persistence algorithm that we will encounter in Section 5.2. Figure 15 graphs the Betti numbers β_n for V_ϵ, $0 \leq \epsilon \leq 0.4$ and $0 \leq n \leq 2$. The Betti number β_3 is identically zero and is not plotted. We also do not consider β_4, as homology requires 5-simplices to determine if a 4-cycle is a boundary, but we only provide the 4-skeleton. As expected, the complex becomes connected starting at $\epsilon = 0.18$, the maximum interpoint distance of closest points. The Betti numbers of the complex match those of the conformation space surface for all $\epsilon \geq 0.3391$. Since we do not know the correct scale, however, we would not be able to determine the Betti numbers of the conformation space from this graph alone. But the complexes at different scales are related to each other. Our success in topological analysis requires analysis across scale to determine the topological features of the unknown space.

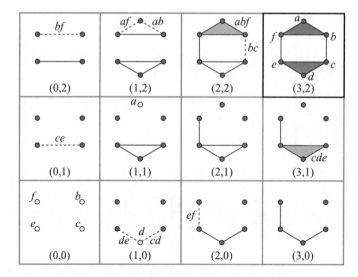

FIGURE 16. A bifiltration of the complex in Figure 14, now with coordinate $(3,2)$. Simplices are highlighted and named at their critical coordinates.

5. Multiscale Invariants

In this section, we provide multiscale solutions for the second step of the analysis pipeline. We extend our combinatorial representation to approximate the unknown space at multiple scales. We then introduce three modern multiscale invariants that analyze the topology of the resulting multiscale structure, identifying robust features that persist across scale. We conclude, as usual, by applying our multiscale tools to the analysis of cyclooctane dataset.

5.1. Multifiltration Model. Our first model is based on notions from *Morse theory* [52]. Let $\mathbb{N} \subseteq \mathbb{Z}$ be the set of non-negative integers. For vectors in $\mathbb{N}^d$ or $\mathbb{R}^d$, we say $u \leq v$ if $u_i \leq v_i$ for all $1 \leq i \leq d$. The relation $\leq$ forms a partial order on $\mathbb{N}^d$ and $\mathbb{R}^d$. A cell complex K is *multifiltered* if we are given a family of subcomplexes $\{K_u\}_u$, where $u \in \mathbb{R}^d$, so that $K_u \subseteq K_v$ whenever $u \leq v$. Intuitively, a multifiltered complex only *grows* with increasing coordinate, so the model describes a monotonically growing space. We call the family of subspaces $\{K_u\}_u$ a *multifiltration*. A one-dimensional multifiltration is a *filtration*. All the simplicial methods in Section 3 give rise to filtrations with increasing ϵ, but cubical complexes do not as the vertices change at different scales.

A *critical coordinate* u for cell $\sigma \in K$ is a minimal coordinate, with respect to the partial order $\leq$, such that $\sigma \in K_u$. A multifiltered complex K where each cell σ has a unique critical coordinate u_σ is *one-critical* [9].

EXAMPLE 5.1. Figure 16 shows a two-dimensional multifiltration, a *bifiltration*, of the complex in Figure 14. The bifiltration is one-critical, with each simplex being highlighted and named at its unique critical coordinate.

A one-critical multifiltration is a natural model for scientific data. Suppose a sampled dataset $S \subseteq \mathbb{Y}$ is augmented with $d - 1$ real-valued functions $f_j \colon S \to \mathbb{R}$

with $d > 1$. The functions measure information about the unknown space $\mathbb{X}$ at each point.

EXAMPLE 5.2 (graphics). In *computer graphics*, one approach to rendering surfaces is to construct a digitized model. A three-dimensional object is sampled by a range scanner that employs multiple cameras to sense the surface position as well as normals and textures [**71**]. Here, S is the set of positions, while the functions f_j are surface attributes, such as normal and texture, sampled at S.

We begin by approximating S with a filtered complex K, using any method that yields a filtration, such as the simplicial methods in Section 3. Suppose each cell $\sigma \in K$ enters the complex at scale $\epsilon(\sigma)$. To incorporate the functions f_j into topological analysis, we first extend them to the cells in the complex. For $\sigma \in K$ and f_j, let $f_j(\sigma)$ be the maximum value f_j takes on σ's vertices; that is, $f_j(\sigma) = \max_{v \in \sigma} f_j(v)$, where $v \in S$. This extension defines $d - 1$ functions on the complex, $f_j \colon K \to \mathbb{R}$. We combine all filtration functions into a d-variate function $F \colon K \to \mathbb{R}^d$, where

$$F(\sigma) = (f_1(\sigma), f_2(\sigma), \ldots, f_{d-1}(\sigma), \epsilon(\sigma)).$$

We multifilter K via the *sublevel sets* $\{K_u\}_u$ of F for $u \in \mathbb{R}^d$:

$$K_u = \{\sigma \in K \mid F(\sigma) \leq u\}.$$

Each simplex σ enters K_u at $u = F(\sigma)$ and remains in the complex for all $u \geq F(\sigma)$. Equivalently, $F(\sigma)$ is the unique critical coordinate at which σ enters the filtered complex. That is, the multifiltrations built by this process are always one-critical.

Finally, since complex K is finite, there are a finite number of critical coordinates in each dimension where the complex grows in the multifiltration. Restricting to the Cartesian product of these critical values, we parameterize the resulting discrete grid using $\mathbb{N}$ in each dimension. This parameterization gives us coordinates in $\mathbb{N}^d$ for a multifiltration, as shown for the bifiltration in Figure 16 [**10**].

5.2. Persistent Homology. We are now interested in the homology of our multiscale model for representing data. That is, we want to know the homology of the complexes at all scales, as well as the relationship between their homologies. Suppose we are given a multifiltration $\{K_u\}_u$, $u \in \mathbb{N}^d$. For each pair $u, v \in \mathbb{N}^d$ with $u \leq v$, $K_u \subseteq K_v$ by definition, so $K_u \hookrightarrow K_v$. Since homology is a functor, this inclusion induces a linear map

$$\iota_n(u, v) \colon H_n(K_u) \to H_n(K_v)$$

that maps an n-dimensional homology class within K_u to the one that contains it within K_v [**38**]. The *nth persistent homology* is $\operatorname{im} \iota_n$, the image of ι_n for all pairs $u \leq v \in \mathbb{N}^d$ [**10**].

For characterization and computation, we begin with one-parameter multifiltrations (filtrations). We follow the same algebraic approach we used for characterizing homology in Section 4.3. For a filtration, we have [**81**]:

(1) Correspondence: The nth persistent homology of a filtration over ring R is a graded $R[t]$-module, where $R[t]$ is the ring of polynomials with indeterminate t over R.

(2) Classification: Over fields k, $k[t]$ is a PID, and any graded $k[t]$-module decomposes uniquely into:

$$\bigoplus_{i=1}^{\ell} \Sigma^{\alpha_i} k[t] \;\oplus\; \bigoplus_{j=1}^{m} \Sigma^{\gamma_j} k[t]/(t^{\delta_j}),$$

where Σ^{α} denotes an α-shift upward in grading, and $\alpha_i, \gamma_j, \delta_j \in \mathbb{N}$.

(3) Parameterization: The classification gives us ℓ half-infinite intervals $[\alpha_i, \infty)$ and m finite intervals $[\gamma_j, \gamma_j + \delta_j)$. The *persistence barcode* is the multiset of these $\ell + m$ intervals and forms the parameterization [**11**].

There is a one-to-one correspondence between the parameterization and finitely generated graded $k[t]$-modules, so this parameterization is a complete invariant, up to isomorphism. Note that while $\mathbb{Z}$ is a PID, $\mathbb{Z}[t]$ is not, so the classification above does not extend to integer coefficients.

The barcode intervals have a natural interpretation. By the theory of persistent homology, each n-simplex either *creates* an n-dimensional homology class, or *destroys* an $(n-1)$-dimensional class by merging it with a class created earlier. For each class, persistence pairs its *creator* $\sigma \in K$ with the *destroyer* $\tau \in K$, if one exists. We may also pair the grades at which σ and τ enter the filtration to get an interval representing each class. This barcode interval is the *lifetime* of the homology class within the filtration. A half-infinite interval $[\alpha_i, \infty)$ represents a class that is created at α_i and still exists within the completed complex. A finite interval $[\gamma_j, \gamma_j + \delta_j)$ represents a class that is created at γ_j and lives only δ_j grades in the filtration, at which point it merges with the boundary class. Alternatively, we may form intervals using the ϵ at which the two simplices enter the filtration, getting a barcode that describes homology with respect to scale.

EXAMPLE 5.3. In a multifiltration, any path with monotonically increasing coordinates is a filtration, such as the bottom row in the bifiltration in Figure 16. The filtered complex at coordinate $(3, 0)$ has 5 vertices and 3 edges. Figure 17 graphs β_0 for this filtration above the x-axis, where the unit is filtration grade. Below the axis, we see the β_0 barcode. Each interval is the lifetime of a connected component in this filtration. The left endpoint is labeled with the simplex that created the component. The right endpoint is labeled with the simplex that destroyed the component, if such a simplex exists. The component created by simplex d and destroyed by simplex cd immediately has zero lifetime, so we do not draw it. The barcode deconstructs the β_0 graph into a set of intervals. We may recover the β_0

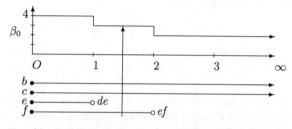

FIGURE 17. Above the x-axis, β_0 is plotted for the bottom row filtration of Figure 16. Below is the labeled β_0 barcode. Since the vertical arrow intersects three intervals at $x = 1.5$, $\beta_0 = 3$ at that x. The axis unit is filtration grade.

graph by sweeping a vertical line from left to right and counting the number of intervals that the line intersects, as demonstrated by the vertical arrow at $x = 1.5$ in the figure.

Persistence barcodes have been quite useful in topological data analysis. Suppose that a geometric process constructs a filtration so that the lifetime of a homology class denotes its significance. Then, we may use barcodes to separate topological noise from features. We have applied barcodes successfully in a number of areas, including shape description [17], biophysics [43], and computer vision [8].

Having characterized persistent homology, we next turn to its computation. Since $k[t]$ is a PID, the reduction algorithm for homology extends naturally to persistent homology. The matrix for a boundary operator now has polynomial entries from $k[t]$ that encode the filtration ordering. While the field homology of any single complex in a filtration is a vector space, the persistent homology of the filtration has torsion. This means that the reduction algorithm will require both row and column operations to reduce each matrix to Smith normal form.

EXAMPLE 5.4. For the filtration in Example 5.3, we use the n-simplices at their critical coordinates as a basis for the chain group C_n. Over $\mathbb{Z}_2[t]$, the matrix for ∂_1 of the filtration is:

$$\left[\begin{array}{c|ccc} & cd & de & ef \\ \hline b & 0 & 0 & 0 \\ c & t & 0 & 0 \\ d & 1 & 1 & 0 \\ e & 0 & t & t^2 \\ f & 0 & 0 & t^2 \end{array} \right].$$

For instance, $\partial_1(cd) = t \cdot c + 1 \cdot d$, as c enters the filtration one grade earlier than d. We may now reduce this matrix with the reduction algorithm as for regular homology.

Alternatively, we may utilize the *persistence algorithm*, which computes directly on matrices with field entries [81]. The algorithm takes advantage of the filtration ordering to require only elementary column operations, allowing it to represent matrices as columns and reduce them to column echelon form. There is a wonderful relationship between the reduction and persistence algorithms [81]. The latter has been refined over the years. For its most recent distillation, see [78]. The algorithm is implemented in several publicly available software packages [56, 66].

Finally, both the reduction and persistence algorithms may also generate descriptions of generators for homology classes, as we do in Example 4.4 by augmenting the boundary matrix. Traditionally, these descriptions are not generated as the focus is on the algebraic characterization of the homology groups. The generators are useful, however, within geometric applications of computational topology [30, 82]. We recommend Jeff Erickson's chapter in this volume for an introduction to current results on geometrically optimal generators.

5.3. Multidimensional Persistence. Our success in characterizing homology of filtrations motivates us to move to higher dimensional multifiltrations. Once again, we follow our algebraic approach. For a multifiltration, we have [10]:

(1) Correspondence: The nth homology of a multifiltration over field k is an n-graded A_n-module M, where $A_n = k[x_1, \ldots, x_n]$ is the n-graded module of polynomials with n indeterminates over k.
(2) Classification: Unlike its one-dimensional counterpart, A_n is not a PID and A_n-modules have no structure theorem. Nevertheless, we establish a full classification of this structure in terms of three invariants. The first invariant, $\xi_0(M)$ is the multiset of generators for the free approximation of M. The second invariant, $\xi_1(M)$ is the multiset of generators for the *free hull* of M. These invariants have intuitive meaning as analogs of the left and right endpoints of intervals in a barcode, respectively. Unfortunately, there is no way to *match* these endpoints consistently as the remaining invariant corresponds to the set of orbits in a set under group action.
(3) Parameterization: The third invariant corresponds to the set of orbits of an algebraic group action on an algebraic variety. Unfortunately, such a set is not, in general, an algebraic variety. The number of orbits may be uncountable, giving us a *continuous* invariant.

To summarize, no complete invariant exists for persistent homology of multifiltrations, in dimension higher than 1. The discrete invariants ξ_0, ξ_1 above do not capture persistence information, which is contained in the intervals, not their endpoints.

Instead, we may use an incomplete invariant. Recall that persistence is the image of the map $\iota_n(u,v)\colon H_n(K_u) \to H_n(K_v)$. The *$n$th rank invariant* is

$$\rho_n(u,v) = \operatorname{rank} \iota_n(u,v),$$

for all pairs $u \leq v \in \mathbb{N}^d$ [10]. The rank invariant is equivalent to the persistent barcode in the one-dimensional case, so it is complete when it can be. Unlike the barcode, the rank invariant extends to higher dimensions as an incomplete invariant.

We next turn to the computation of multidimensional persistence. We assume we are given a d-dimensional multifiltration of a cell complex K with m cells. Any pair $u \leq v \in \mathbb{N}^d$ defines a two-level one-dimensional filtration, where we map u to 0 and v to 1. We may compute the barcodes for this filtration in $\Theta(m^3)$ time using the persistence algorithm in Section 5.2. We then read $\rho_n(u,v)$ directly from the β_n-barcode: It is the number of intervals that contain both 0 and 1. To compute the full rank invariant, we need to consider all distinct pairs of complexes in a multifiltration that are comparable by the partial order $\leq$. Unfortunately, there are constructions with $\Theta(m^d)$ distinct complexes, implying $\Theta(m^{2d})$ comparable pairs, and a $\Theta(m^{2d+3})$ running time. To store the rank invariant, we also require $\Theta(m^{2d})$ space [9]. This is clearly not a feasible method.

For one-critical multifiltrations, described in Section 5.1, we can use more sophisticated algorithms. The n-graded chain modules C_n for one-critical multifiltrations are free, as each cell enters the complex only once. The boundary operator $\partial_n\colon C_n \to C_{n-1}$, in turn, is a homomorphism between free multigraded modules and may be written as a matrix with polynomial entries.

EXAMPLE 5.5. For the bifiltration in Figure 16, we use n-simplices in their critical coordinates as a basis for chain group C_n. Over $A_2 = \mathbb{Z}_2[x_1, x_2]$, the matrix for ∂_1 of the multifiltration is:

$$\begin{bmatrix}
 & ab & bc & cd & de & ef & af & bf & ce \\
\hline
a & x_2 & 0 & 0 & 0 & 0 & x_1 & 0 & 0 \\
b & x_1 x_2^2 & x_1^2 x_2^2 & 0 & 0 & 0 & 0 & x_2^2 & 0 \\
c & 0 & x_1^2 x_2^2 & x_1 & 0 & 0 & 0 & 0 & x_2 \\
d & 0 & 0 & 1 & 1 & 0 & 0 & 0 & 0 \\
e & 0 & 0 & 0 & x_1 & x_2^2 & 0 & 0 & x_2 \\
f & 0 & 0 & 0 & 0 & x_1^2 & x_1 x_2^2 & x_2^2 & 0
\end{bmatrix}.$$

Recall from Equation 4.2 that the nth homology group is $H_n = \ker \partial_n / \operatorname{im} \partial_{n+1}$. To compute homology, we have three tasks, all of which may be recast into problems in computational commutative algebra [19].

(1) Compute the boundary module ($\operatorname{im} \partial_{n+1}$): This is the *submodule membership problem*. We first compute a Gröbner basis using the Buchberger algorithm. We then check membership using the division algorithm for multivariate polynomials.
(2) Compute the cycle module ($\ker \partial_n$): The problem is equivalent to computing the *syzygy submodule* using Schreyer's algorithm.
(3) Compute the quotient H_n: We need to test whether the generators of the syzygy submodule are in the boundary submodule. But this is simply an instance of our first task.

While the above solution is theoretically sound, it is practically infeasible. The SMP problem is a generalization of the *polynomial ideal membership problem* at the ring level, and PIMP is already EXPSPACE-complete, requiring exponential space and time [54]. The Buchberger algorithm is doubly-exponential, although impractical singly-exponential versions do exist.

We can exploit the structure provided by a multigrading, however, to derive polynomial-time algorithms. The key insight is that we ensure that the matrix entries are always homogeneous monomials, as in Example 5.5. The resulting multigraded algorithms run in worst-case $O(m^3)$ space and $O(m^7)$ time, where m is the size of the multifiltration [9]. Empirically, the time bound seems to be tight. While the reduction in complexity is theoretically significant, this time bound still implies that multiparameter topological analysis is out of reach for large datasets.

5.4. Zigzag Persistence. We end our discussion of topological invariants with recent developments for extracting persistent information for yet another model for scientific data. A primary characteristic of our model in the section so far is that it is monotonically increasing as in a multifiltration $\{K_u\}_u$, subcomplexes nest: $K_u \subseteq K_v$ whenever $u \leq v$. But we have nonmonotonicity in a number of application areas.

EXAMPLE 5.6 (molecular rigidity). *Flexible docking* models biological macromolecules, such as protein-protein complexes, by allowing flexibility near active sites. We would like to identify flexible regions of a protein algorithmically. Based on the molecular conjecture [70], now a theorem [44], we model a protein by a multigraph, where covalent bonds, hydrogen bonds, salt bridges, and hydrophobic contacts or tethers are represented as edges [39]. The program FIRST partitions this multigraph into flexible and rigid regions by extending the *pebble game* to three-dimensions [31]. Covalent bonds, however, have picosecond vibrations that cause noncovalent bonds to be unstable, resulting in the *flickering* of edges in the associated multigraph [**47**]. Flickering edges can be modeled by adding and deleting edges from a dynamically changing cell complex. The resulting history of the complex, however, is no longer a filtration.

Instead, we model nonmonotinicity as a sequence of topological spaces $\{Y_i\}_i \in \mathbb{N}$ that are not necessarily nested. Since any pair of spaces Y_i, Y_j include into their union $Y_i \cup Y_j$, we use unions of consecutive spaces from the sequence to build the following diagram:

$$\begin{array}{ccccccc} & Y_0 \cup Y_1 & & Y_1 \cup Y_2 & & \cdots \\ & \nearrow \quad \nwarrow & \nearrow \quad \nwarrow & \nearrow \\ Y_0 & & Y_1 & & Y_2 & \end{array}$$

where all maps are inclusions. Let $X_{2n} = Y_n$ and $X_{2n+1} = Y_n \cup Y_{n+1}$ for $n \in \mathbb{N}$ to rewrite the diagram as:

$$X_1 \hookrightarrow X_2 \hookleftarrow X_3 \hookrightarrow X_4 \hookleftarrow \cdots X_m,$$

where we have assumed that the resulting sequence has m terms. If our spaces are cell complexes, the right arrows here indicate cell *addition*, and the left arrows indicate cell *deletion*. We generalize this model further by allowing the maps to be homomorphisms:

(5.1) $$X_1 \to X_2 \leftarrow X_3 \to X_4 \leftarrow \cdots \to X_m,$$

Since some maps could be identities, the general model is a family of complexes with forward or backward homomorphisms in any order. Due to the alternating directions of the maps, Diagram (5.1) is called a *zigzag* [**6**]. Note that if we only have arrows to the right in this diagram, and the arrows are inclusions, we get a diagram for a filtration:

(5.2) $$X_1 \to X_2 \to X_3 \to X_4 \to \cdots \to X_m,$$

Over a field k, the homology of each complex is a k-vector space. Applying the nth homology functor to Diagram (5.1), we get

(5.3) $$V_1 \to V_2 \leftarrow V_3 \to V_4 \leftarrow \cdots \to V_m,$$

where $V_j = H_n(X_j)$ is the nth homology of the jth space and the maps are induced maps at the homology level. Diagrams such as (5.3) are the objects of study in *representation theory* [**22, 34**]. A *quiver* is a pair $Q = (Q_0, Q_1)$, where Q_0 is a finite set of *vertices* and Q_1 is a finite set of *arrows (directed edges)* between them. That is, a quiver is a directed graph. The quiver for Diagram (5.3) is

(5.4) $$\bullet \longrightarrow \bullet \longleftarrow \bullet \longrightarrow \bullet \longleftarrow \cdots \longrightarrow \bullet$$

A quiver has an *underlying undirected graph*. This graph is a *path* for our quiver:

(5.5) $$\bullet \longrightarrow \bullet \longrightarrow \bullet \longrightarrow \bullet \longrightarrow \cdots \longrightarrow \bullet$$

A *representation* V for a quiver Q is a collection $\{V_x \mid x \in Q_0\}$ of finite-dimensional k-vector space together with a collection $\{V_{ab}: V_a \to V_b \mid ab \in Q_1\}$ of k-linear maps. Diagram (5.3) is a representation for quiver (5.4). Given representation V, the *dimension vector* $d_V: Q_0 \to \mathbb{N}$ is $d_V(x) = \dim_k(V_x)$ for all $x \in Q_0$; that is, $d_V(x)$ is the dimension of the vector space at node x.

As with finitely-generated modules in Section 4.3 or finitely-generated graded $R[t]$-modules in Section 5.2, quivers have a classification theorem stipulating that every representation has a unique decomposition into a direct sum of *indecomposable* representations, up to isomorphism and permutations of components. A quiver has *finite type* if it decomposes into a finite number of indecomposables. *Gabriel's Theorem* states that a quiver is of finite type iff the underlying undirected graph is a disjoint union of the following graphs [**33**]:

(5.6)
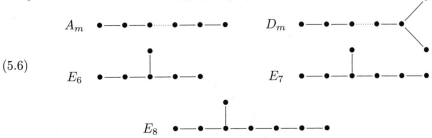

Kac's Theorem states that the set of dimension vectors of indecomposable representations of a quiver Q does not depend on the orientation of the arrows [**41**]. Following our algebraic approach, we have [**6**]:

(1) Correspondence: The nth homology of a zigzag of length m over a field is a representation of a quiver in Diagram (5.4).
(2) Classification: The zigzag quiver has an underlying undirected graph in Diagram (5.5), which is a path of length m, equivalent to type A_m in Diagram (5.6). By Gabriel's theorem, the zigzag quiver has finite type.
(3) Parameterization: By Kac's theorem, the invariant depends only on the underlying undirected graph. The model for one-dimensional persistence in Diagram (5.2) also gives a quiver of finite type A_m. We already know that persistent homology has barcodes as invariants. Therefore, as a corollary of Kac's theorem, zigzag persistence is parameterizable by barcodes.

Having characterized zigzag persistence, we next turn to its computation. From representation theory, one may derive a general scheme for computing zigzag persistence from of a representation, that is, given the set of vector spaces and linear maps as input. Given a topological space $\mathbb{X}$ and a Morse function $f \colon \mathbb{X} \to \mathbb{R}$ defined on it, one may compute the persistent homology of the level sets $f^{-1}(c)$ of this space using zigzag persistence [**7**]. Finally, a recent result gives an algorithm for computing zigzag persistence of a simplicial complex with a sequence of n additions and deletions in $O(M(n) + n^2 \log n)$ time, where $M(n)$ denotes the cost of multiplying two $n \times n$ matrices [**55**]. In this approach, a simplex with multiplicity is treated as multiple different simplices.

5.5. Multiscale Analysis. We have now looked at three multiscale invariants for the second step of the analysis pipeline. We end this section by completing this step for the cyclooctane dataset S using the 4-dimensional VR complex built in Section 3.6. The complex is already filtered by ϵ, as described in Section 3.3 and we have a filtration without any additional computation. Given a filtration, the natural model for analysis is persistent homology from Section 5.2.

We have already computed persistence barcodes in Section 4.4, as we used the persistence algorithm to compute homology, reading off the Betti numbers using the technique in Figure 17. Once again, there is no need for further computation. Figure 18 draws the 3,475 non-empty intervals of the β_1-barcode of this filtration. Most 1-cycles have short lifetimes: The average lifetime is 0.0382 with a standard deviation of 0.0260. But there is one cycle with a half-infinite interval, meaning

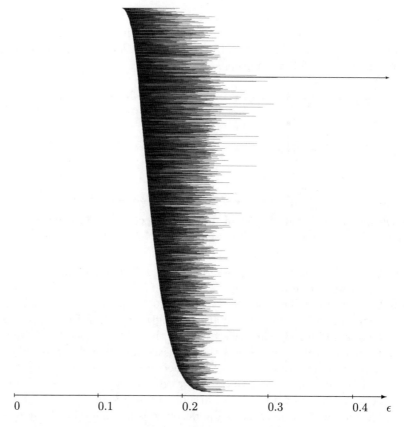

FIGURE 18. Cyclooctane β_1-barcode for the filtered VR complex V_ϵ, $0 \leq \epsilon \leq 0.4$, built in Section 3.6. There are 3,475 non-empty intervals, one of which is half-infinite.

that the cycle's homology class still exists at the maximum scale $\hat{\epsilon} = 0.4$. Even at this scale, this 1-cycle has had a life time of 0.2545, more than 8 standard deviations away from the average. Recall that the VR complex method from Section 3.3 is a geometric process that is based on local distances. Therefore, we are confident that this outlier is a topological feature and $\beta_1 = 1$ for this complex. By a similar procedure, we determine $\beta_0 = 1$ and $\beta_2 = 2$. Our results match the Betti numbers computed from the reconstructed surface of this dataset, as listed in Table 2.

We have now successfully completed a multiscale analysis of the 24-dimensional cyclooctane conformation space dataset containing 6,400 samples. Topological data analysis, using the two-step pipeline in Figure 6, only required 24.47 seconds on a desktop machine. By comparison, geometric reconstruction of the surface using a specialized algorithm is numerically challenging and requires domain knowledge, such as the intrinsic dimension of the unknown space and its types of non-manifold structure [**50**].

6. Reduced Representations

So far, this chapter has been organized around the full implementation of the two-step pipeline in Figure 6. The division of topological analysis into two steps, while useful, stems from a geometric point of view and is somewhat artificial. From an algebraic point of view, the two steps are very much interrelated. Homology is defined on a chain complex, as described in Section 4.3. In our pipeline, the chain complex is always derived from a cell complex built in step one, but this derivation is not a requirement. If we can obtain a chain complex without building a cell complex explicitly, we may still compute homology. Such an approach is desirable given the massive size of the cell complexes that we are now able to build with the methods in Section 3. For example, the simplicial complex representing the cyclooctane dataset has more than three million simplices defined on only 6,400 points.

In this section, we attempt to reduce the size of our representations by combining the two steps of the pipeline. We begin by describing reduction methods that maintain the category of a cell complex, yielding complexes with fewer cells. For even further reduction, we switch category to the simplicial set, a combinatorial structure that allows for collapsed simplices. We next use simplicial sets to define tidy sets, a method for computing homology of any clique complex without its full construction. We end the section by analyzing the cyclooctane dataset using tidy sets.

6.1. Reductions. The traditional approach to dealing with massive complexes has been to reduce the size of the complex before computing homology. This approach is somewhat justified since the reduction algorithm in Section 4.3 has supercubical complexity in the size of the complex over integers, retaining quadratic space and cubic time complexity over fields [**68**]. It is a reasonable to search for heuristics that reduce the size of the complex, while preserving its topology. To be useful, the heuristics must be simple and fast.

There have been a number of reduction techniques proposed for different categories of complexes. For simplicial complexes, the earliest, and perhaps simplest method, is elementary contraction, proposed by Whitehead in defining the simple homotopy type [**75**]. Let K be a simplicial complex. A simplex $\sigma \in K$ has a *free face* $\tau \subseteq \sigma$ if τ has no other cofaces in K. The pair (σ, τ), then, is a *free pair*. An *elementary contraction* removes a free pair (σ, τ) from a complex K. Since an elementary contraction is a *deformation retraction*, a type of homotopy equivalence, the resulting smaller complex $K - \{\sigma, \tau\}$ has the same homotopy type as K [**38**]. But while deformation retractions are continuous, elementary contractions are combinatorial, involving only the deletion of simplices. Elementary contractions may be extended to other cell complexes, such as cubical complexes. Another heuristic is the recent *LC-reduction* [**16**] that produces not only a homotopic, but isomorphic complex [**51**].

EXAMPLE 6.1 (contraction). Figure 19 displays six elementary contractions in two steps for the complex in (a). In the first step, the three highlighted triangles and their dashed edges form free pairs and are removed in (b). In the second step, the three dashed edges and their highlighted vertices form free pairs and are removed. The final complex in (c) has the same homotopy type as the original complex in (a). It is minimal with respect to elementary contraction.

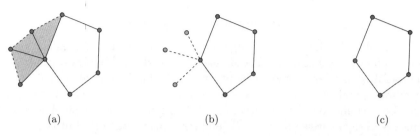

(a)   (b)   (c)

FIGURE 19. Elementary contractions. The three highlighted triangles and dashed edges (a) form free pairs and are removed in (b). Then, the dashed edges and highlighted vertices form free pairs and are removed, resulting in a minimal complex (c), homotopy equivalent to (a).

The key property of both heuristics is that the reduced complex remains in the initial category. For instance, elementary contractions on a simplicial complex always yield a simplicial complex. Due to the popularity of simplicial complexes in computational topology, the techniques have been used widely. For *cubical complexes*, the CHomP project has examined a large number of heuristics over the years, resulting in an array of homology engines [**15, 63**].

A stronger reduction is to collapse a cell into a point, as all cells are contractible by design. Such collapses, however, may change the category type of the structure.

EXAMPLE 6.2 (collapse). Figure 20 shows that collapsing edge bc in triangle abc (a) results in a 2-gon ad (b), which is not a simplicial complex as its two edges are both named ad.

(a) 3-gon abc (b) 2-gon ad

FIGURE 20. The simplicial collapse of edge bc of the triangle abc (a) yields a 2-gon ad (b) that is no longer a simplicial complex.

6.2. Simplicial Sets. To allow for simplicial collapses, we move into the category of simplicial sets. Intuitively, a simplicial set models a well-behaved topological space. One such space is a simplicial complex within which any simplex may be collapsed, and any subset of vertices may be identified.

Let K be a simplicial complex with n-simplices K_n. We define the *simplicial set* X corresponding to K to be a collection of sets $\{X_n\}_n$ together with maps:

$$d_i \colon X_n \to X_{n-1},$$
$$s_i \colon X_n \to X_{n+1},$$

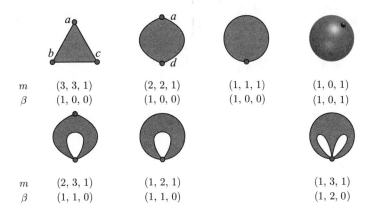

FIGURE 21. The 7 possible 2-simplices in a simplicial set. The triangle abc is the only one allowed in a complex. The rest have collapsed edges (top row), identified vertices (bottom row), or both. The vector m counts the non-degenerate simplices and the vector β holds the Betti numbers.

where d_i is the *ith face operator* and s_i is the *ith degeneracy operator* defined as follows:

$$d_i([v_0, \ldots, v_n]) = [v_0, \ldots, \hat{v}_i, \ldots, v_n],$$
$$s_i([v_0, \ldots, v_n]) = [v_0, \ldots, v_i, v_i, \ldots, v_n].$$

That is, the ith face operator d_i deletes the ith vertex, and the ith degeneracy operator s_i repeats it. We now define X_n inductively using the degeneracy operators:

$$X_0 = K_0,$$
$$X_n = K_n \cup \cup_i^n s_i(X_{n-1}), \quad n > 0.$$

It is easy to verify that $\{X\}_n$ together with these operators satisfy the axioms for a simplicial set [**53**]. A simplex $\sigma \in X$ such that $\sigma = s_i(\tau)$ for some $\tau \in X$ is *degenerate* and $\sigma \notin K$. Otherwise, σ is *non-degenerate* and $\sigma \in K$.

EXAMPLE 6.3 (triangle). Figure 21 gives the seven possible 2-simplices in a simplicial set, in contrast to the only possible 2-simplex in a simplicial complex, namely the triangle abc on the top left. We now represent the 2-gon from Example 6.2 as well as spaces with different topological types, such as the 2-sphere on the top right corner. For each simplex, the vectors m and β hold the number of non-degenerate simplices and the Betti numbers, respectively. As a simplicial complex K, abc has

$$K_0 = \{a, b, c\},$$
$$K_1 = \{ab, bc, ac\},$$
$$K_2 = \{abc\}.$$

As a simplicial set X, abc has

$$X_0 = \{a, b, c\},$$
$$X_1 = \{ab, bc, ac, aa, bb, cc\},$$
$$X_2 = \{abc, aab, abb, bbc, bcc, aac, acc, aaa, bbb, ccc\},$$

where the set X_n is K_n augmented with degenerate simplices, such as abb, a triangle with one collapsed edge.

Since simplicial sets are capable of representing collapsed simplices, we now incorporate this reduction into the model. Given a simplicial set X and an n-simplex $\sigma \in X$, the *collapse of* σ identifies σ to a single point, giving us a new simplicial set $X' = X/\sigma$. To construct X', we introduce a new vertex v and replace σ, its faces, and its degeneracies, with appropriate degeneracies of v. We first gather σ's non-degenerate k-faces inductively for $k \geq 0$:

$$\bar{F}_k(\sigma) = \begin{cases} \emptyset, & \text{if } k > n, \\ \{\sigma\}, & \text{if } k = n, \\ \cup_{i=0}^{k+1} d_i(\bar{F}_{k+1}(\sigma)), & \text{if } k < n. \end{cases}$$

By adding the degenerate faces, we get all the faces of σ:

$$F_k(\sigma) = \begin{cases} \bar{F}_0(\sigma), & \text{if } k = 0, \\ \bar{F}_k(\sigma) \cup \bigcup_{i=0}^{k-1} s_i(F_{k-1}(\sigma)), & \text{if } k > 0. \end{cases}$$

We replace these faces with degeneracies of v:

$$X'_k = (X_k - F_k(\sigma)) \cup \{s_0^k(v)\},$$

where s_0^k denotes applying the degeneracy operator k times. To complete the definition of X' as a simplicial set, we now define the operators for any $\tau \in X'$:

$$d'_i(\tau) = \begin{cases} d_i(\tau), & \text{if } d_i(\tau) \notin F_{i-1}(\sigma), \\ s_0^{i-1}(v), & \text{otherwise.} \end{cases}$$

$$s'_i(\tau) = \begin{cases} s_i(\tau), & \text{if } s_i(\tau) \notin F_{i+1}(\sigma), \\ s_0^{i+1}(v), & \text{otherwise.} \end{cases}$$

EXAMPLE 6.4 (2-gon). In Example 6.3, we listed the n-simplices of the triangle abc in Figure 21 as a simplicial set X. We now collapse edge bc to a new vertex d to get the 2-gon $X' = ad$ in the figure. We have

$$\bar{F}_2(bc) = \emptyset,$$
$$\bar{F}_1(bc) = \{bc\},$$
$$\bar{F}_0(bc) = F_0(bc) = \{b, c\},$$
$$F_1(bc) = \{bc, bb, cc\},$$
$$F_2(bc) = \{bbc, bcc, bbb, ccc\},$$
$$X'_0 = \{a, d\},$$
$$X'_1 = \{ab, ac, aa, dd\},$$
$$X'_2 = \{abc, aab, abb, aac, acc, aaa, ddd\}.$$

The operators follow easily, e.g. $d'_0(abc) = dd$.

To extend simplicial homology from Section 4.3 to simplicial sets, we just need a chain complex. Let X be a simplicial set. The *nth chain group* $C_n(X)$ of X is the free Abelian group on K's set of oriented, non-degenerate, n-simplices. The *boundary homomorphism* $\partial_n \colon C_n \to C_{n-1}$ is the linear extension of

$$\partial_n = \sum_{i=0}^{n} (-1)^i d_i,$$

where d_i are the face operators and a degenerate face is treated as 0. The boundary homomorphism connects the chain groups into a chain complex, and homology follows.

EXAMPLE 6.5 (collapsed boundary). The face operators for our collapsed set in Example 6.4 give us the correct boundary. For instance, we have $d_0(abc) = dd$, $d_1(abc) = ac$, and $d_2(abc) = ab$, giving us $\partial_2(abc) = -ac + ab$, as dd is degenerate. Taking another boundary, we have

$$\partial_1 \partial_2(abc) = \partial_1(-ac) + \partial_1(ab) = -(d-a) + (d-a) = 0.$$

After the collapse, the simplex abc represents the 2-gon, and its boundary is still a 1-cycle.

We may now collapse simplices in a simplicial complex to get a smaller simplicial set to represent the unknown space of our point set. In practice, however, computing homology of geometric complex with the persistence algorithm exhibits linear time behavior [77]. For the cyclooctane dataset, constructing the complex of more than 3 million simplices takes 11.12 seconds in Section 3.6, while computing homology takes 13.35 seconds in Section 4.4. For larger complexes, we spend most of the time building and storing the complex, not computing its homology. We need reduction *during* construction, not *after*.

6.3. Tidy Sets. Tidy sets are clique complexes that are reduced during construction. Recall from Sections 3.3 and 3.4 that the VR and witness complexes are both clique complexes and are popular in topological analysis. Clique complexes, therefore, present an excellent model for reduction using simplicial sets.

Generally, we construct skeletons of clique complexes using *bottom-up* algorithms, as we did for the cyclooctane dataset in Section 3.6. Alternatively, we may compute the maximal cliques directly as they become maximal simplices in the clique complex [79]. Maximal simplices are a minimal description for a simplicial complex as their closure under the subset operation enumerates the full complex. We would need the full description for computing homology, but we may also reduce the complex via *top-down* reduction first.

Let Q and C be disjoint sets of maximal sets, and $\mathcal{X}(Q, C)$ be the simplicial set having the sets in Q as maximal simplices and the sets in C as collapsed maximal simplices. We use the tuple (Q, C) to denote $\mathcal{X}(Q, C)$. Initially, a clique complex is the simplicial set $K = \mathcal{X}(Q, \emptyset)$. A simplicial set X is *acyclic* if $H_0(X) \cong \mathbb{Z}$ and $H_n(X) \cong \{0\}$ for $n > 0$. Contractible spaces, such as simplices in a simplicial complex, are acyclic.

Given a clique complex represented as a tuple (Q, C), we perform two types of reductions that we will describe informally. A *leaf* is a simplex in a simplicial complex that has an acyclic intersection with the rest of the complex, the intersection being its "stem". The notion of a leaf may also be extended naturally to simplicial

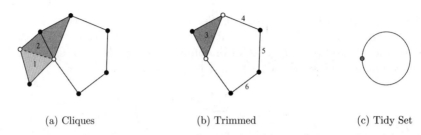

FIGURE 22. Starting with the cliques (a), we trim cliques 1 and 2, and thin cliques 3 through 6 (b) to get the tidy set (c), which is homotopy equivalent to the clique complex (a).

sets. Our first reduction is removing leaves, which preserves homology while keeping a simplicial complex in its category. For cubical complexes, removing leaves is called *shaving* for full-dimensional cubes [62]. Our second reduction is collapsing acyclic simplices, which also preserves homology, but may change the category to a simplicial set. Given tuple (Q, C) with $\sigma \in Q$, we have two reductions that we may perform easily.

$$\begin{array}{lll} \textit{trim:} & (Q, C) \mapsto (Q - \{\sigma\}, C) & \sigma, \text{ a leaf} \\ \textit{thin:} & (Q, C) \mapsto (Q - \{\sigma\}, C \cup \{\sigma\}) & \sigma, \text{ acyclic} \end{array}$$

Thinning is closely related to the construction of acyclic subspaces for homology computation [58]. A *tidy set* is a trimmed, then thinned, simplicial complex [80]. A tidy set is minimal with respect to trimming and thinning.

EXAMPLE 6.6. Figure 22 illustrates the construction of the tidy set for a small complex. We start with the set of maximal cliques (a), rendered as maximal simplices. The intersection of clique 1 with the rest of the cliques is the dashed edge with white vertices. Since this intersection is acyclic, clique 1 is a leaf and is removed. Clique 2 is similarly trimmed. Clique 3 (b), however, intersects the remaining cliques in the two white vertices, so it is not a leaf and cannot be trimmed. Instead, we thin cliques 3 through 6 in order to get the tidy set (c), a loop with one vertex and one edge. The tidy set has the homotopy type of the complex in (a). Compare with elementary contractions in Figure 19.

EXAMPLE 6.7. Figure 23 shows projections of the 1-skeletons of three homologous structures: A clique complex defined by 331 maximal cliques (a), its trimmed complex (b) with 88 remaining cliques, and its tidy set with 23 uncollapsed cliques (c).

We have algorithms for computing tidy sets, based on greedy trimming, and thinning in both simplicial complex and set categories [80]. Testing whether a simplex is acyclic or a leaf involves homology, and much of the algorithmic design involves postponing homology computation as long as it is possible.

6.4. Tidy Analysis. Having described tidy sets as an alternative to our two-step pipeline, we now analyze our cyclooctane dataset $\mathcal{S}$ once again. We begin with the neighborhood graph built in Section 3.6. Recall that the maximum scale is $\hat{\epsilon} = 0.4$ and the graph has 76,657 edges.

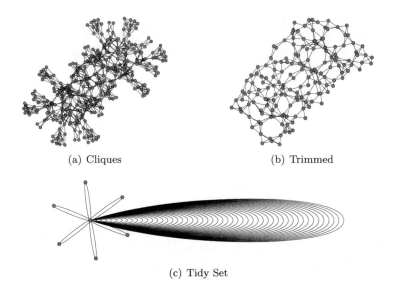

FIGURE 23. A clique complex (a) is trimmed (b) and thinned (c), resulting in its tidy set. We project 1-skeletons only.

We enumerate the maximal cliques in the neighborhood graph in 1.17 seconds using the IK-GX algorithm [**13**]. There are 23,279 maximal cliques, with the average size of a clique being about 8.68 and the maximum being 16, implying that the full VR complex is 15-dimensional.

Since we only built a 4-skeleton in Section 3.6, we first construct the 4-skeleton of the tidy set for comparison. From the maximal cliques, the tidy set algorithm trims 20,246 (87%) and collapses another 1,860 (8%). The remaining 1,173 cliques (5%) give rise to a 4-dimensional tidy set with 155,202 simplices, as compared to the 4-dimensional VR complex with more than 3 million simplices constructed earlier. The entire construction process, including clique enumeration, is 11.73 seconds. In another 0.36 seconds, we compute homology using the persistence algorithm. We find $\beta_0 = 1$, $\beta_1 = 1$, $\beta_2 = 2$, matching our earlier analysis results in Sections 4.3 and Section 5.5. But the tidy set is about 5% of the size of the VR complex, and our total analysis time drops from 24.47 to 12.09 seconds.

The reduction in size motivates us to construct the full tidy set instead of a low-dimensional skeleton. The largest uncollapsed clique in the tidy set has size 12, and we construct the full 11-dimensional tidy set in 11.99 seconds with 202,406 simplices. That is, we only require 0.26 seconds to construct an additional 47,204 simplices. By comparison, the 15-dimensional VR complex has more than 13 million simplices, requiring a computer with large memory for its construction, and even larger memory for its homology computation. Since our tidy set is 66 times smaller, we compute homology groups in all 11 dimensions in only another 0.52 seconds. We find $\beta_0 = 1$, $\beta_1 = 1$, $\beta_2 = 2$, as before, but also $\beta_n = 0$ for all $3 \leq n \leq 11$. The triviality of homology in all higher dimensions is a strong indication that the cyclooctane dataset has intrinsic dimension 2, which indeed is the case [**50**]. Although our only assumption is that the unknown space is topological, our analysis is yielding information about the intrinsic dimension of the dataset.

Due to the size of simplicial complexes, topological analysis has been limited to low dimensional features, such as components [17], tunnels [8], and voids [43]. The tidy set is the first method that enables topological analysis in higher dimensions. On the other hand, this method does not yield filtrations, so we cannot analyze tidy sets at multiple scales using persistent homology. There are several approaches, however, for multiscale analysis using tidy sets, such as zigzag persistence.

7. Conclusion

With its focus on qualitative information, topological data analysis is the first step toward a robust understanding of data. In this chapter, we looked at current multiscale structures and invariants for computing the topology of data. Swift advances in technology allow us to acquire high-resolution data, transmit it through fast networks, and store it on distributed cloud infrastructure. We are engulfed in heterogeneous scientific data without theory or algorithms for its analysis. To extract information from these massive datasets, we need multiscale, nonmonotonic, probabilistic models to represent their structure, as well as randomized and streaming algorithms for their analysis.

Acknowledgments

The author thanks Shawn Martin for generously providing the cyclooctane dataset S as well as its reconstructed surface.

References

[1] F. Aurenhammer, *Power diagrams: Properties, algorithms and applications*, SIAM Journal on Computing **16** (1987), 78–96.
[2] J.-D. Boissonnat, O. Devillers, and S. Hornus, *Incremental construction of the Delaunay triangulation and the Delaunay graph in medium dimension*, Proc. ACM Symposium on Computational Geometry, 2009, pp. 208–216.
[3] R. Bott and L. W. Tu, *Differential forms in algebraic topology*, Springer-Verlag, New York, NY, 1982.
[4] W. M. Brown, S. Martin, S. N. Pollack, E. A. Coutsias, and J.-P. Watson, *Algorithmic dimensionality reduction for molecular structure analysis*, Journal of Chemical Physics **129** (2008), no. 064118.
[5] G. Carlsson, *Topology and data*, Bulletin of the American Mathematical Society (New Series) **46** (2009), no. 2, 255–308.
[6] G. Carlsson and V. de Silva, *Zigzag persistence*, Foundations of Computational Mathematics **10** (2010), 367–405.
[7] G. Carlsson, V. de Silva, and D. Morozov, *Zigzag persistent homology and real-valued functions*, Proc. ACM Symposium on Computational Geometry, 2009, pp. 247–256.
[8] G. Carlsson, T. Ishkhanov, V. de Silva, and A. Zomorodian, *On the local behavior of spaces of natural images*, International Journal of Computer Vision **76** (2008), no. 1, 1–12.
[9] G. Carlsson, G. Singh, and A. Zomorodian, *Computing multidimensional persistence*, Journal of Computational Geometry **1** (2010), no. 1, 72–100.
[10] G. Carlsson and A. Zomorodian, *The theory of multidimensional persistence*, Discrete & Computational Geometry **42** (2009), no. 1, 71–93.
[11] G. Carlsson, A. Zomorodian, A. Collins, and L. J. Guibas, *Persistence barcodes for shapes*, International Journal of Shape Modeling **11** (2005), no. 2, 149–187.
[12] F. Cazals, J. Giesen, M. Pauly, and A. Zomorodian, *The conformal alpha shape filtration*, The Visual Computer **22** (2006), no. 8, 531–540.
[13] F. Cazals and C. Karande, *Reporting maximal cliques: new insights into an old problem*, Research Report 5642, INRIA, 2005.
[14] CGAL, *Computational Geometry Algorithms Library*, http://www.cgal.org.
[15] CHomP, *Computational Homology Project*, 2011, http://chomp.rutgers.edu/.

[16] Y. Civan and E. Yalçin, *Linear colorings of simplicial complexes and collapsing*, Journal of Combinatorial Theory Series A **114** (2007), no. 7, 1315–1331.
[17] A. Collins, A. Zomorodian, G. Carlsson, and L. Guibas, *A barcode shape descriptor for curve point cloud data*, Computers & Graphics **28** (2004), no. 6, 881–894.
[18] T. H. Cormen, C. E. Leiserson, R. L. Rivest, and C. Stein, *Introduction to algorithms*, third ed., The MIT Press, Cambridge, MA, 2009.
[19] D. A. Cox, J. Little, and D. O'Shea, *Using algebraic geometry*, second ed., Graduate Texts in Mathematics, vol. 185, Springer-Verlag, New York, 2005.
[20] M. de Berg, O. Cheong, M. van Kreveld, and M. Overmars, *Computational geometry: Algorithms and applications*, third ed., Springer-Verlag, New York, 2008.
[21] V. de Silva and G. Carlsson, *Topological estimation using witness complexes*, Proc. IEEE/Eurographics Symposium on Point-Based Graphics, 2004, pp. 157–166.
[22] H. Derksen and J. Weyman, *Quiver representations*, Notices of the American Mathematical Society **52** (2005), no. 2, 200–206.
[23] T. K. Dey, *Curve and surface reconstruction*, Cambridge Monographs on Applied and Computational Mathematics, vol. 23, Cambridge University Press, New York, NY, 2007.
[24] D. L. Donoho and C. Grimes, *Hessian eigenmaps: Locally linear embedding techniques for high-dimensional data*, Proc Natl Acad Sci USA **100** (2003), no. 10, 5591–5596.
[25] J.-G. Dumas, F. Heckenbach, B. D. Saunders, and V. Welker, *Computing simplicial homology based on efficient Smith normal form algorithms*, Algebra, Geometry, and Software Systems, 2003, pp. 177–207.
[26] D. Dummit and R. Foote, *Abstract algebra*, third ed., John Wiley & Sons, Inc., New York, NY, 2004.
[27] H. Edelsbrunner, D. G. Kirkpatrick, and R. Seidel, *On the shape of a set of points in the plane*, IEEE Transactions on Information Theory **29** (1983), 551–559.
[28] H. Edelsbrunner and E. P. Mücke, *Three-dimensional alpha shapes*, ACM Transactions on Graphics **13** (1994), 43–72.
[29] H. Edelsbrunner and N. R. Shah, *Incremental topological flipping works for regular triangulations*, Proc. ACM Symposium on Computational Geometry, 1992, pp. 43–52.
[30] H. Edelsbrunner and A. Zomorodian, *Computing linking numbers in a filtration*, Homology, Homotopy and Applications **5** (2003), no. 2, 19–37.
[31] FIRST *(Floppy Inclusions and Rigid Substructure Topography)*, http://flexweb.asu.edu/software/.
[32] *Folding@Home: Distributed Computing*, http://folding.stanford.edu/.
[33] P. Gabriel, *Unzerlegbare Darstellungen I*, manuscripta mathematica **6** (1972), no. 1, 71–103, (German).
[34] P. Gabriel and A. V. Roiter, *Representations of finite-dimensional algebras*, Springer-Verlag, New York, NY, 1997.
[35] R. Ghrist, *Barcodes: the persistent topology of data*, Bulletin of the American Mathematical Society (New Series) **45** (2008), no. 1, 61–75.
[36] R. Ghrist and A. Muhammad, *Coverage and hole-detection in sensor networks via homology*, Proc. International Symposium on Information Processing in Sensor Networks, 2005.
[37] M. Gromov, *Hyperbolic groups*, Essays in Group Theory (S. Gersten, ed.), Springer-Verlag, New York, NY, 1987, pp. 75–263.
[38] A. Hatcher, *Algebraic topology*, Cambridge University Press, New York, NY, 2002, http://www.math.cornell.edu/~hatcher/AT/ATpage.html.
[39] D. J. Jacobs, A. J. Rader, L. A. Kuhn, and M. F. Thorpe, *Protein flexibility prediction using graph theory*, Proteins: Structure, Function, and Genetics **44** (2001), 150–165.
[40] I. T. Jolliffe, *Principle component analysis*, second ed., Springer-Verlag, New York, NY, 2002.
[41] V. G. Kac, *Infinite root systems, representations of graphs and invariant theory*, Inventiones Mathematicae **56** (1980), no. 1, 57–92.
[42] T. Kaczynski, K. Mischaikow, and M. Mrozek, *Computational homology*, Springer-Verlag, New York, NY, 2004.
[43] P. M. Kasson, A. Zomorodian, S. Park, N. Singhal, L. J. Guibas, and V. S. Pande, *Persistent voids: a new structural metric for membrane fusion*, Bioinformatics **23** (2007), no. 14, 1753–1759.
[44] N. Katoh and S-i. Tanigawa, *A proof of the molecular conjecture*, Proc. ACM Symposium on Computational Geometry, 2009, pp. 296–305.

[45] F. Klein, *A comparative review of recent researches in geometry*, Bull. New York Math. Soc. **2** (1892–1893), 215–249, Translated by M. W. Haskell.
[46] D. Kozlov, *Combinatorial algebraic topology*, Springer-Verlag, New York, NY, 2008.
[47] T. Mamonova, B. Hespenheide, R. Straub, M. F. Thorpe, and M. Kurnikova, *Protein flexibility using constraints from molecular dynamics simulations*, Physical Biology **2** (2005), S137–S147.
[48] A. A. Markov, *Insolubility of the problem of homeomorphy*, Proc. International Congress of Mathematics, 1958, pp. 14–21.
[49] S. Martin, A. Thompson, E. A. Coutsias, and J.-P. Watson, *Topology of cyclo-octane energy landscape*, Journal of Chemical Physics **132** (2010), no. 234115.
[50] S. Martin and J.-P. Watson, *Non-manifold surface reconstruction from high-dimensional point cloud data*, Computational Geometry: Theory & Applications **44** (2011), no. 8, 427–441.
[51] J. Matoušek, *LC reductions yield isomorphic simplicial complexes*, Contributions to Discrete Mathematics **3** (2008), no. 2, 37–39.
[52] Y. Matsumoto, *An introduction to Morse theory*, Iwanami Series in Modern Mathematics, vol. 208, American Mathematical Society, Providence, RI, 2002.
[53] J. P. May, *Simplicial objects in algebraic topology*, D. Van Nostrand Co., Inc., Princeton, NJ, 1967.
[54] E. W. Mayr, *Some complexity results for polynomial ideals*, Journal of Complexity **13** (1997), no. 3, 303–325.
[55] N. Milosavljević, D. Morozov, and P. Skraba, *Zigzag persistent homology in matrix multiplication time*, Proc. ACM Symposium on Computational Geometry, 2011, pp. 216–225.
[56] D. Morozov, *Dionysus*, http://www.mrzv.org/software/dionysus/.
[57] D. M. Mount and S. Arya, *ANN: A library for approximate nearest neighbor searching*, version 1.1.1, http://www.cs.umd.edu/~mount/ANN/.
[58] M. Mrozek, P. Pilarczyk, and N. Żelazna, *Homology algorithm based on acyclic subspace*, Computers and Mathematics with Applications **55** (2008), no. 11, 2395–2412.
[59] National Library of Medicine, *The Visible Human Project*, http://www.nlm.nih.gov/research/visible/visible_human.html/.
[60] P. Niyogi, S. Smale, and S. Weinberger, *Finding the homology of submanifolds with high confidence from random samples*, Discrete & Computational Geometry **39** (2008), no. 1, 419–441.
[61] NOAA Satellite and Information Service, *National Geophysical Data Center*, http://www.ngdc.noaa.gov/.
[62] P. Pilarczyk, *Computer assisted method for proving existence of periodic orbits*, Topological Methods in Nonlinear Analysis **13** (1999), 365–377.
[63] *RedHom*, http://redhom.ii.uj.edu.pl/.
[64] J. J. Rotman, *An introduction to algebraic topology*, Springer-Verlag, New York, NY, 1988.
[65] S. Roweis and L. K. Saul, *Nonlinear dimensionality reduction by locally linear embedding*, Science **290** (2000), no. 5500, 2323–2326.
[66] H. Sexton and M. V. Johansson, *JPlex*, http://comptop.stanford.edu/programs/jplex/.
[67] *The Stanford 3D Scanning Repository*, http://www-graphics.stanford.edu/data/3Dscanrep/.
[68] A. Storjohann, *Near optimal algorithms for computing Smith normal forms of integer matrices*, Proc. International Conference on Symbolic and Algebraic Computation, 1996, pp. 267–274.
[69] _____, *Computing Hermite and Smith normal forms of triangular integer matrices*, Linear Algebra and Its Applications **282** (1998), no. 1–3, 25–45.
[70] T. S. Tay and W. Whiteley, *Recent advances in the generic rigidity of structures*, Structural Topology **9** (1984), 31–38.
[71] G. Turk and M. Levoy, *Zippered polygon meshes from range images*, Proc. SIGGRAPH, 1994, pp. 311–318.
[72] F. Uhlig, *Transform linear algebra*, Prentice Hall, Upper Saddle River, NJ, 2002.
[73] R. Vidal, Y. Ma, and S. Sastry, *Generalized principal component analysis*, IEEE Trans. Pattern Anal. Mach. Intell. **27** (2005), no. 12, 1945–1959.
[74] L. Vietoris, *Über den höheren zusammenhang kompakter Räume und eine Klasse von zusammenhangstreuen Abbildungen*, Mathematische Annalen **97** (1927), no. 1, 454–472.

[75] J. H. C. Whitehead, *Simplicial spaces, nuclei, and m-groups*, Proceedings of the London Mathematical Society **s2-45** (1939), no. 1, 243–327.

[76] C. K. Yap, *Robust geometric computation*, Handbook of Discrete and Computational Geometry (J. E. Goodman and J. O'Rourke, eds.), CRC Press, LLC, Boca Raton, FL, second ed., 2004, pp. 927–952.

[77] A. Zomorodian, *Topology for computing*, paperback ed., Cambridge University Press, New York, NY, 2009.

[78] _____, *Computational topology*, Algorithms and Theory of Computation Handbook (M. Atallah and M. Blanton, eds.), vol. 2, Chapman & Hall/CRC Press, Boca Raton, FL, second ed., 2010.

[79] _____, *Fast construction of the Vietoris-Rips complex*, Computers & Graphics **34** (2010), no. 3, 263 – 271.

[80] _____, *The tidy set: A minimal simplicial set for computing homology of clique complexes*, Proc. ACM Symposium of Computational Geometry, 2010, pp. 257–266.

[81] A. Zomorodian and G. Carlsson, *Computing persistent homology*, Discrete & Computational Geometry **33** (2005), no. 2, 249–274.

[82] _____, *Localized homology*, Computational Geometry: Theory & Applications **41** (2008), no. 3, 126–148.

THE D. E. SHAW GROUP, NEW YORK, NY

Topological Dynamics: Rigorous Numerics via Cubical Homology

Marian Mrozek

ABSTRACT. We review techniques of computer assisted proofs in dynamics based on topological tools, rigorous numerics and computational homology.

1. Introduction

Analysis and design in science and engineering heavily depends on mathematical models of natural and technical processes. The use of models provides a cost-effective alternative to experimentation and prototyping. However, analysis of the mathematical models is often not possible due to the complexity of the problems and consequently digital computers are used to study the models. This works perfectly when the models are based on discrete mathematics, because then integer arithmetic is fully sufficient. But most of the models take the form of a differential equation which requires the use of numerical approximation schemes and floating-point arithmetic. In consequence, the adequateness of approximation depends only on limit theorems. Such theorems, constituting the essence of classical numerical analysis, are often not easy to prove and usually apply only to finite time spans. However, in most problems the asymptotic behaviour of the system is of interest and there are examples of asymptotic behaviour present in numerical schemes which may be proved to be non existent in the system approximated. Moreover, in the standard floating-point arithmetic used in traditional numerical schemes rounding errors cannot be controlled. This makes the approach far from being fully reliable.

A method of controlling rounding errors, known as interval arithmetic (see [47, 28]), is almost as old as digital computers. However, the size of intervals grows exponentially with the number of operations and consequently the method is extremely greedy with respect to the computational power. But, the constant progress in the efficiency of digital computers together with the development of

---

2010 *Mathematics Subject Classification.* Primary 37B30, 37M99, 37D45, 55-04, 55N35; Secondary 34C28, 34C45, 52B99.

Key words and phrases. computer assisted proof, rigorous numerics, interval arithmetic, chaotic dynamics, symbolic dynamics, Conley index, index pair, cubical set, homology algorithm, reduction methods.

Partially supported by Polish MNSzW, Grant N N201 419539.

new, non-standard techniques in numerical analysis bring more and more rigorous numerical studies of continuous mathematical models. In particular, theorems about the models are obtained by means of computer assisted proofs. Among many problems addressed by the new methods is the asymptotic study of dynamical systems. In this review we present one of the streams of the modern rigorous numerical analysis of dynamical systems based on topological tools, with particular emphasis on homology algorithms.

Throughout the paper we denote the sets of reals, rationals, integers and natural numbers respectively by $\mathbb{R}, \mathbb{Q}, \mathbb{Z}, \mathbb{N}$. We use the superscript $+$ or $-$ in this notation to denote numbers which are respectively non-negative or non-positive.

2. Dynamical Systems

2.1. Preliminaries. Let X be a topological space and let T be an element of $\{\mathbb{R}, \mathbb{R}^+, \mathbb{Z}, \mathbb{Z}^+\}$. We consider the elements of T as time: continuous if $T \in \{\mathbb{R}, \mathbb{R}^+\}$ or discrete if $T \in \{\mathbb{Z}, \mathbb{Z}^+\}$.

A *(semi)dynamical system* is a continuous map $\varphi : X \times T \to X$ such that for any $x \in X$ and $s, t \in T$

$$\varphi(\varphi(x,t),s) = \varphi(x, s+t),$$
$$\varphi(x, 0) = x.$$

In the case when $T = \mathbb{R}$ the dynamical system is often referred to as a *flow* and when $T = \mathbb{Z}^+$ as a *discrete semidynamical system* (dsds). The map $\varphi_t : X \ni x \to \varphi(x, t) \in X$ is called the *t-translation map*. In the case of discrete time the (semi)dynamical system φ may be identified with the *generator* of φ, i.e. $f := \varphi_1$. This is because in this case $\varphi(x, n) = f^n(x)$, the nth iterate of f at x.

One of the important sources of flows are autonomous ordinary differential equations while their discretizations such as Euler or Runge-Kutta methods induce discrete dynamical systems.

The *trajectory (orbit)* of $x \in X$ is the set $\varphi(x) := \{\varphi(x,t) \mid t \in T\}$. A point $x \in X$ is called a *stationary* point if $\varphi(x) = x$. It is called a *periodic* point if there exists a $t \in T^+$ such that $\varphi(x, t) = x$. The *invariant part* of $N \subset X$ is defined as the set

$$\operatorname{Inv}(N, \varphi) := \{x \in N \mid \varphi(x) \subset N\}.$$

A subset $A \subset X$ is called *invariant* if $\operatorname{Inv}(A, \varphi) = A$. The *alpha and omega limit sets* of an $x \in X$ are defined by

$$\alpha(x) := \{y \in X \mid \exists t_n \to -\infty \text{ s.t. } y = \lim \varphi(x, t_n)\},$$
$$\omega(x) := \{y \in X \mid \exists t_n \to +\infty \text{ s.t. } y = \lim \varphi(x, t_n)\}.$$

Note that alpha and omega limit sets are invariant.

Given a flow $\varphi : X \times \mathbb{R}^+ \to X$ and a closed subset A of X put

$$A_0 := \{x \in A \mid \exists t > 0 \; \varphi(x, (0,t)) \cap A = \emptyset \text{ and } \varphi(x,t) \in A\}.$$

One defines the *return time map* of A by

$$t_{\varphi,A} : A_0 \ni x \to \sup\{t \in \mathbb{R}^+ \mid \varphi(x, (0,t]) \cap A = \emptyset\}.$$

It is then straightforward to verify that the following map

$$P_{\varphi,A} : \{x \in A_0 \mid t_{\varphi,A}(x) < \infty\} \ni x \to \varphi(x, t_{\varphi,A}(x)) \in A.$$

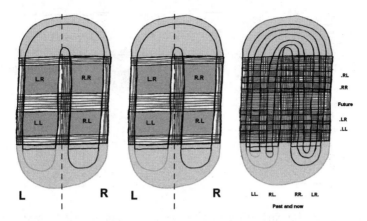

FIGURE 1. The domain (left), the first iterate (middle) and the second iterate (right) of the Smale horseshoe map. The hatched vertical strips indicate the parts of the square N in the domain which do not belong to the image of the map. Similarly, the hatched horizontal strips indicate the parts of N which do not belong to the inverse image of the map. What is left consists of four squares (middle) and sixteen squares (right). The trajectory of each point in the invariant part of N is in one-to-one correspondence with a sequence of symbols L and R indicating whether the respective iterate of the point is in the left or right half-plane.

is well defined. It is called the *Poincaré map* associated with A. We say that A is a *Poincaré section* if $P_{\varphi,A}$ has non-empty domain and is continuous.

2.2. Chaotic dynamics. Consider
$$\Sigma_k := \{0, 1, 2, \ldots k-1\}^{\mathbb{Z}}$$
as a metric space with the metric
$$d(\alpha, \beta) := \sum_{i=-\infty}^{\infty} \frac{1 - \delta_{\alpha(i),\beta(i)}}{2^{|i|}},$$
where δ_{mn} stands for the Kronecker delta. A *full shift* on k symbols is the discrete dynamical system generated on Σ_k by
$$\sigma : \Sigma_k \ni \alpha \to \sigma(\alpha) := (n \to \alpha(n+1)) \in \Sigma_k.$$

It is straightforward to observe that every finite sequence of symbols in Σ_k is in one-to-one correspondence with a periodic point of σ. This is one of the characteristic features of chaotic dynamics. Another feature is positive topological entropy (see [1] for the definition of topological entropy). In the case of full shift on k symbols the topological entropy is $\log k$. In 1967 Stephen Smale observed that chaotic dynamics may be also present in case of a plane homeomorphism. He considered a map which squishes a square, then stretches the result into a long strip, and finally folds the strip into the shape of a horseshoe (see Fig. 1). Such a map may be extended to a homeomorphism of the plane.

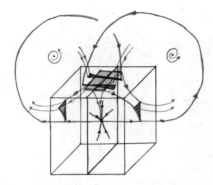 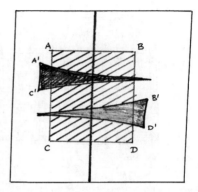

FIGURE 2. Linearization of the Lorenz equations (left) and the Poincaré map of the top face of a cubical neighborhood of the origin (right).

THEOREM 2.1. [40] *Let N denote the square part of the domain of the horseshoe map h. Then there exists a homeomorphism $\rho : \mathrm{Inv}(N, h) \to \Sigma_2$ such that $\sigma\rho = \rho h$.*

The map satisfying the assertion of Theorem 2.1 is called a *conjugacy*. Conjugacy preserves topological entropy and maps periodic points to periodic points.

In 1950s Edward Norton Lorenz, an American mathematician and meteorologist, became skeptical of the appropriateness of the mathematical models used in meteorology. In a 1963 paper [20] he presented a set of differential equations, nowadays known as the Lorenz equations:

(2.1) $$\begin{cases} \dot{x} = \sigma(y - x), \\ \dot{y} = Rx - y - xz, \\ \dot{z} = xy - bz. \end{cases}$$

He studied numerically these equations and observed very sensitive dependence on initial conditions: trajectories originating even very closely quickly diverged showing no resemblance in the future.

The linearization of the Lorenz equations at the origin combined with numerical estimates of the trajectories outside a cubical neighborhood at the origin (see Fig. 2) leads to the observation that the trajectories of the flow return to the top face of the cube in the pattern reminiscent of the behaviour of Smale horseshoe map [41]. This raises several questions: is horseshoe dynamics present in the Lorenz system? or maybe it is only present in the numerical scheme? or the observed chaotic behaviour of the trajectories is only the consequence of the rounding errors?

Lorenz ran many numerical experiments to get understanding of the unusual behaviour of equations (2.1). On the basis of these experiments he conjectured that the observed phenomenon is an inherent feature of the equations and not a consequence of approximations and roundings present in the numerical simulation.

2.3. Ghost solutions. The need to distinguish between the dynamics of the original system and the numerical scheme used to do numerical simulations is justified by the following simple example. The differential equation in the complex plane

$$z' = (\alpha i - |z|)z, \quad z \in \mathbb{C}$$

has only one periodic trajectory: the stationary point at the origin. However, its Euler discretization
$$\Phi_h(z) := z(1 + h(\alpha i - |z|))$$
for every $h > 0$ exhibits invariant circles of radius
$$r_\pm := \frac{1 \pm \sqrt{1 - h^2\alpha^2}}{h}.$$
Another example is provided by the logistic equation given by
$$y' = y(1 - y).$$
It may be solved explicitly and has very simple dynamics, far from any chaotic behaviour. However, Hale and Koçak [16] proved that the two step numerical scheme
$$\Phi_{h,\lambda}\begin{pmatrix} y_1 \\ y_2 \end{pmatrix} := \begin{pmatrix} \frac{1-\lambda}{1+\lambda}y_2 + \frac{2\lambda}{1+\lambda}y_1 + 2hy_1(1 - y_2) \\ y_1 \end{pmatrix}$$
contains an invariant subset conjugate to the Smale horseshoe for every $h > 0$.

The trajectories which are present in the dynamics of a discretization of a differential equation but disappear in the dynamics of the flow induced by the equations are called *ghost solutions* [43].

The first rigorous result showing that the horseshoe dynamics observed in numerical experiments with the Lorenz system is not a ghost solution was the following theorem.

THEOREM 2.2. [23, 24] *Consider the Lorenz equations (2.1) and put*
$$P := \{(x, y, z) \in \mathbb{R}^3 \mid z = 53\}.$$
For all parameter values in a sufficiently small neighborhood of $(\sigma, R, b) = (45, 54, 10)$, *there exists a Poincaré section* $N \subset P$ *such that the Poincaré map g induced by (1) is Lipschitz and well defined. Furthermore, there exists a $d \in \mathbb{N}$ and a continuous surjection* $\rho : \mathrm{Inv}(N, g) \to \Sigma_2$ *such that*
$$\rho \circ g^d = \sigma \circ \rho$$
where $\sigma : \Sigma_2 \to \Sigma_2$ *is the full shift dynamics on two symbols.*

The proof of Theorem 2.2 is based on rigorous numerics and topological tools. The scheme of the proof is presented in Figure 3. The goal of rigorous numerics is to perform numerical simulations in such a way that rigorous conclusions can be drawn about the features of the output of simulations with respect to the studied dynamical system. This requires exact bounds for the errors resulting from the time and space discretization. Such an approach is in contrast to classical numerics, which concentrates only around limit theorems about the numerical schemes.

Rigorous numerics itself is not sufficient to draw rigorous conclusions about the studied dynamical system. For this, another tool is needed, which allows to tie the features of the output of rigorous numerical computations with the features of the original system. The tool used in the proof of Theorem 2.2 is the topological invariant of dynamical systems called Conley index. We present a brief summary of the ideas of rigorous numerics in Section 2.4 and a review of the Conley index theory in Section 2.5.

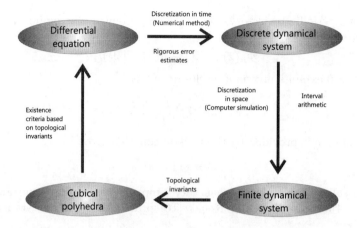

FIGURE 3. Computer assisted proofs in qualitative analysis of differential equations.

2.4. Rigorous numerics. The fundamental tool of rigorous numerics is *interval arithmetic*, first proposed by M. Warmus [**47**] in 1956 and later rediscovered and popularised by R.E. Moore [**28**] in 1959. It is defined as follows. Let $\hat{\mathbb{R}} \subset \bar{\mathbb{R}}$ be a fixed, finite set of *representable numbers*. We define the set of *representable intervals* by

$$\mathcal{I} := \{[a,b] \mid a, b \in \hat{\mathbb{R}}, a \leq b\}.$$

For $\diamond \in \{+, -, *, /\}$ and $I, J \in \mathcal{I}$ we denote by $I \diamond J$ the smallest representable interval that contains

$$\{a \diamond b \mid a \in I, b \in J\}.$$

Note that given the endpoints of I and J, one can easily provide an algorithm constructing the endpoints of $I \diamond J$. Intervals provide rigorous bounds for roundings which inevitably show up when arithmetic operations are performed on a finite subset of real numbers available in the digital processor. More precisely, let $f : \mathbb{R}^m \multimap \mathbb{R}^n$ be a rational function. The *interval extension* of f, is a map $[f] : \mathcal{I}^m \multimap \mathcal{I}^n$ obtained by replacing the arithmetic operations in f with their interval counterparts. The interval extension is used in the following result.

THEOREM 2.3. [**28**] *Assume* $D \subset \mathbb{R}^m$ *and* $f : D \to \mathbb{R}^n$ *is a rational function. Then for any intervals* $\mathbf{x}_1, \ldots, \mathbf{x}_n \in \operatorname{dom}[f]$ *and* $x_i \in \mathbf{x}_i$ *we have*

$$f(x_1, \ldots, x_m) \in [f](\mathbf{x}_1, \ldots, \mathbf{x}_m).$$

Now, if $f : D \to \mathbb{R}^n$ is an arbitrary function for which we have a rational approximation $g : D \to \mathbb{R}^n$ such that for $(x_1, \ldots x_m) \in D$ and some $\mathbf{w} \in \mathcal{I}^n$

$$f(x_1, \ldots, x_m) - g(x_1, \ldots, x_m) \in \mathbf{w},$$

then we have the following estimate

$$f(\mathbf{x}) \subset [g](\mathbf{x})[+]\mathbf{w}.$$

The simplest case when the outcome of interval evaluations of a function may be translated into a rigorous conclusion about the properties of the function is the question of the existence of a zero of a continuous rational function $f : [a, b] \to \mathbb{R}$. If the interval evaluation of f at the endpoints does not contain zero then the signs

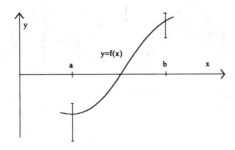

FIGURE 4. Interval estimates of the values of a continuous rational function at the endpoint of a an interval leading to the conclusion of the existence of a zero.

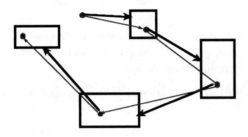

FIGURE 5. Interval enclosures as a multivalued map.

of $f(a)$ and $f(b)$ may be rigorously determined. Thus, if the signs turn out to be opposite, the function has a zero by the Darboux property (see Fig. 4).

Unfortunately, not many questions may be reduced just to the existence of a zero of a continuous function. To extend the range of applicability of methods based on rigorous numerics, it is convenient to view the interval enclosure of a map $f : \mathbb{R}^n \to \mathbb{R}^m$ as a multi-valued map (see Fig. 5).

Let X, Y be topological spaces. A *multivalued map* $F : X \rightrightarrows Y$ from X to Y is a function $F : X \to 2^Y$ from X to subsets of Y. The *image* of $A \subset X$ is

$$F(A) := \bigcup_{x \in A} F(x).$$

The *weak preimage* of $B \subset Y$ under F is

$$F^{-1}(B) := \{x \in X \mid F(x) \cap B \neq \emptyset\}.$$

F is *upper semicontinuous* if $F^{-1}(B)$ is closed for any closed set $B \subset Y$, and it is *lower semicontinuous* if the set $F^{-1}(U)$ is open for any open set $U \subset Y$.

2.5. Topological invariants of dynamical systems. One of the first topological tools used in the qualitative study of dynamical systems was Ważewski's Theorem [48]. Recall that $A \subset X$ is a *deformation retract* of X if there exists a map $r : X \to A$ homotopic to the identity on X such that $r_{|A} = \text{id}_{|A}$. Given a flow φ on X we define the *exit set* of $N \subset X$, denoted N^-, by

$$\{x \in N \mid \exists \epsilon > 0 : \varphi(x, t) \notin N \text{ for } 0 < t < \epsilon\}.$$

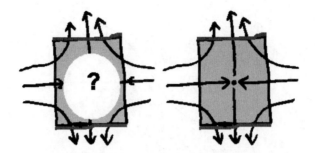

FIGURE 6. An isolating neighborhood with the exit set at the bottom and top side of the box (left). Since the assumptions of Ważewski's Theorem are satisfied, whatever we do to complete the missing dynamics we need to add at least one trajectory entirely contained in the isolating neighborhood (right).

A compact set N is called an *isolating block* if N^- is closed.

In terms of the concept of isolating block and exit set, Ważewski's Theorem (see Fig. 6) may be formulated as follows.

THEOREM 2.4. [48] *If N is an isolating block and N^- is not a deformation retract of N, then there exists an $x \in N$ such that $\varphi(x) \subset N$.*

Charles Conley [9] developed the concepts from Ważewski's Theorem into an index theory. We say that a compact set N is an *isolating neighborhood* iff
$$x \in \operatorname{bd} N \;\Rightarrow\; \varphi(x) \not\subset N.$$
One easily verifies that N is an isolating neighborhood iff $\operatorname{Inv}(N,\varphi) \subset \operatorname{int} N$. A compact set $S \subset X$ is called an *isolated invariant set* if there exists an isolating neighborhood N such that $S = \operatorname{Inv}(N,\varphi)$.

THEOREM 2.5. [9] *For every isolating neighborhood N of S there exists an isolating block M such that $S \subset M \subset N$. If M_1 and M_2 are two such blocks, then $(M_1/M_1^-, [M_1^-])$ and $(M_2/M_2^-, [M_2^-])$ are homotopy equivalent and, in particular,*
$$H^*(M_1, M_1^-) \cong H^*(M_2, M_2^-),$$
where H^* stands for the functor of Alexander-Spanier cohomology.

Theorem 2.5 allows us to define the *homotopy Conley index* of S and N as the homotopy type of $(M_1/M_1^-, [M_1^-])$ for any isolating block satisfying $S \subset M \subset N$ (see Fig. 7) and the *cohomological Conley index* of S and N as
$$\operatorname{Con}^*(N,\varphi) := \operatorname{Con}^*(S,\varphi) := H^*(M, M^-).$$

In order to define the Conley index for a discrete time dynamical system denote by $f : X \to X$ the *generator* of φ, i.e. $f(x) := \varphi(x,1)$. We say that a function $\sigma : \mathbb{Z} \to N$ is a *solution* to f through x in $N \subset X$ if $\sigma(0) = x$ and $f(\sigma(n)) = \sigma(n+1)$ for all $n \in \mathbb{Z}$. The *invariant part* of $N \subset X$ is defined by
$$\operatorname{Inv}(N, f) := \{\, x \in N \mid \exists \sigma : \mathbb{Z} \to N \text{ a solution to } f \text{ through } x \,\}$$
A compact set $S \subset X$ is called an *isolated invariant set* for f if there exists a compact neighborhood N of S such that $S = \operatorname{Inv}(N,\varphi) \subset \operatorname{int} N$. Then, N is called an *isolating neighborhood* (for S).

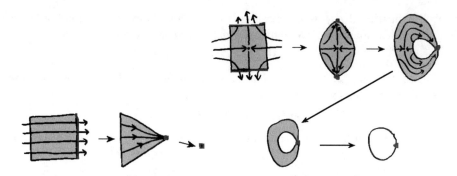

FIGURE 7. Sample computation of the homotopy Conley index: homotopy type of a point (left) and a circle (right)

A pair of compact sets $P = (P_1, P_2)$ is called an *index pair* for f and an isolated invariant set S if the following three conditions are satisfied

$$f(P_2) \cap P_1 \subset P_2,$$
$$P_1 \cap \operatorname{cl}(f(P_1) \setminus P_1) \subset P_2,$$
$$S = \operatorname{Inv}(\operatorname{cl}(P_1 \setminus P_2), f) \subset \operatorname{int}(P_1 \setminus P_2).$$

An index pair in the discrete case plays a similar role to an isolating block in the continuous case. However, $H^*(P_1, P_2)$ is not an invariant as one might expect. To get an invariant we need to introduce one more concept. To simplify the presentation, from now on we consider the cohomology with rational coefficients. and we denote by $\mathcal{V}_0$ the category of finite dimensional vector spaces over $\mathbb{Q}$. Let $\operatorname{Endo}(\mathcal{V}_0)$ denote the *category of endomorphisms* of $\mathcal{V}_0$, i.e. the category whose objects are pairs (V, v) consisting of a finite dimensional vector space and a distinguished endomorphism $v : V \to V$ and whose morphisms are morphisms in $\mathcal{V}_0$ which commute with the distinguished endomorphisms. An important full subcategory of $\operatorname{Endo}(\mathcal{V}_0)$ is $\operatorname{Auto}(\mathcal{V}_0)$ consisting of objects whose distinguished endomorphisms are isomorphisms (automorphisms). Now, let V be an object of $\mathcal{V}_0$. We define the *generalized kernel* of an endomorphism $\alpha \in \mathcal{V}_0(V, V)$ by

$$\operatorname{gker} \alpha := \bigcup_{n \in \mathbb{N}} \ker \alpha^n$$

and the *Leray functor* $L : \operatorname{Endo}(\mathcal{V}_0) \to \operatorname{Auto}(\mathcal{V}_0)$ by

$$L(V, v) := (V/\operatorname{gker} v, v')$$

where v' denotes the quotient endomorphism

$$v' : V/\operatorname{gker} \ni [x] \mapsto [v(x)] \in V/\operatorname{gker}.$$

A quadruple $P = (P_1, P_2, \bar{P}_1, \bar{P}_2)$ is an *index quadruple* for f and S if (P_1, P_2) is an index pair for f and S and $(\bar{P}_1, \bar{P}_2)$ is a topological pair such that the maps

$$f_{P\bar{P}} : (P_1, P_2) \ni x \to f(x) \in (\bar{P}_1, \bar{P}_2)$$
$$\iota_{P\bar{P}} : (P_1, P_2) \ni x \to x \in (\bar{P}_1, \bar{P}_2)$$

are well defined and $\iota_{P\bar{P}}$ is an excision (induces an isomorphism in cohomology). Given an index quadruple, we define the *index map* as the composition

$$I_P := H^*(f_{P\bar{P}}) \circ H^*(\iota_{P\bar{P}})^{-1}.$$

THEOREM 2.6. **[29, 30, 31]** *For every isolating neighborhood N of f there exists an index quadruple P such that*

$$\operatorname{Inv}(N, f) \subset P_1 \subset \bar{P}_1 \subset N.$$

Moreover, if P and Q are two such quadruples, then the objects $L(H^*(P_1, P_2), I_P)$ and $L(H^*(Q_1, Q_2), I_Q)$ in $\operatorname{Auto}(\mathcal{V}_0)$ are isomorphic in this category.

Now, we are able to define the cohomological *Conley index* of f in N by

$$\operatorname{Con}(N, f) := (CH^*(N, f), \chi(N, f)) := L(H^*(P_1, P_2), I_P),$$

where P is any index quadruple for f and S. Note that the choice of Alexander-Spanier cohomology is just to simplify this definition, because otherwise extra assumptions are needed to guarantee that $\iota_{P\bar{P}}$ induces an isomorphism. By modifying slightly the definition of index pair an analogous Conley index theory may be built on the basis of any homology or cohomology theory. One can show that index pairs satisfying the extra assumptions exist and actually are not particularly more difficult to find.

Historically, the first definition of Conley index for discrete dynamical system, based on shape theory, was proposed by Robbin and Salamon **[39]** in 1988.

A pictorial sketch how the Conley index of the horseshoe map may be computed is presented in Figure 8. We take an index pair (P_1, P_2) where P_1 is the square in the left top par of the figure and P_2 is a subset of P_1 consisting of the three strips marked with horizontal lines. The homology $H_1(P_1, P_2)$ has two generators α and β. The map f stretches α across α and β and stretches β across α and β reversing orientation. Therefore, the index map has the matrix

$$A := \begin{bmatrix} 1 & -1 \\ 1 & -1 \end{bmatrix}.$$

One easily verifies that $A^2 = 0$, therefore the generalized kernel of the index map is $\mathbb{Q}^2$ and consequently the Conley index is trivial.

Another horseshoe map is presented in Figure 9. A similar calculations shows that in this case the Conley index in dimension one is the endomorphism of $\mathbb{Q}$ multiplying by factor 2.

The question is whether the Conley index is sufficient to detect the full shift dynamics as in the horseshoe map. For this we need the following definitions. Given a compact set $N = N_0 \cup N_1$ with $N_0 \cap N_1 = \emptyset$ and a sequence $\alpha \in \{0,1\}^n$ put

$$N_\alpha := \bigcap_{i=0}^{n-1} f^i(N_{\alpha_i})$$

and for $\bar{\alpha} = (\alpha^1, \alpha^2, \ldots \alpha^m)$ with $\alpha^j \in \{0,1\}^n$ put

$$N_{\bar{\alpha}} := \bigcup_{j=1}^m N_{\alpha^j}.$$

One can show that if N is an isolating neighborhood for f then so is N_α and $N_{\bar{\alpha}}$.

The following theorem shows that the full shift dynamics may be regained up to a semiconjugacy from the knowledge of the Conley index.

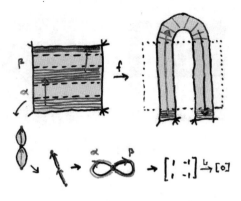

FIGURE 8. Computation of the homological Conley index of the horseshoe map.

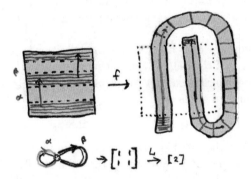

FIGURE 9. Computation of the cohomological Conley index of the G-horseshoe map.

THEOREM 2.7. [22] *Assume $N = N_0 \cup N_1$ is an isolating neighbourhood for f such that N_0 and N_1 are disjoint compact polyhedra. If for $k = 0, 1$*

$$\mathrm{Con}^n(N_k, f) = \begin{cases} (\mathbf{Q}, \mathrm{Id}) & \text{if } n = 1 \\ 0 & \text{otherwise} \end{cases}$$

and the map parts $\chi^*(N_{00,01,11}, f)$, $\chi^*(N_{00,10,11}, f)$ of the Conley indexes $\mathrm{Con}^*(N_{00,01,11}, f)$, $\mathrm{Con}^*(N_{00,10,11}, f)$ are different from the identity then there exists a $d \in \mathbf{N}$ and a continuous surjection $\rho : \mathrm{Inv}(N, f) \to \Sigma_2$ such that $\rho \circ f^d = \sigma \circ \rho$, where $\sigma : \Sigma_2 \to \Sigma_2$ is the full shift dynamics on two symbols. Moreover, for each periodic sequence $\alpha \in \Sigma_2$ there exists a periodic point $x \in N$ such that $\rho(x) = \alpha$.

Some generalizations of this theorem may be found in [**45, 44**].

3. Discretizing dynamics

To use theorems characterizing dynamics in rigorous numerics we need first to discuss how spaces and maps may be discretized to be stored in computer memory and analysed algorithmically.

FIGURE 10. A two dimensional (left) and three dimensional (right) cubical set.

3.1. Discretization of space. By the discretization of space we mean a countable family of subsets of $\mathbb{R}^d$ together with some way of identifying each member of the family by some finite code. A classical example is the collection of simplicial complexes. A useful alternative, particularly convenient in rigorous numerics (due to the use of interval arithmetic) and imaging is the family of cubical sets (see Fig. 10).

To define this family we need the following definitions. An *elementary interval* is an interval $[k, l] \subset \mathbb{R}$ such that k, l are integers satisfying $l = k$ (degenerate case) or $l = k + 1$ (nondegenerate case). An *elementary cube* Q in $\mathbb{R}^d$ is a Cartesian product of elementary intervals

$$I_1 \times I_2 \times \cdots \times I_d \subset \mathbb{R}^d.$$

The *dimension* of Q is the number of nondegenerate intervals in the product. We denote the set of all elementary cubes in $\mathbb{R}^d$ by $\mathcal{K}^d$ or briefly by $\mathcal{K}$ if d is clear from the context. We write $\mathcal{K}_k$ for the set of all elementary cubes in $\mathcal{K}$ of dimension k. An elementary cube is said to be *full* if its dimension is d. For $\mathcal{A} \subset \mathcal{K}$ we use notation $|\mathcal{A}| := \bigcup \mathcal{A}$ and for $A \subset \mathbb{R}^d$ we write $\mathcal{K}(A) := \{ Q \in \mathcal{K} \mid Q \subset A \}$.

The set $A \subset \mathbb{R}^d$ is *cubical* if there exists a finite family $\mathcal{A} \subset \mathcal{K}$ such that $A = |\mathcal{A}|$. The family $\mathcal{A}$ is referred to as the *representation* of A. The unique minimal representation, the *minimal representation* of A, is denoted by $\mathcal{K}_{\min}(A)$. A cubical set is a *full cubical set* if its minimal representation consists only of full elementary cubes.

The class of cubical sets is not smaller than the class of polytopes in the sense of the following theorem.

THEOREM 3.1. [4] *Every polytope is homeomorphic to a cubical set.*

Given an arbitrary subset $A \subset \mathbb{R}^d$ we define the family of elementary cubes with non-empty intersection with A by

$$o_d(A) := \{ Q \in \mathcal{K} \mid Q \cap A \neq \emptyset \}.$$

For $\mathcal{N} \subset \mathcal{K}$ we also define discrete analogues of the interior and boundary by

$$\operatorname{int} \mathcal{N} := \{ Q \in \mathcal{N} \mid o_d(Q) \subset \mathcal{N} \},$$
$$\operatorname{bd} \mathcal{N} := \mathcal{N} \setminus \operatorname{int}(\mathcal{N}).$$

3.2. Discretization of maps. Let $\mathcal{X}$ be a finite subfamily of $\mathcal{K}^d$ and let $\mathcal{F} : \mathcal{X} \rightrightarrows \mathcal{X}$ be a multivalued map. We refer to such a map as a *combinatorial map*

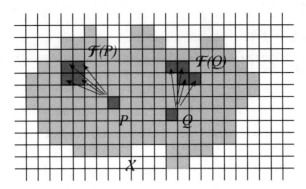

FIGURE 11. A combinatorial multivalued map.

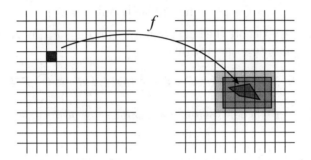

FIGURE 12. The construction of a combinatorial enclosure of a continuous map f. The cube in the left grid is mapped by f into an unknown region (dark gray) in the right grid. The methods of interval arithmetic and rigorous numerics enable the computation of an upper estimate of the value in the form of a product of intervals (mid gray). The estimate is covered by the boxes from the grid (light gray) to obtain the value of the combinatorial enclosure.

(see Fig. 11). The *associated digraph* of $\mathcal{F}$ has $\mathcal{X}$ as the set of vertices and an edge from P to Q whenever $Q \in \mathcal{F}(P)$.

A combinatorial multivalued map $\mathcal{F} : \mathcal{X} \rightrightarrows \mathcal{X}$ is a *combinatorial enclosure* of $f : X \to X$ [**46, 18**] if for every $Q \in \mathcal{X}$

$$o_d(f(Q)) \subset \mathcal{F}(Q).$$

In this case we say that f is a *selector* of $\mathcal{F}$. If $\mathcal{F} : \mathcal{X} \to \mathcal{X}$ is a combinatorial enclosure of $f : X \to X$, then for every $Q \in \mathcal{X}$

$$f(Q) \subset \operatorname{int} |\mathcal{F}(Q)|.$$

An example of a combinatorial multivalued map is presented in Fig. 11 and of a combinatorial enclosure in Fig. 12.

3.3. Discretization of topological dynamics. Let $\mathcal{F} : \mathcal{K} \to \mathcal{K}$ be a combinatorial multivalued map and let I be an interval in $\mathbb{Z}$ containing 0. A *solution* through $Q \in \mathcal{K}$ under $\mathcal{F}$ is a function $\Gamma : I \to \mathcal{K}$ satisfying the following two properties:

(1) $\Gamma(0) = Q$,
(2) $\Gamma(n+1) \in \mathcal{F}(\Gamma(n))$ for all n such that $n, n+1 \in I$.

Note that in the language of the associated digraph a solution is just a path in the digraph.

Assume $\mathcal{N} \subset \mathcal{K}$ is finite. The *invariant part* of $\mathcal{N}$ under $\mathcal{F}$ is

$$\mathrm{Inv}(\mathcal{N}, \mathcal{F}) := \{\, Q \in \mathcal{N} \mid \text{there exists a full solution } \Gamma : \mathbb{Z} \to \mathcal{N} \,\}.$$

The *positively invariant part* and the *negatively invariant part* of $\mathcal{N}$ under $\mathcal{F}$ are defined respectively by

$$\mathrm{Inv}^+(\mathcal{N}, \mathcal{F}) := \{\, Q \in \mathcal{N} \mid \text{there exists a solution } \Gamma : \mathbb{Z}^+ \to \mathcal{N} \,\}$$
$$\mathrm{Inv}^-(\mathcal{N}, \mathcal{F}) := \{\, Q \in \mathcal{N} \mid \text{there exists a solution } \Gamma : \mathbb{Z}^- \to \mathcal{N} \,\}$$

We have the following obvious formula

$$\mathrm{Inv}(\mathcal{N}, \mathcal{F}) = \mathrm{Inv}^-(\mathcal{N}, \mathcal{F}) \cap \mathrm{Inv}^+(\mathcal{N}, \mathcal{F}).$$

Let $\mathcal{F}_\mathcal{N} : \mathcal{N} \rightrightarrows \mathcal{N}$ denote the map given by $\mathcal{F}_\mathcal{N}(Q) := \mathcal{F}(Q) \cap \mathcal{N}$. One can show that there exists an integer n such that

$$\mathrm{Inv}^-(\mathcal{N}, \mathcal{F}) = \bigcap_{i=0}^n \mathcal{F}_\mathcal{N}^i(\mathcal{N}) \quad \text{and} \quad \mathrm{Inv}^+(\mathcal{N}, \mathcal{F}) = \bigcap_{i=0}^n \mathcal{F}_\mathcal{N}^{-i}(\mathcal{N}).$$

A finite subset $\mathcal{N}$ of $\mathcal{K}$ is an *isolating neighborhood* for $\mathcal{F}$ if

$$\mathrm{Inv}(\mathcal{N}, \mathcal{F}) \subset \mathrm{int}\, \mathcal{N}.$$

We say that $(\mathcal{P}_1, \mathcal{P}_2)$ is a *combinatorial index pair* for $\mathcal{F}$ in $\mathcal{N}$ if $\mathcal{P}_2 \subset \mathcal{P}_1 \subset \mathcal{N}$ and the following three conditions are satisfied

$$\mathcal{F}(\mathcal{P}_i) \cap \mathcal{N} \subset \mathcal{P}_i,$$
$$\mathcal{F}(\mathcal{P}_1) \cap \mathrm{bd}\, \mathcal{N} \subset \mathcal{P}_2,$$
$$\mathrm{Inv}(\mathcal{N}, \mathcal{F}) \subset \mathcal{P}_1 \setminus \mathcal{P}_2.$$

The following theorem shows that the construction of an index pair of a continuous map may be replaced by the computation of a combinatorial index pair of some its combinatorial enclosure.

THEOREM 3.2. [30, 46, 31] *Assume $\mathcal{N}$ is an isolating neighborhood for $\mathcal{F}$ and $(\mathcal{P}_1, \mathcal{P}_2)$ is a combinatorial index pair for $\mathcal{F}$ in $\mathcal{N}$. Then for any selector f of $\mathcal{F}$ the set $|\mathcal{N}|$ is an isolating neighborhood for f and $(|\mathcal{P}_1|, |\mathcal{P}_2|)$ is a index pair for f.*

The question is if we can compute a combinatorial index pair algorithmically. For this, we need the following theorem.

THEOREM 3.3. [31] *Assume $\mathcal{N}$ is an isolating neighborhood for $\mathcal{F}$. Let*

$$\mathcal{P}_1 := \mathrm{Inv}^-(\mathcal{N}, \mathcal{F}) \quad \text{and} \quad \mathcal{P}_2 := \mathrm{Inv}^-(\mathcal{N}, \mathcal{F}) \setminus \mathrm{Inv}^+(\mathcal{N}, \mathcal{F}).$$

Then $(\mathcal{P}_1, \mathcal{P}_2)$ is a combinatorial index pair for $\mathcal{F}$ in $\mathcal{N}$ and

$$|\mathcal{P}_1| \setminus |\mathcal{P}_2| \subset \mathrm{int}\, |\mathcal{N}|.$$

Moreover, if

$$\bar{\mathcal{P}}_1 := \mathcal{P}_1 \cup \mathcal{F}(\mathcal{P}_1) \quad \text{and} \quad \bar{\mathcal{P}}_2 := \mathcal{P}_2 \cup (\mathcal{F}(\mathcal{P}_1) \setminus \mathcal{P}_1),$$

then for any selector f of $\mathcal{F}$ the quadruple $(|\mathcal{P}_1|, |\mathcal{P}_2|, |\bar{\mathcal{P}}_1|, |\bar{\mathcal{P}}_2|)$ is an index quadruple.

```
function combinatorialIndexPair(set N, combinatorialMap F)
S⁺ := positiveInvariantPart(N, F);
S⁻ := negativeInvariantPart(N, F);
if S⁻ ∩ S⁺ ⊂ int(N) then
    P₁ := S⁻;
    P₂ := S⁻ \ S⁺;
    P̄₁ := P₁ ∪ F(P₁);
    P̄₂ := P₂ ∪ (F(P₁) \ P₁);
    return (P₁, P₂, P̄₁, P̄₂);
else
    return "Failure";
endif;
```

TABLE 1. Algorithm computing index quadruple of a combinatorial multivalued map.

```
function negativeInvariantPart(set N, combinatorialMap F)
F := restrictedMap(F, N);
S := C := N;
repeat
    S' = S;
    C := evaluate(F, C);
    S := S ∩ C;
until (S = S');
return S;

function positiveInvariantPart(set N, combinatorialMap F)
Finv := evaluateInverse(F);
return negativeInvariantPart(N, Finv);
```

TABLE 2. Algorithms computing negative and positive invariant parts of a combinatorial multivalued map.

Theorem 3.3 leads to an algorithm presented in Table 1. One can show that if this algorithm is called with a collection of cubes $\mathcal{N}$ and a combinatorial enclosure of f on input and it does not fail, then it returns a representation of an index quadruple of f. The algorithm uses algorithmic computation of the negative and positive invariant parts of a combinatorial enclosure. The respective algorithms are presented in Table 2. It is straightforward to observe that these algorithms, when called with a collection of cubes $\mathcal{N}$ and a combinatorial multivalued map $\mathcal{F}$ on input, always stop and return respectively the negative and positive invariant parts of $\mathcal{F}$ in $\mathcal{N}$. Note that the algorithms in Table 1 and in Table 2 are not optimised to keep them simple but it is not difficult to optimise them using the standard techniques of graph algorithms.

4. Homology algorithms

The results of the previous section show that the methods of rigorous numerics combined with topological invariants of dynamical systems may provide rigorous results about the qualitative features of dynamical systems like the existence of chaotic invariant sets as long as we have efficient algorithms to compute the homology of spaces and maps.

The classic way of computing homology of cubical sets consists in immediate algebraization: we triangulate the set, construct the matrices of boundary maps, compute their Smith diagonalization and use it to read Betti numbers [38]. The obvious advantage of such an approach is that it is based on standard and widely implemented linear algebra tools. However, building triangulation of cubical sets increases data size, particularly in high dimension (see [5]). Moreover, the complexity of Smith diagonalization is supercubical, which restricts the applicability of the method to sets with relatively small triangulation. The matrices of boundary maps are sparse but the gain from it is not substantial because of the well known fill-in process. Moreover, the use of sparse matrices in general requires dynamic storage allocation, which causes a substantial overhead.

In the case of cubical sets the homology may be computed directly from the list of elementary cubes without the need to triangulate the set. Moreover, an essential speed up may be obtained by preprocessing the Smith diagonalization with geometric reductions. The general idea is to first reduce the set so that the representation used to store the set is preserved and the homology is not changed in course of the reductions. If the resulting set is substantially smaller, we gain by circumventing the need to construct a big chain complex and applying the algebraization to a smaller set. However, the benefits show up only when the complexity of reduction is Cn with a small C and the set after the reductions is indeed significantly smaller. An additional gain comes from the use of bitmaps. Bitmaps provide a very compact and extremely efficient way of representing cubical sets in computer memory.

In this section we review some recent ideas concerning speeding up algorithmic homology computations by means of geometric reductions.

4.1. Cubical homology. As we already mentioned, the homology of a cubical set may be computed directly from the list of elementary cubes without the need to triangulate the set. We begin with briefly recalling the respective theory as presented in [18]. Given an elementary cube Q we define the associated *elementary chain* by

$$\widehat{Q}(P) = \begin{cases} 1 & \text{if } P = Q, \\ 0 & \text{otherwise.} \end{cases}$$

A *cubical chain* is a finite linear combination of elementary chains of the same dimension, called the *dimension* of the chain. All cubical chains of dimension q form an Abelian group, denoted C_q and called the *group of q-chains*. Given two elementary chains $\widehat{P}, \widehat{Q}$, we define their *cubical product* by

$$\widehat{P} \diamond \widehat{Q} := \widehat{P \times Q}$$

and we extend this definition linearly to arbitrary chains.

The *boundary operator* is defined as a homomorphism $\partial : C_q \to C_{q-1}$ given on generators by

$$\partial \widehat{Q} := \begin{cases} 0 & \text{if } Q = [l], \\ \widehat{[l+1]} - \widehat{[l]} & \text{if } Q = [l, l+1], \\ \partial \widehat{I} \diamond \widehat{P} + (-1)^{\dim I} \widehat{I} \diamond \partial \widehat{P} & \text{if } Q = I \times P. \end{cases}$$

All these definitions are not related to a particular cubical set yet. To restrict the attention to a cubical set we define the *support* of a cubical chain $c = \sum_{i=1}^{n} \alpha_i \widehat{Q}_i$ by $|c| := \bigcup \{ Q_i \mid \alpha_i \neq 0 \}$. Now, given a cubical set X we define the group of q-chains of X by

$$C_q(X) := \{ c \in C_q \mid |c| \subset X \}.$$

Is is easy to verify that we have the induced boundary operator

$$\partial_q^X : C_q(X) \to C_{q-1}(X)$$

which satisfies $\partial_{q-1}^X \circ \partial_q^X = 0$. This allows us to define the qth *homology group* of X by

$$H_q(X) := Z_q(X)/B_q(X)$$

where $Z_q(X)$ stands for the kernel of ∂_q^X and $B_q(X)$ for the image of ∂_{q+1}^X. One can prove that this construction provides homology theory isomorphic to simplicial or singular homology of cubical sets [18].

4.2. Geometric reductions. The simplest way to perform geometric reductions, called *shaving*, is based on the elementary observation that if $X = \bigcup \mathcal{X}$ is cubical and $Q \in \mathcal{X}$ is an elementary cube such that $Q \cap X$ is acyclic then Q may be removed from X without changing its homology (see Fig. 13). This method is particularly useful when X is a full cubical set, because then, in particular, acyclicity tests may be performed via lookup tables containing acyclicity information of all configurations of neighbours of a full cube. The lookup table requires 2^{3^d-1} entries with d denoting the embedding dimension. The lookup tables may be computed and stored for $d \in \{2,3\}$ and in this case provide extremely fast acyclicity tests. Unfortunately, the amount of memory needed for dimension 4 and higher is prohibitive for this method. However, some partial acyclicity tests may be used in higher dimensions.

Unfortunately, in many applications shaving does not provide reduction substantial enough for the algebraic methods applied to the reduced set to succeed. A method proposed in [33], which usually leads to far deeper reductions is based on the fact that if X is cubical and $A \subset X$ is acyclic then

$$H_n(X) \cong \begin{cases} H_n(X, A), & \text{for } n \geq 1 \\ \mathbb{Z} \oplus H_n(X, A), & \text{for } n = 0. \end{cases}$$

Thus, instead of shaving we construct a possibly large acyclic subset and then remove its cubical interior and compute algebraically the relative homology (see Fig. 14).

In the representation of a cubical set in which all faces and not only the full elementary cubes are stored, a method similar in spirit to shaving but not requiring lookup tables is available. It consists in reducing the so called free faces [18]. In the setting of cubical sets or simplicial complexes a *free face* is a face, i.e. an elementary cube or a simplex, which has exactly one face in its coboundary. If P

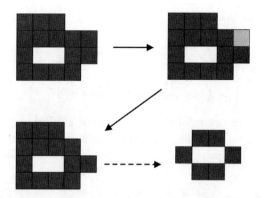

FIGURE 13. Shaving of a cubical set. Given a cubical set (top left) we select a cube whose intersection with the remaining cubes is acyclic (light gray cube in top right) and remove it from the set (bottom left). We continue this process as long as such cubes exist, arriving at a set with the same homology but often significantly smaller (bottom right).

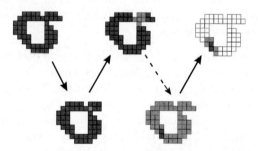

FIGURE 14. Geometric reduction by acyclic subspace construction. We start with the top left dark gray cubical set, select a square (light gray) and start adding more squares to it as long as the constructed light gray set remains acyclic. When the acyclic subset cannot be grown anymore (bottom right) we compute the homology of the whole set relative the light gray acyclic set but first we use excision to get rid of as many light gray squares as possible (top right).

is a free face and Q is the unique face in its coboundary, then we can remove P and Q without changing the homology. This may be viewed as a combinatorial counterpart of deformation retraction. On the algebraic level, it corresponds to removing a row of the boundary matrix which is zero except one entry which is an invertible element. A sample process of reducing a simplicial complex via a sequence of free face reductions is presented in Figure 15.

The free face reductions often do not go sufficiently deep to give some gains. Much better results may be obtained by applying the dual process of free coface reductions. A *free coface* is a face with exactly one face in its boundary. The formal justification of this method is based on the one space homology theory with compact supports for locally compact sets [**42**] and its combinatorial version proposed in [**34**].

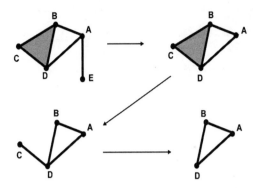

FIGURE 15. Reduction by free face collapses.

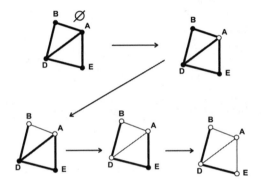

FIGURE 16. Reduction by free coface collapses.

Unfortunately, a cubical set or simplicial complex does not have any free cofaces. But, by moving from standard homology to reduced homology one resolves the limitation (see Fig. 16). The respective algorithm, called the coreduction homology algorithm, is presented in Table 3.

Another tool which has some potential in speeding up homology computations is discrete Morse theory. However, in order to build a Morse complex on an arbitrary cubical set or simplicial complex one needs first a Morse function. The coreduction homology algorithm may be adapted for this task [17].

4.3. Computing homology of continuous maps. To apply topological methods in rigorous numerics of dynamical systems, in particular to compute the Conley index, one needs not only algorithms computing homology of sets but also algorithms computing homology of continuous maps. In principle, computing homology of a map $f : X \to Y$ consists in finding a finite representation of f suitable to build the associated chain map. However, this raises some questions: what constitutes an acceptable finite representation? how one can build it? An acceptable finite representation must at least be constructible from an algorithm approximating values of f and carry all the information necessary to build a chain map with the same homology as f.

A natural candidate for the chain map is the chain map of a simplicial approximation of f. Unfortunately, building a simplicial approximation when only

```
input cubical set S;
Q := empty queue;
enqueue(Q,s);
while Q ≠ ∅ do
    s:=dequeue(Q);
    if bd_S s = {t} then
        remove s from S;
        remove t from S;
        foreach u ∈ cbd_S t do
        if u ∉ Q then
            enqueue(Q, u);
        endif;
    else if bd_S s = ∅ then
        foreach u ∈ cbd_S s do
        if u ∉ Q then
            enqueue(Q, u);
        endif;
    endif;
endwhile ;
output reduced cubical set S;
```

TABLE 3. Coreduction homology algorithm.

approximations of values of f are available is not easy. Moreover, in the case of cubical homology there is no natural counterpart of the simplicial approximation. We present some alternative approaches below.

For an elementary interval I put

$$\overset{\circ}{I} := \begin{cases} (l, l+1) & \text{if } I = [l, l+1], \\ [l] & \text{if } I = [l, l]. \end{cases}$$

and for an elementary cube $Q = I_1 \times I_2 \times \ldots \times I_d \subset \mathbf{R}^d$ define the associated *elementary cell* as $\overset{\circ}{Q} := \overset{\circ}{I}_1 \times \overset{\circ}{I}_2 \times \ldots \times \overset{\circ}{I}_d$.

Let X and Y be cubical sets. A multivalued map $F : X \rightrightarrows Y$ is *cubical* [18] if $F(x)$ is a cubical set for every $x \in X$ and for all $Q \in \mathcal{K}(X)$ and $x, x' \in \overset{\circ}{Q}$ we have $F(x) = F(x')$.

Let $\mathcal{X}, \mathcal{Y} \subset \mathcal{K}$ and let $\mathcal{F} : \mathcal{X} \rightrightarrows \mathcal{Y}$ be a multivalued combinatorial map. Define the multivalued maps $\lfloor \mathcal{F} \rfloor, \lceil \mathcal{F} \rceil : |\mathcal{X}| \rightrightarrows |\mathcal{Y}|$ by

$$\lfloor \mathcal{F} \rfloor(x) := \bigcap \{ |\mathcal{F}(Q)| \mid x \in Q \in \mathcal{X} \},$$
$$\lceil \mathcal{F} \rceil(x) := \bigcup \{ |\mathcal{F}(Q)| \mid x \in Q \in \mathcal{X} \}.$$

THEOREM 4.1. [30] *The maps $\lfloor \mathcal{F} \rfloor$ and $\lceil \mathcal{F} \rceil$ are cubical. The map $\lfloor \mathcal{F} \rfloor$ is lower semicontinuous and the map $\lceil \mathcal{F} \rceil$ is upper semicontinuous.*

A chain map $\varphi : C(X) \to C(Y)$ is a *chain selector* of $F : X \rightrightarrows Y$ if the following two conditions are satisfied

$$|\varphi(\widehat{Q})| \subset F(\overset{\circ}{Q}) \text{ for all } Q \in \mathcal{K}(X),$$
$$\varphi(\widehat{Q}) \in \widehat{\mathcal{K}}_0(F(Q)) \text{ for any vertex } Q \in \mathcal{K}_0(X).$$

We say that $F : X \rightrightarrows Y$ is *acyclic-valued* if for every $x \in X$ the set $F(x)$ is acyclic i.e. its homology is isomorphic to the homology of a singleton.

THEOREM 4.2. **[2]** *Every lower semicontinuous, acyclic-valued cubical map admits a chain selector and any two such chain selectors are chain homotopic.*

For a lower semicontinuous, acyclic-valued cubical map $F : X \rightrightarrows Y$ we put $H_*(F) := H_*(\varphi)$ for any chain selector φ of F.

The proof of Theorem 4.2 is constructive and an implementation of an algorithm based on the proof was proposed by Mazur, Szybowski in **[21]**. Unfortunately, the algorithm is very slow, because huge linear systems need to be solved.

For a bounded set $A \subset \mathbb{R}^d$ let $\operatorname{ch}(A)$ be the smallest cubical set containing A. A map $F : X \rightrightarrows Y$ is a *representation* of a continuous map $f : X \to Y$ if F is a lower semicontinuous multivalued cubical map and $f(x) \in F(x)$ for every $x \in X$. One can show that the map

$$M_f : X \ni x \to \operatorname{ch}(f(\operatorname{ch}(x))) \subset Y$$

is a representation of $f : X \to Y$. Unfortunately, in general M_f need not be acyclic-valued. To get an acyclic-valued representation we use the concept of rescalling, playing a role analogous to barycentric subdivisions. For $\alpha \in \mathbb{N}^d$ put

$$\Lambda^\alpha : \mathbb{R}^d \ni x \to (\alpha_1 x_1, \alpha_2 x_2, \ldots, \alpha_d x_d) \in \mathbb{R}^d,$$
$$\Omega^\alpha : \mathbb{R}^d \ni x \to (\alpha_1^{-1} x_1, \alpha_2^{-1} x_2, \ldots, \alpha_d^{-1} x_d) \in \mathbb{R}^d,$$
$$X^\alpha := \Lambda^\alpha(X),$$
$$\Lambda_X^\alpha : X \ni x \to \Lambda^\alpha(x) \in X^\alpha,$$
$$f^\alpha := f \circ \Omega^\alpha.$$

THEOREM 4.3. **[18]** *The following properties are satisfied:*
- $M_{\Lambda_X^\alpha}$ *is acyclic-valued,*
- M_{f^α} *is acyclic-valued for sufficiently large α,*
- $H_*(M_{f^\alpha}) \circ H_*(M_{\Lambda_X^\alpha})$ *does not depend on α for large α.*

For a continuous map $f : X \to Y$ we put

$$H_*(f) := H_*(M_{f^\alpha}) \circ H_*(M_{\Lambda_X^\alpha})$$

for a sufficiently large α. One can show that this definition coincides with the definition in simplicial or singular theory (see **[18]**).

THEOREM 4.4. **[18]** *If $F : X \rightrightarrows Y$ is an upper semicontinuous cubical map, then*

$$G(F) := \{ (x, y) \in X \times Y \mid y \in F(x) \}$$

is a cubical set.

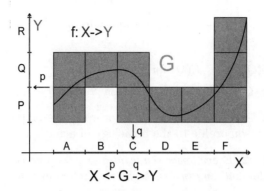

FIGURE 17. Computing homology of a map $f : X \to Y$ by projecting from the graph G of an acyclic-valued representation (marked gray) via projections $p : G \to X$ and $q : G \to Y$.

Let $p : G(F) \to X$ and $q : G(F) \to Y$ denote projections. Then p and q induce chain maps and by the Vietoris-Begle Mapping Theorem $H_*(p)$ is an isomorphism if F is acyclic-valued. Moreover, the following diagram commutes [15]

$$\begin{array}{ccc} & G(F) & \\ {}^p\swarrow & & \searrow^q \\ X & \overset{F}{\rightrightarrows} & Y. \end{array}$$

This leads to the following theorem.

THEOREM 4.5. [25] *If $F : X \rightrightarrows Y$ is an upper semicontinuous, acyclic-valued representation of $f : X \to Y$, then*

$$H_*(f) = H_*(q) H_*(p)^{-1}.$$

The homology map algorithm based on Theorem 4.5 consists of the following steps.

(1) Construct a combinatorial enclosure $\mathcal{F}$ of $f : X \to Y$.
(2) Construct the graph G of $F := \lceil \mathcal{F} \rceil$.
(3) If the homologies of the values of F are not trivial, refine the grid and go to 1.
(4) Find the homologies of the projections $p : G \to X$ and $q : G \to Y$.
(5) Return $H_*(q) H_*(p)^{-1}$.

The algorithm was implemented by P. Pilarczyk [8]. The implementation is satisfactorily fast for a class of practical problems but too slow for many other problems.

In order to further speed up computations of the homology of continuous maps one can combine the graph method with coreductions. For this we need to discuss first how to decompose homology classes on generators. Let $\mathcal{K}_q(X) = \{Q_1^q, Q_2^q, \ldots, Q_{r_q}^q\}$ and let $\{[u_1], [u_2], \ldots, [u_n]\}$ be generators of the homology group $H_q(X)$. Take a homology class $[z] \in H_q(X)$. When computing the homology of maps one faces the problem of finding a representation of $[z]$ in terms of the given generators

$$[z] = \sum_{i=1}^{n} x_i [u_i].$$

The problem reduces to solving the equation

$$z = \sum_{i=1}^{n} x_i u_i + \partial c$$

for some unknown variables $x_1, x_2, \ldots, x_n \in \mathbb{Z}$ and $c \in C_{q+1}(X)$.

Note that every chain involved in this system may be represented as a linear combination of elementary chains $\widehat{Q}_j^q$ with respective integer coefficients

$$z = \sum_{j=1}^{r_q} z_j \widehat{Q}_j^q, \quad u_i = \sum_{j=1}^{r_q} u_{ij} \widehat{Q}_j^q, \quad c = \sum_{k=1}^{r_{q+1}} y_k \widehat{Q}_k^{q+1},$$

$$\partial \widehat{Q}_k^{q+1} = \sum_{j=1}^{r_q} a_{kj} \widehat{Q}_j^q, \quad \partial c = \sum_{j=1}^{r_q} \left(\sum_{k=1}^{r_{q+1}} a_{kj} y_k \right) \widehat{Q}_j^q.$$

Therefore the equation reduces to

$$z_j = \sum_{i=1}^{n} u_{ij} x_i + \sum_{k=1}^{r_{q+1}} a_{kj} y_k \text{ for } j = 1, 2, \ldots, r_q,$$

which is a system of r_q linear equations with $n + r_{q+1}$ unknowns. Unfortunately, when the number of cubes is huge, then also the number of equations and unknowns is huge. Therefore, it is desirable to replace the system of equations with an essentially smaller system.

This leads us to the concept of a homology model. There are explicitely given chain maps $\pi^f : \mathcal{C} \to \mathcal{C}^f$ and $\iota^f : \mathcal{C}^f \to \mathcal{C}$ between the chain complex $\mathcal{C}$ of the set to which the elementary reduction and coreduction homology algorithm is applied and the chain complex $\mathcal{C}^f$ of the set after the reductions. The maps π^f and ι^f may be used to easily transport homology classes between $H_*(\mathcal{C})$ and $H_*(\mathcal{C}^f)$, because the cost of transporting one generator through π^f or ι^f, though in general quadratic, in this case is linear. Therefore, the chain complex $\mathcal{C}^f$ may serve as a convenient model to solve the problems of decomposing homology classes on generators, because instead of solving the problem in $H_*(\mathcal{C})$ we may transport it to $H_*(\mathcal{C}^f)$, solve it there and then transport the solution back to $H_*(\mathcal{C})$. The method may be applied to compute homology of inclusion maps and persistent homology [35], projections, and via the projections of the graph of an acyclic representation to a general continuous map [37].

The respective algorithm for general continuous maps consists of the following steps.

(1) Using coreductions construct homology models of X, Y, and G
(2) Using homology models find the homology of the projections $p : G \to X$ and $q : G \to Y$
(3) Compute the inverse of p_* and return $q_* p_*^{-1}$

One of the advantages of the algorithm is the fact that there is no need to preserve the acyclicity under reductions as in the case of the graph approach. In consequence the algorithm is significantly faster.

Finally let us mention that a very promising but based on entirely different ideas and not yet implemented approach to computing homology of continuous maps is presented in [32].

	$T \times S^1$	$(S^1)^3$	$S^1 \times K$	$T \times T$
dim	5	6	6	6
size in millions	0.07	0.10	0.40	2.36
H_0	$\mathbb{Z}$	$\mathbb{Z}$	$\mathbb{Z}$	$\mathbb{Z}$
H_1	$\mathbb{Z}^3$	0	$\mathbb{Z}^2 + \mathbb{Z}_2$	$\mathbb{Z}^4$
H_2	$\mathbb{Z}^3$	0	$\mathbb{Z} + \mathbb{Z}_2$	$\mathbb{Z}^6$
H_3	$\mathbb{Z}$	$\mathbb{Z}$		$\mathbb{Z}^4$
H_4				$\mathbb{Z}$
Linbox::Smith	130	350	> 600	> 600
CHomP::homcubes	1.3	1.7	10	56
RedHom::CR	0.03	0.04	0.26	2.5
RedHom::CR+DMT	0.02	0.08	0.5	1.1

TABLE 4. Running times in seconds for various homology algorithms applied to some cubical manifolds.

	$T \times S^1$	$(S^1)^3$	$S^1 \times K$	$T \times T$
original size	73728	110592	402144	2359296
size after free face reductions	4716	1774	11916	151820
size after free coface reductions	860	400	1874	31810
reduction factor	98.83%	99.64%	99.53%	98.6517%

TABLE 5. The original size, the size after reductions and coreductions and the reduction factor for the examples in Table 4.

	P0001	P0050	P0100
dim	3	3	3
size in millions	75.56	73.36	71.64
H_0	$\mathbb{Z}^7$	$\mathbb{Z}^2$	$\mathbb{Z}$
H_1	$\mathbb{Z}^{6554}$	$\mathbb{Z}^{2962}$	$\mathbb{Z}^{1057}$
H_2	$\mathbb{Z}^2$		
Linbox::Smith	> 600	> 600	> 600
CHomP::homcubes	400	360	310
RedHom::CR	18	16	15
RedHom::CR+DMT	8	7	6
RedHom::AS	10	5	3.5

TABLE 6. Running times in seconds for various homology algorithms applied to some cubical sets from numerical simulation of Cahn-Hillard equation.

4.4. CAPD and CAPD::RedHom software projects. CAPD [6] is a software project developing tools for rigorous numerics and computer assisted proofs in dynamical systems. CAPD::RedHom [7] is a subproject providing implementations of homology algorithms. Originally CAPD::RedHom was developed as a set

	d4s8f50	d4s12f50	d4s16f50	d4s20f50
dim	4	4	4	4
size in millions	0.07	0.34	1.04	2.48
H_0	$\mathbb{Z}^2$	$\mathbb{Z}^2$	$\mathbb{Z}^2$	$\mathbb{Z}^2$
H_1	$\mathbb{Z}^2$	$\mathbb{Z}^{17}$	$\mathbb{Z}^{30}$	$\mathbb{Z}^{51}$
H_2	$\mathbb{Z}^{174}$	$\mathbb{Z}^{1389}$	$\mathbb{Z}^{5510}$	$\mathbb{Z}^{15401}$
H_3	$\mathbb{Z}^2$	$\mathbb{Z}^{15}$	$\mathbb{Z}^{71}$	$\mathbb{Z}^{179}$
Linbox::Smith	120	> 600	> 600	> 600
CHomP::homcubes	1	8.3	41	170
RedHom::CR	0.08	1.4	15	140
RedHom::CR+DMT	0.03	0.16	0.58	2.9

TABLE 7. Running times in seconds for various homology algorithms applied to some random cubical sets.

	random set	Björner set	S^5
dim	4	2	5
size in millions	4.8	1.9	4.3
H_0	$\mathbb{Z}$	$\mathbb{Z}$	$\mathbb{Z}$
H_1	$\mathbb{Z}^{39}$	0	0
H_2	$\mathbb{Z}^{84}$	$\mathbb{Z}$	0
H_3			0
H_4			$\mathbb{Z}$
CHomP::homsimpl	830	310	2100
RedHom::CR+DMT	65	11	100

TABLE 8. Running times in seconds for various homology algorithms applied to some simplicial complexes.

of implementations of homological algorithms needed in CAPD. However, since the range of possible applications of these implementations turned out to go far beyond dynamics and currently encompasses such areas as image analysis [36], material science [35], electromagnetic engineering [14] and sensor networks [12], the software is now developed as a stand-alone project for general applications.

CAPD::RedHom is based on geometric reduction algorithms. The software implements several homology algorithms, in particular the acyclic subspace homology algorithm AS [33], the coreduction homology algorithm CR [34] and an algorithm based on discrete Morse theory DMT [17]. The software computes Betti and torsion numbers, homology generators, homology maps and persistence intervals over $\mathbb{Z}$ and $\mathbb{Z}_p$ coefficients. The implementation is based on C++ templates. It is generic and very efficient. It accepts as input cubical sets, simplicial complexes, cubical CW complexes, as well as general regular CW complexes [13].

In order to compare the efficiency of CAPD::RedHom homology software with other homology software, in particular homology software available from [8] and [19], we present in Tables 4, 6, 7, 8 computation times for some cubical manifolds,

Size	2048	3200	4608	6272	8192
CHomP::homcubes	24.562	117.625	453.563	600.14	2679.98
RedHom::CR	0.063	0.109	0.141	0.187	0.235

TABLE 9. Running times in seconds for the graph approach and coreduction approach homology map algorithms applied to inclusion maps.

Size in millions	3.65	7.75	14.13	23.29
CHomP::homcubes	10.3	25.6	54.5	121.0
RedHom::CR	1.5	3.2	6.8	11.7

TABLE 10. Running times in seconds for the graph approach and coreduction approach homology map algorithms applied to a continuous self-map on a torus.

cubical sets arising from numerical studies of PDE's, random cubical sets, as well as some simplicial complexes.

The geometric reductions (acyclic subspace construction, free face reductions, free coface reductions) are performed in linear time. However, the worst case performance of these reduction algorithms is the same as the worst case performance of the Smith diagonalization algorithm, i.e. cubical. This is because so far nothing is proved about the depth of reductions. However, numerical experiments indicate that the depth of reductions is significant. Table 5 presents the original size, the size after free face reductions, the size after free coface reductions and the reduction factor (the number of cells removed as the percentage of the original size) for cubical sets in Table 4. The table shows that almost everything is reduced. Consequently, the observed complexity is close to linear.

A comparison of some map examples is given in Table 9 and 10.

5. Applications

In this final section we briefly review some applications. In all the examples the procedure for obtaining the results consisted in applying the following scheme

(1) Choose a theoretical result concluding the requested thesis when the Conley indexes of certain isolating neighborhoods are as required.
(2) Using the methods of Section 2.4 in the case of a dynamical system given explicitly or experimental result in the case of a time series construct a sufficiently good combinatorial enclosure $\mathcal{F}$ of the generator of the dynamical system
(3) Choose candidates for the isolating neighborhoods
(4) Using the methods of Section 3.3 verify the isolation and construct the combinatorial index pairs for the selected candidates.
(5) Using the methods od Section 4 compute the space and map part of the Conley indexes of the isolating neighborhoods.
(6) If the computed Conley indexes verify the assumptions of the theoretical result selected in the first step, proclaim success.

The choices were made either by trial and error or by utilizing some auxiliary algorithms based on classical numerics.

5.1. Lorenz equations.
For a $k \times k$ matrix $A = (A_{ij})$ over $\mathbb{Z}_2$ put
$$\Sigma(A) := \{\alpha \in \Sigma_k \mid \forall i \in \mathbb{Z}\ A_{\alpha(i)\alpha(i+1)} = 1\}.$$
The set $\Sigma(A)$ is invariant under the shift map σ, so $(\Sigma(A), \sigma_{|\Sigma(A)})$ is a discrete dynamical system called *subshift of finite type*. One can prove that the topological entropy $h(\Sigma(A))$ of a subshift of finite type satisfies $h(\Sigma(A)) = \log(\lambda(A))$, where $\lambda(A)$ stands for the spectral radius of A.

THEOREM 5.1. [26] *Consider the Lorenz equations and the plane $P := \{(x, y, z) \mid z = 27\}$. For all parameter values in a sufficiently small neighborhood of $(\sigma, R, b) = (28, 10, 8/3)$ there exists a Poincaré section $N \subset P$ such that the associated Poincaré map g is Lipschitz and well defined. Furthermore, for*

$$A := \begin{bmatrix} 0 & 1 & 1 & 0 & 0 & 0 \\ 0 & 0 & 0 & 1 & 1 & 0 \\ 0 & 0 & 0 & 0 & 0 & 1 \\ 1 & 0 & 0 & 0 & 0 & 0 \\ 0 & 1 & 1 & 0 & 0 & 0 \\ 0 & 0 & 0 & 1 & 1 & 0 \end{bmatrix}$$

there is a continuous surjection $\rho : \mathrm{Inv}(N, g) \to \Sigma(A)$ such that
$$\rho \circ g = \sigma \circ \rho.$$
In particular $h(\mathrm{Inv}(N, g)) \geq 0.48$. Moreover, for every $\alpha \in \Sigma(A)$ which is periodic there exists an $x \in \mathrm{Inv}(N, g)$ on a periodic trajectory such that $\rho(x) = \alpha$.

5.2. Hénon map.
Consider the Hénon map $h : \mathbb{R}^2 \to \mathbb{R}^2$ given by the formula
$$h(x, y) := (1 + y/5 - ax^2, 5bx)$$
at the classical parameter values $a = 1.4$ and $b = 0.2$.

THEOREM 5.2. [18] *The discrete dynamical system induced by the Hénon map admits an invariant set S semiconjugate with a subshift of finite type on 8 symbols and topological entropy $h = 0.28$ Moreover, if*

$$A := \begin{bmatrix} 0 & 0 & 0 & 0 & 0 & 0 & 0 & 1 \\ 0 & 0 & 0 & 0 & 0 & 0 & 1 & 0 \\ 0 & 0 & 0 & 0 & 0 & 1 & 0 & 0 \\ 0 & 0 & 0 & 0 & 1 & 0 & 0 & 0 \\ 1 & 0 & 0 & 0 & 0 & 0 & 0 & 0 \\ 0 & 1 & 0 & 0 & 0 & 0 & 0 & 0 \\ 0 & 1 & 0 & 1 & 0 & 0 & 0 & 0 \\ 0 & 1 & 1 & 1 & 0 & 0 & 0 & 0 \end{bmatrix},$$

then for each periodic sequence $\theta \in \Sigma(A)$ with period p the set $\rho^{-1}(\theta)$ contains a periodic orbit with period p. In particular $h(S) \geq 0.28$.

Recently, Day, Frongilo and Trevion [11] improved the above result by providing a computer assisted proof that the classical Hénon map admits an isolated invariant set S semiconjugated to a subshift on 129 symbols and $h(S) \geq 0.42$.

5.3. Infinite dimensional dynamical systems.
Kot-Schaffer growth-dispersal model for plants
$$\Phi : L^2([-\pi,\pi]) \to L^2([-\pi,\pi])$$
is a discrete dynamical system in infinite dimensional space given by
$$\phi(a)(y) := \frac{1}{2\pi} \int_{-\pi}^{\pi} \mu b(x,y) a(x) \left(1 - \frac{a(x)}{c(x)}\right) dx$$
where $\mu > 0$. Via Fourier expansion the map may be replaced by a map acting on an infinite sequence of Fourier coefficients. Thus, the method of self-consistent apriori bounds [49] may be applied so that the study of dynamics reduces to a discrete dynamical system in $\mathbb{R}^{11}$, which may be treated by the methods reviewed in this paper. This, in particular, leads to the following theorem.

THEOREM 5.3. [10] *The Kot-Schaffer system admits a stationary solution p and a 2-periodic solution q in the ball of radius 10^{-11} in L^2 and C^0 norms respectively around certain functions p^* and q^* whose all non-zero Fourier coefficients are explicitly given. Moreover, there exists a heteroclinic connection running from a neighborhood of p to a neighborhood of q.*

THEOREM 5.4. [10] *The Kot-Schaffer system admits an invariant set with entropy not less than 0.18.*

□

5.4. Time series analysis of experiments.
Let X be a compact submanifold of $\mathbb{R}^{d_0}$ and let $f : X \to X$ describe a physical process governed by a smooth discrete dynamical system with unknown f. Let $\mu : X \to \mathbb{R}$ be a smooth form on X accessible by measurement. For $d > 2d_0$ define
$$\mu^d : X \ni x \to (\mu(x), \mu(f(x)), \ldots, \mu(f^{d-1}(x))) \in \mathbb{R}^d.$$
A *time series* of f is a sequence $\tau = \{\tau_i\}_{i=1}^n \subset \mathbb{R}^n$ with large n such that $\tau_i = \mu(f^i(x))$ for some $x \in X$. Let $\mathcal{K}_\delta := \delta\mathcal{K}$ denote the δ-*rescaling* of $\mathcal{K}$ for some $\delta > 0$. Put $\tau_i^d := (\tau_i, \tau_{i+1}, \ldots, \tau_{i+d-1})$ and $\Theta_d^\tau := \{\tau_i^d \mid 1 \leq i \leq n-d+1\}$. We take $\mathcal{X}^\tau := \{Q \in \mathcal{K}_\delta^d \mid Q \cap \Theta_d^\tau \neq \emptyset\}$ and define a multivalued map $\mathcal{F} : \mathcal{X}^\tau \rightrightarrows \mathcal{X}^\tau$ given by
$$Q \to \{P \in \mathcal{X}^\tau \mid \tau_i^d \in Q \text{ and } \tau_{i+1}^d \in P \text{ for some } 1 \leq i \leq n-d\}.$$
We say that $B \subset \mathbb{R}^d$ is δ-*stiff* if for every $\mathcal{C} \subset \mathcal{K}_\delta^d$ the set $|\mathcal{C}| \cap B$ is a deformation retract of $|\mathcal{C}|$.

Assume $\mu^d(X)$ is δ-stiff, $\mathcal{F}$ is acyclic-valued with a continuous selector f and $\mathcal{N} \subset \mathcal{K}_\delta^d$ is an isolating neighborhood for $\mathcal{F}$. Then one can prove that $|\mathcal{N}|$ is an isolating neighborhood for f and
$$\text{Con}(\mathcal{N}, \mathcal{F}) = \text{Con}((\mu^n)^{-1}(|\mathcal{N}|), f).$$

This allows to study the dynamics of f via the experimental information gathered in the map $\mathcal{F}$. As a test of this method, an experiment with a magnetoelastic ribbon, i.e. a thin strip of material with varying modulus of elasticity was set up [27]. Such a ribbon clamped from the bottom buckles under its own weight in a magnetic field. The motion of the ribbon was measured by means of a photonic

FIGURE 18. The isolating neighborhood N with an index pair (P_1, P_2) consisting of $P_1 = N$ (gray and dark gray) and P_2 (dark gray), constructed from the time series of a mangnetoelastic ribbon.

sensor providing about 100000 consecutive data points. The resulting multivalued map $\mathcal{F}$ turned out to have an isolating neighborhood (see Fig. 18) with index map

$$\begin{bmatrix} 0 & 0 & 0 & 1 \\ 0 & 1 & 1 & 0 \\ 1 & 0 & 0 & 0 \\ 0 & 1 & 1 & 0 \end{bmatrix},$$

which led us to the conclusion that the dynamics is chaotic with topological entropy of the dynamics of the ribbon at least 0.38.

5.5. Databases for multiparameter systems. As our last example we briefly summarize results of [**3**], a recent major project whose goal was to show that the methods outlined in this review may be successfully applied to parametrized dynamical systems with a wide range of parameters.

Let X be a locally compact subset of $\mathbb{R}^d$ and Λ a compact, locally contractible, connected subset of $\mathbb{R}^q$. A parametrized family of discrete semidynamical systems is given by

$$f\colon X \times \Lambda \ni (x, \lambda) \mapsto f(x, \lambda) =: f_\lambda(x) \in X.$$

The general problem is to determine regions of parameter λ where the global asymptotic dynamics is common. The problem of this type is important in biology and social sciences where it is often difficult or even impossible to obtain the numerical values of parameters from experiments.

In the study of parametrized dynamical systems it is often convenient to consider the *unified semidynamical* system $F\colon X \times \Lambda \to X \times \Lambda$ given by

$$F(x, \lambda) = (f_\lambda(x), \lambda) = (f(x, \lambda), \lambda).$$

For $A \subset X \times \Lambda$ and $\Lambda_0 \subset \Lambda$ we write $A_{\Lambda_0} := A \cap X \times \Lambda_0$ and $F_{\Lambda_0} := F_{|X \times \Lambda_0}$.

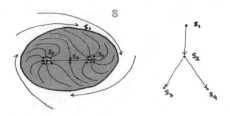

FIGURE 19. Morse decomposition (left) and the Morse graph (right) of an isolated invariant set S.

Let S be an isolated invariant set of F. A *Morse decomposition* of S is a finite collection
$$\mathbf{M}(S) = \{ M(p) \subset S \mid p \in \mathcal{P} \}$$
of disjoint isolated invariant sets of F, called *Morse sets*, which are indexed by the set $\mathcal{P}$ on which there exists a partial order $>$, called an *admissible order*, such that for every $(x, \lambda) \in S \setminus \bigcup_{p \in \mathcal{P}} M(p)$ and any trajectory γ of F through (x, λ) in S there exist indices $p > q$ such that under F
$$\omega(\gamma) \subset M(q) \quad \text{and} \quad \alpha(\gamma) \subset M(p).$$

The *Morse graph* of $\mathbf{M}(S)$ is an directed graph with vertices in $\mathcal{P}$ and edges given by minimal relations $p > q$. The *Conley-Morse graph* is a Morse graph labeled at vertices by the Conley indices of the respective isolated invariant sets.

Let $\mathcal{X} \subset \mathcal{K}_\delta^d$ be a finite subfamily. Fix a combinatorial enclosure $\mathcal{F}$ of F and an invariant set $\mathcal{S}$ of $\mathcal{F}$. A *combinatorial Morse decomposition* of $\mathcal{S}$ is a collection
$$\{ \mathcal{M}(p) \subset \mathcal{S} \mid p \in \mathcal{P} \}$$
of disjoint, non-empty and minimal under inclusion invariant sets of $\mathcal{F}$, called *combinatorial Morse sets*. Note that if $\mathcal{F}$ is a combinatorial enclosure of F then there is a natural partial order $>$ on $\mathcal{P}$ given by $p > q$ iff there exists a finite trajectory from an element of $\mathcal{M}(p)$ to an element of $\mathcal{M}(q)$.

In the sequel assume that $X = |\mathcal{X}|$ for some $\mathcal{X} \subset \mathcal{K}_\delta^d$ and $\Lambda = |\mathcal{Q}|$ for some $\mathcal{Q} \subset \mathcal{K}_\delta^q$. Let $S := \operatorname{Inv}(B, F)$ be an isolated invariant set of F for some isolating neighborhood $B \subset X \times \Lambda$. Then, for each $Q \in \mathcal{Q}$ the set $B_Q := \mathcal{X}(\bigcup_{\lambda \in Q} B_\lambda)$ satisfies $S_Q = \operatorname{Inv}\left(|B_Q| \times Q, F_Q\right)$.

THEOREM 5.5. [3] *Let $Q \in \mathcal{Q}$ and let $\{ \mathcal{M}_Q(p) \mid p \in \mathcal{P}_Q \}$ be a combinatorial Morse decomposition for $\mathcal{F}_Q := \mathcal{F}_{|X \times Q}$. If $\mathcal{F}_Q(\mathcal{M}_Q(p)) \subset \mathcal{B}_Q$ for all $p \in \mathcal{P}_Q$, then the Morse graph for the combinatorial Morse decomposition is the Morse graph for the Morse decomposition of S_Q defined by*
$$\mathbf{M}(S_Q) := \{ \operatorname{Inv}\left(|\mathcal{M}_Q(p)| \times Q, F_Q\right) \mid p \in \mathcal{P}_Q \}.$$
Moreover, each $|\mathcal{M}_Q(p)|$ is an isolating neighborhood for $\operatorname{Inv}|\mathcal{M}_Q(p)|$.

For $P, Q \in \mathcal{Q}$ such that $P \cap Q \neq \emptyset$ put
$$R := \{ (p, q) \in \mathcal{P}_P \times \mathcal{P}_Q \mid \mathcal{M}_P(p) \cap \mathcal{M}_Q(q) \neq \emptyset \}.$$

If R is a graph isomorphism, then we say that P and Q share a Morse graph. The equivalence classes with respect to the transitive closure of this relation are called *continuation classes*.

THEOREM 5.6. [3] *The Conley-Morse graphs in the same continuation class coincide.*

The *continuation graph* consists of the continuation classes as vertices with edges joining two classes $[P]$ and $[Q]$ iff there exist $P_0 \in [P]$ and $Q_0 \in [Q]$ such that $P_0 \cap Q_0 \neq \emptyset$. The continuation graph constitutes a database which captures the types of dynamics detectable in the parametrized dynamical system at a given level of resolution.

The following discrete dynamical systems in known as *Leslie population model*

$$(x, \lambda) = \left(\begin{bmatrix} x_1 \\ x_2 \end{bmatrix}, \begin{bmatrix} \theta_1 \\ \theta_2 \\ p \end{bmatrix} \right) \mapsto f(x, \lambda) = \begin{bmatrix} (\theta_1 x_1 + \theta_2 x_2) e^{-0.1(x_1 + x_2)} \\ p x_1 \end{bmatrix}.$$

Take

$$\begin{aligned} \Lambda &:= \{ (\theta_1, \theta_2) \in [8, 37] \times [3, 50] \}, \\ X &:= [-0.001, 320.056] \times [-0.001, 224.040]. \end{aligned}$$

The methods presented in these notes allow rigorous analysis of the dynamics of the Leslie population model in the whole parameter range Λ providing a database which may be inquired for the existence of particular types of dynamics. Details are presented in [3].

References

[1] R. L. ADLER, A. G. KONHEIM AND M. H. MCANDREW, Topological entropy, *Trans. Amer. Math. Soc.* **114**(1965) 309–319.

[2] M. ALLILI AND T. KACZYNSKI, An algorithmic approach to the construction of homomorphisms induced by maps in homology, *Trans. Amer. Math. Soc.* **352**(2000) 2261–2281.

[3] Z. ARAI, H. KOKUBU, W. KALIES, K. MISCHAIKOW, H. OKA, P. PILARCZYK, A Database Schema for the Analysis of Global Dynamics of Multiparameter Systems, *SIAM J. App. Dyn. Sys.* **8**(2009) 757–789.

[4] J. BLASS, W. HOŁSZTYŃSKI, Cubical polyhedra and homotopy, I, II, III, IV, V, *Atti Accad. Naz. Lincei Rend. Cl. Sci. Fis. Mat. Natur.* **50**(2)(1971), 131–138; **50**(6)(1971), 703–708; **53**(8)(1972), 275–279; **53**(8)(1972), 402–409; **54**(1973), 416–425.

[5] A. BLISS, F.E. SU, Lower Bounds for Simplicial Covers and Triangulations of Cubes, *Discrete Comput. Geom.* **33**(2005) 669–686.

[6] CAPD: Computer assisted proofs in dynamics,
http://capd.ii.uj.edu.pl/.

[7] CAPD::RedHom: Reduction homology algorithms,
http://redhom.ii.uj.edu.pl/.

[8] CHOMP: Computational homology project,
http://chomp.rutgers.edu/.

[9] C. CONLEY, Isolated invariant sets and the Morse index, *CBMS Regional Conference Series in Math* **38**(1978) .

[10] S. DAY, O. JUNGE, K. MISCHAIKOW, A rigorous numerical method for the global analysis of infinite-dimensional discrete dynamical systems, *SIAM J. App. Dyn. Sys.* **3**(2004) 117–160.

[11] S. DAY, R. FRONGILO, R. TREVINO, Algorithms for rigorous entropy bounds and symbolic dynamics, *SIAM J. App. Dyn. Sys.* **7**(2008) 1477–1506.

[12] P. DŁOTKO, R. GHRIST, M. JUDA, M. MROZEK, Distributed computing of homological coverage in sensor networks, *Applicable Algebra in Engineering, Communication and Computing*, accepted .

[13] P. Dłotko, T. Kaczynski, M. Mrozek, T. Wanner, Coreduction Homology Algorithm for Regular CW-Complexes, *Discrete and Computational Geometry* **46**(2011) 361–380; (DOI: 10.1007/s00454-010-9303-y).

[14] P. Dłotko, R. Specogna, F. Trevisan, Automatic generation of cuts on large-sized meshes for the T-Omega geometric eddy-current formulation, *Computer Methods in Applied Mechanics and Engineering* **198**(2009) 3765–3781.

[15] A. Granas and L. Górniewicz, Some general theorems in coincidence theory, *J. Math. Pure Appl.* **60**(1981) 661–373.

[16] J.K. Hale and H. Koçak, Dynamics and Bifurcations, *Springer-Verlag* (1991).

[17] S. Harker, K. Mischaikow, M. Mrozek, V. Nanda, H. Wagner, M. Juda, P. Dłotko, The Efficiency of a Homology Algorithm based on Discrete Morse Theory and Coreductions in Proceedings of the 3rd International Workshop on Computational Topology in Image Context, Rocío González Díaz and Pedro Real Jurado (Eds.), *Image A* **1**(2010) 41–47, Chipiona, Spain, November 2010, (ISSN: 1885-4508).

[18] T. Kaczynski, K. Mischaikow, M. Mrozek, Computational Homology, *Springer Verlag* (2004).

[19] Linbox: Exact computational linear algebra, http://www.linalg.org/.

[20] E.N. Lorenz, Deterministic Nonperiodic Flow, *Journal of the Atmospheric Science* **20**(1963) 130–141.

[21] M. Mazur, J. Szybowski, Algebraic construction of a coboundary of a given cycle, *Opuscula Mathematica* **27**(2007) 291–300.

[22] K. Mischaikow, M. Mrozek, Isolating neighborhoods and chaos, *Jap. J. Ind. & Appl. Math.* **12**(1995) 205–236.

[23] K. Mischaikow, M. Mrozek, Chaos in Lorenz equations: a computer assisted proof, *Bull. Amer. Math. Soc. (N.S.)* **33**(1995) 66–72.

[24] K. Mischaikow, M. Mrozek, Chaos in the Lorenz equations: a computer assisted proof. Part II: details, *Mathematics of Computation* **67**(1998) 1023-1046.

[25] K. Mischaikow, M. Mrozek, P. Pilarczyk, Graph Approach to the Computation of the Homology of Continuous Maps, *Foundations of Computational Mathematics* **5**(2005) 199–229.

[26] K. Mischaikow, M. Mrozek, A. Szymczak, Chaos in the Lorenz equations: a computer assisted proof. Part III: the classical parameter values, *J. Diff. Equ.* **169**(2001) 17-56.

[27] K. Mischaikow, M. Mrozek, J. Reiss, A. Szymczak, Construction of Symbolic Dynamics from Experimental Time Series, *Physical Review Letters* **82**(1999) 1144.

[28] R.E. Moore, Interval analysis, *Prentice-Hall* (1966).

[29] M. Mrozek, Leray functor and the cohomological Conley index for discrete time dynamical systems, *Trans. Amer. Math. Soc.* **318**(1990) 149–178.

[30] M. Mrozek, Topological invariants, multivalued maps and computer assisted proofs in Dynamics, *Computers Math. Applic.* **32**(1996) 83–104.

[31] M. Mrozek, Index Pairs Algorithms, *Foundations of Computational Mathematics* **6**(2006) 457-493.

[32] M. Mrozek, Čech Type Approach to Computing Homology of Maps, *Discrete and Computational Geometry* **44**(2010) 546–576.

[33] M. Mrozek, P. Pilarczyk, N. Żelazna, Homology algorithm based on acyclic subspace, *Computers and Mathematics with Applications* **55**(2008) 2395–2412.

[34] M. Mrozek, B. Batko, Coreduction homology algorithm, *Discrete and Computational Geometry* **41**(2009) 96–118.

[35] M. Mrozek, T. Wanner, Coreduction Homology Algorithm for Inclusions and Persistence, *Computers and Mathematics with Applications* **60.10**(2010) 2812-2833 (DOI: 10.1016/j.camwa.2010.09.036).

[36] M. Mrozek, M. Żelawski, A. Gryglewski, S. Han, A. Krajniak, Homological methods for extraction and analysis of linear features in multidimensional images, *Pattern Recognition* **45**(2012) 285–298 (DOI:10.1016/j.patcog.2011.04.020).

[37] M. Mrozek, H. Wagner, Coreduction Homology Algorithm for Continuous Maps, in preparation.

[38] J.R. Munkres, Elements of Algebraic Topology, *Addison-Wesley* (1984).

[39] J. ROBBIN AND D. SALAMON, Dynamical systems, shape theory, and the Conley index, *Ergodic Theory and Dynamical Systems* **8***(1988) 375–393.
[40] S. SMALE, Differentiable dynamical systems, *Bull. Amer. Math. Soc.* **73**(1967) 747–817.
[41] C. SPARROW, The Lorenz Equations: Bifurcations, Chaos and Strange Attractors, *Springer-Verlag* (1982).
[42] N.E. STEENROD, Regular cycles of compact metric spaces, *Ann. Math.* **41**(1940) 833–851.
[43] A.M. STUART, A.R. HUMPHRIES, Dynamical Systems and Numerical Analysis, *Cambridge Univ. Press* (1998).
[44] A. SZYMCZAK, The Conley index and symbolic dynamics, *Topology* **35**(1996) 287–299.
[45] A. SZYMCZAK, The Conley index for decompositions of isolated invariant sets, *Fundamenta Mathematicae* **148**(1995) 71–90.
[46] A. SZYMCZAK, A combinatorial procedure for finding isolating neighborhoods and index pairs, *Proc. Royal Soc. Edinburgh, Ser. A* **127A**(1997) 1075–1088.
[47] M. WARMUS, Calculus of approximations, *Bull. de l'Académie Polonaise des Sciences Cl. III)* **4**(1956) 253–259.
[48] T. WAŻEWSKI, Une méthode topologique de l'examen du phénomène asymptotique relativement aux équations différentielles ordinaires, *Rend. Accad. Nazionale dei Lincei, Cl. Sci. fisiche, mat. e naturali* **8.3**(1947) 210–215.
[49] P. ZGLICZYŃSKI, K. MISCHAIKOW, Rigorous numerics for partial differential equations: The Kuramoto-Sivashinsky equation, *Foundations of Computational Mathematics* **1**(2001) 255–288.

INSTITUTE OF COMPUTER SCIENCE, JAGIELLONIAN UNIVERSITY, UL. PROF. STANISŁAWA L OJASIEWICZA 6, 30-348 KRAKÓW, POLAND AND CHAIR OF COMPUTATIONAL MATHEMATICS, WSB-NLU, UL. ZIELONA 27, 33-300 NOWY SĄCZ, POLAND

E-mail address: `Marian.Mrozek@ii.uj.edu.pl`

Euler Calculus with Applications to Signals and Sensing

Justin Curry, Robert Ghrist, and Michael Robinson

ABSTRACT. This article surveys the Euler calculus — an integral calculus based on Euler characteristic — and its applications to data, sensing, networks, and imaging.

1. Introduction

This work surveys the theory and applications of EULER CALCULUS, an integral calculus built with the Euler characteristic, χ, as a measure. While the theory engages an ethereal swath of topology (complexes, sheaves, and cohomology), the applications (to signal processing, data aggregation, and sensing) are concrete. These notes are meant to be read by both pure and applied mathematicians.

In the mid-1970s, MacPherson [60] and Kashiwara [53] independently published seminal works on constructible sheaves. Their respective motivations appeared quite different. MacPherson was interested in answering a conjecture of Deligne and Grothendieck on the theory of Chern classes for complex algebraic varieties with singularities. Kashiwara had been following up on work from his 1970 thesis on the algebraic study of partial differential equations via D-modules. In each setting — singularities, solutions, and obstructions — were understood using sheaf theory. Both MacPherson and Kashiwara made use of constructible sheaves and functions to provide algebraic characterizations of the local nature of singularities. Both provided index-theoretic formulae and developed a calculus relying on Euler characteristic. If it was in sheaf theory that Euler calculus was born, it had an earlier conception in geometry, going back at least to work of Blaschke [13] and perhaps before, though neither he nor those who followed (Hadwiger, Groemer, Santalo, Federer, Rota, etc.) developed the full calculus that arose from sheaves.

The language of Euler calculus was slow to form and be appreciated. The short survey paper of Viro [78] cited MacPherson and mentioned simple applications of the Euler integral to algebraic geometry. The short survey paper of

---

2010 *Mathematics Subject Classification.* Primary 55N30; Secondary 53C65, 28E99.

Key words and phrases. Euler characteristic, sheaves, cohomology, integral geometry, signal processing.

All authors supported by DARPA DSO - HR0011-07-1-0002, AFOSR FA9550-09-1-0643, and ONR N000140810668.

Schapira [**71**] relied more upon Kashiwara [**53, 54**] and was full of interesting formulations and applications — it was explicitly motivated by the work of Guibas, Ramshaw, and Stolfi [**47**] in computational geometry. In addition, the followup paper of Schapira [**72**] contained a prescient application of Euler integral transforms to problems of tomography and reconstruction of images from the Euler characteristics of slicing data. In the decade following these works, the language of Euler integration was used infrequently, hiding mostly in works on real-algebraic geometry and constructible sheaves and paralleled in the combinatorial geometry literature [**43, 68, 63, 58, 22, 70**].

There has been a recent renaissance of appreciation for Euler calculus. Some of this activity comes from the role of Euler characteristic as an elementary type of MOTIVIC MEASURE in motivic integration [**24, 45**]. Applications to algebraic geometry seem to be the primary impetus for interest in the subject [**78, 61, 56, 45**]. Parallel applications to integral geometry also have recently emerged. This survey comes out of the recent applications [**5, 4**] to problems in sensing, networks, signal processing, and data aggregation. These applications, presaged by Schapira [**72**], are poised to impact a number of problems of contemporary relevance.

It is somewhat remarkable that Schapira's deep insights saw little-to-no followup. One explanation is that in this, as in many other things, Schapira is ahead of his time. However, the language in which his results were couched — sheaf theory — was and is beyond the grasp of nearly all researchers in the application areas to which his results were directed. In the intervening decades, algebraic-topological methods have become more numerous and palatable to scientists and engineers, and the basics of homology and cohomology are now not so foreign outside of Mathematics departments. The same cannot yet be said of sheaf theory. It is with this in mind that this article exposits the Euler calculus from both explicit/applied and implicit/sheaf-theoretic perspectives.

The article begins with a concrete presentation of the Euler integral (§2-4); continues with a gentle if brief introduction to the topology undergirding the Euler characteristic (§5-8); then advances to the sheaf-theoretic view (§9-13). With this full span of concepts established, this article turns to the many applications of the Euler calculus to engineering systems and data aggregation (§14-15). We emphasize issues connected with implementation of the Euler calculus, including numerical approximation (§16-18), the use and inversion of integral transforms (§19-23), and the extension to a real-valued theory (§24-28). This last development, motivated by the need to build an honest numerical analysis for the Euler calculus, flows back to the abstract sea from which Euler calculus was born, by yielding fundamental connections to Morse theory. The article concludes (§29) with a collection of open problems and directions for further research.

The all-encompassing title of this work is a misnomer: our applications of the Euler calculus focus primarily on problems inspired by engineering systems and data. This is by no means the sole — or even most important — application of the Euler calculus. Other survey articles on applications of the Euler calculus (*e.g.*, [**45**]) detail applications that have no overlap with those of this paper at all: it is a broad subject. Several classical results in topology/geometry (the Gauss-Bonnet and Riemann-Hurwitz theorems) are both simplified and illuminated by an application of the Euler calculus. More applications are either implicit or emerging in the literature:

(1) A careful reading of work by R. Adler and others on Gaussian random fields [1, 3], in which one wants to compute the expected Euler characteristic of an excursion set of a random smooth distribution over a domain, reveals the generous use of Euler calculus, without the language. Recent preprints [2, 14] incorporate this language.
(2) The exciting work on persistent homology and associated barcodes for data [82, 20, 36] has a recent connection to Euler calculus. O. Bobrowski and M. Strom Borman [14] define the Euler characteristic of a barcode and relate this quantity to Euler integrals.
(3) The work of S. Gal on configuration spaces [33] and its applications to robotics [31] resonates with similar constructions in the algebraic geometry literature [45] and is clearly awaiting an Euler calculus reinterpretation.
(4) Computational complexity of Euler characteristic computation for semi-algebraic sets has been considered by Basu [10], with recent work of [11] identifying Euler characteristic as an important obstruction in complexity theory related to the classical Toda theorem. This hints at the use of Euler integrals in computational complexity of constructible functions.

It is to be hoped that other problems in Applied Mathematics and Statistics are equally amenable to simplification by means of this elegant and efficacious theory.

The Combinatorial Formulation

2. Euler characteristic

The Euler characteristic is a generalization of counting. Given a finite discrete set X, the Euler characteristic is its cardinality $\chi(X) = |X|$. If one connects two points of X together by means of an edge (in a cellular/simplicial structure), the resulting space has one fewer component and the Euler characteristic is decremented by one. Continuing inductively, the Euler characteristic counts vertices with weight $+1$ and edges with weight -1. This intuition of counting connected components works at first; however, the addition of an edge producing a cycle does not change the count of connected components. To fill in such a cycle with a 2-cell would return to the setting of counting connected components again, suggesting that 2-cells be weighted with $+1$. This intuition of counting with ± 1 weights inspires the following combinatorial definition of χ.

Given a space X and a decomposition of X into a finite number of open cells $X = \coprod_\alpha \sigma_\alpha$, where each k-cell σ_α is homeomorphic to $\mathbb{R}^k$, the EULER CHARACTERISTIC of X is defined as

(2.1) $$\chi(X) = \sum_\alpha (-1)^{\dim \sigma_\alpha}.$$

For an appropriate class of *"tame"* spaces (see §3), this quantity is well-defined and independent of the cellular decomposition of X. This combinatorial Euler characteristic is a homeomorphism invariant, but, as defined, is not a homotopy invariant for non-compact spaces, as, e.g., it distinguishes $\chi((0,1)) = -1$ from $\chi([0,1]) = 1$. Among compact finite cell complexes, it is a homotopy invariant, as will be shown in §7.

Example 2.1. (1) Euler characteristic completely determines the homotopy type of a compact connected graph.
(2) Euler characteristic is also a sharp invariant among closed orientable 2-manifolds: $\chi = 2 - 2g$, where g equals the genus.
(3) Any compact convex subset of $\mathbb{R}^n$ has $\chi = 1$. Removing k disjoint convex open sets from such a convex superset results in a compact space with Euler characteristic $1 - k(-1)^n$.
(4) The n-dimensional sphere $\mathbb{S}^n$ has $\chi(\mathbb{S}^n) = 1 + (-1)^n$.

3. Tame topology

Euler characteristic requires some degree of finiteness to be well-defined. This finiteness, when enlarged to unions, intersections, and mappings of spaces, demands a behavior that is best described as *tameness*. Different mathematical communities have adopted different schemes for imposing tameness on subsets of Euclidean space. Computer scientists often focus on piecewise linear (PL) spaces, describable in terms of affine spaces and matrix inequalities. Combinatorial geometers sometimes use a generalization called POLYCONVEX sets [58, 63]. Algebraic geometers tend to prefer SEMIALGEBRAIC sets — subsets expressible in terms of a finite number of polynomial inequalities. Analysts prefer the use of analytic functions, leading to a class of sets called SUBANALYTIC [75]. Logicians have recently created an axiomatic reduction of these classes of sets in the form of an O-MINIMAL STRUCTURE, a term derived from *order minimal*, in turn derived from model theory. The text of Van den Dries [77] is a beautifully clear reference.

For our purposes, an o-minimal structure $\mathcal{O} = \{\mathcal{O}_n\}$ (over $\mathbb{R}$) denotes a sequence of Boolean algebras $\mathcal{O}_n$ of subsets of $\mathbb{R}^n$ (families of sets closed under the operations of intersection and complement) which satisfies simple axioms:

(1) $\mathcal{O}$ is closed under cross products;
(2) $\mathcal{O}$ is closed under axis-aligned projections $\mathbb{R}^n \to \mathbb{R}^{n-1}$;
(3) $\mathcal{O}$ contains all algebraic sets (zero-sets of polynomials);
(4) $\mathcal{O}_1$ consists of all finite unions of points and open intervals.

Elements of $\mathcal{O}$ are called TAME or, more properly, DEFINABLE sets. Canonical examples of o-minimal systems include semialgebraic sets and subanalytic sets. Note the last axiom: the finiteness imposed there is the crucial piece that drives the theory. (The open intervals need not be bounded, however.)

Given a fixed o-minimal structure, one can work with tame sets with relative ease: *e.g.*, the union and intersection of two tame sets are again tame. A (not necessarily continuous) function between tame spaces is tame (or definable) if its graph (in the product of domain and range) is a tame set. A definable homeomorphism is a tame bijection between tame sets. To repeat: *definable homeomorphisms are not necessarily continuous*. Such a convention makes the following theorem concise:

Theorem 3.1 (Triangulation Theorem [77]). *Any tame set is definably homeomorphic to a subcollection of open simplices in (the geometric realization of) a finite Euclidean simplicial complex.*

The analogue of the Triangulation Theorem for tame mappings is equally important:

Theorem 3.2 (Hardt Theorem [77]). *Given a tame mapping $F: X \to Y$, Y has a definable partition into cells Y_α such that $F^{-1}(Y_\alpha)$ is definably homeomorphic to*

$U_\alpha \times Y_\alpha$ for U_α definable, and F restricted to this inverse image acts as projection to Y_α.

The Triangulation Theorem implies that tame sets always have a well-defined Euler characteristic, as well as a well-defined dimension (the max of the dimensions of the simplices in a triangulation). The surprise is that these two quantities are not only topological invariants with respect to definable homeomorphism; they are complete invariants.

Theorem 3.3 (Invariance Theorem [77]). *Two tame sets in an o-minimal structure are definably homeomorphic if and only if they have the same dimension and Euler characteristic.*

This result reinforces the idea of a definable homeomorphism as a SCISSORS EQUIVALENCE. One is permitted to cut and rearrange a space (or, even, a mapping) with abandon. Recalling the utility of such scissors work in computing areas of planar sets, the reader will not be surprised to learn of a deep relationship between tame sets, the Euler characteristic, and integration.

4. The Euler integral

We now have all the tools at our disposal for an integral calculus based on χ. The *measurable sets* in this theory are the tame sets in a fixed o-minimal structure. From Theorem 3.1 it follows that each such set has a well-defined Euler characteristic. The Euler characteristic, like a measure is ADDITIVE:

Lemma 4.1. *For A and B tame,*

(4.1) $$\chi(A \cup B) = \chi(A) + \chi(B) - \chi(A \cap B).$$

PROOF. The result follows from (1) the Triangulation Theorem, (2) the well-definedness of χ, and (3) counting cells. □

Additivity suggests converting χ into a (signed, integer-valued) measure $d\chi$ against which a indicator function is integrated in the obvious manner:

(4.2) $$\int_X 1_A \, d\chi := \chi(A),$$

The additivity of χ is, crucially, finite; the limiting process used in (standard) measure theory is certainly inapplicable. The natural set of *measurable functions* in this theory are definable functions with finite, discrete image. For the sake of convenience and clear presentation, we use the integers $\mathbb{Z}$ as range and invoke a compactness assumption. In what follows, X will be assumed a tame space in a fixed o-minimal structure.

An integer-valued function $h: X \to \mathbb{Z}$ is CONSTRUCTIBLE if all of its level sets $h^{-1}(n) \subset X$ are tame. Denote by $CF(X)$ the set of bounded compactly supported constructible functions on X. The use of compact support is not strictly needed and is done for convenience. We may sometimes bend this criterion without warning: *caveat lector*. The integrable functions of Euler calculus on X are, precisely, $CF(X)$.

The EULER INTEGRAL is defined to be the homomorphism $\int_X: CF(X) \to \mathbb{Z}$ given by:

(4.3) $$\int_X h \, d\chi = \sum_{s=-\infty}^{\infty} s \, \chi(\{h = s\}).$$

Each level set is tame and thus has a well-defined Euler characteristic. Alternately, one may use triangulation to write $h \in CF(X)$ as $h = \sum_\alpha c_\alpha 1_{\sigma_\alpha}$, where $c_\alpha \in \mathbb{Z}$ and $\{\sigma_\alpha\}$ is a decomposition of X into a disjoint union of open cells, yielding

$$(4.4) \qquad \int_X h\, d\chi = \sum_\alpha c_\alpha \chi(\sigma_\alpha) = \sum_\alpha c_\alpha (-1)^{\dim \sigma_\alpha}$$

That this sum is invariant under the decomposition into definable cells is a consequence of corresponding properties of the Euler characteristic.

Equation (4.3) is deceptive in explicit computations, since the level sets are rarely compact and the Euler characteristics must be computed with care. The following reformulation — a manifestation of the Fundamental Theorem of Integral Calculus — is more easily implemented in practice.

Proposition 4.2. *For any $h \in CF(X)$,*

$$(4.5) \qquad \int_X h\, d\chi = \sum_{s=0}^{\infty} \chi\{h > s\} - \chi\{h < -s\}$$

PROOF. Rewrite h as

$$h = \sum_{s=-\infty}^{\infty} s 1_{\{h=s\}}$$

$$= \sum_{s=0}^{\infty} s(1_{\{h \geq s\}} - 1_{\{h > s\}}) + \sum_{s=0}^{-\infty} s(1_{\{h \leq s\}} - 1_{\{h < s\}})$$

$$= \sum_{s=0}^{\infty} 1_{\{h > s\}} - 1_{\{h < -s\}},$$

where the last equality comes from telescoping sums. □

Example 4.3. In the example illustrated in Figure 1, the integral with respect to Euler characteristic is equal to 6, since the integrand can be expressed as the sum of indicator functions over six (contractible) closed sets with unit Euler characteristics, though such a decomposition is not assumed given.

Euler characteristic is like a measure in another aspect: it is multiplicative under cross products:

Lemma 4.4. *For X and Y definable sets,*

$$(4.6) \qquad \chi(X \times Y) = \chi(X)\chi(Y).$$

PROOF. The product $X \times Y$ has a definable cell structure using products of cells from X and Y. For cells $\sigma \subset X$ and $\tau \subset Y$, the lemma holds via the exponent rule since $\dim(\sigma \times \tau) = \dim \sigma + \dim \tau$. The rest follows from additivity. □

The assertion that $\int d\chi$ is an honest integral is supported by this fact and its corollary: the Euler integration theory admits a Fubini theorem. Calculus students know that $\int f(x,y)\, dA$ is computable via the double integral $\iint f(x,y)\, dx\, dy$. This familiar result is the image of a deeper truth about integrations and projections.

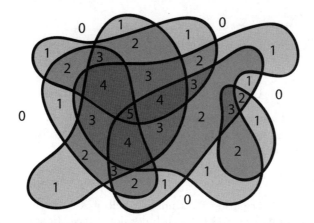

FIGURE 1. A simple example of an integrand $h \in CF(\mathbb{R}^2)$ whose Euler integral equals six: it is the sum of six characteristic functions over contractible discs in the plane.

Theorem 4.5 (Fubini Theorem). *Let $F: X \to Y$ be a tame mapping between tame spaces. Then for all $h \in CF(X)$,*

$$(4.7) \qquad \int_X h\, d\chi = \int_Y \left(\int_{F^{-1}(y)} h(x)\, d\chi(x) \right) d\chi(y).$$

PROOF. If X is homeomorphic to $U \times Y$ and F is projection to the second factor, the result follows from Lemma 4.4. The Hardt Theorem and additivity of the integral complete the proof. □

The (Co)Homological Formulation

As defined, both χ and the integral $\int d\chi$ are explicit, combinatorial, and concrete. Much of the depth and applicability of the theory derives from the pairing of these features with the algebraic-topological formulation. We provide a brief introduction to these methods, referring the interested reader to the better and more in-depth treatment in, *e.g.*, [50] for more details.

5. Homology

The reason for the topological invariance of ostensibly combinatorial χ lies in a particular CATEGORIFICATION — an enrichment of the combinatorial Euler characteristic with (first) linear and (then) homological algebra, yielding an algebraic-topological means of counting and canceling features in a topological space. The simplest such lifting is via cellular homology.

Consider a finite cell complex, X: a space built from standard compact cells (simplices, discs, cubes, or other simple components) assembled by means of attaching maps along cell faces.[1] The mechanics of counting used to define the Euler

---

[1] The reader for whom cell complexes are unfamiliar should consult, *e.g.*, [50, 59]. The reader for whom CW complexes are familiar should replace *cell complex* with *CW complex*.

characteristic of X,

$$\cdots \quad \{k\text{-cells}\} \quad \{(k-1)\text{-cells}\} \quad \cdots \quad \{1\text{-cells}\} \quad \{0\text{-cells}\},$$

may be lifted to a sequence of vector spaces over a field $\mathbb{F}$,

$$\cdots \quad C_k \quad C_{k-1} \quad \cdots \quad C_1 \quad C_0,$$

where each C_k is the $\mathbb{F}$-vector space with basis the k-cells of X. This collection of vector spaces is then enriched to a sequence of linear transformations that detail how the cells are connected together:

$$(5.1) \quad \cdots \longrightarrow C_k \xrightarrow{\partial_k} C_{k-1} \xrightarrow{\partial_{k-1}} \cdots \xrightarrow{\partial_2} C_1 \xrightarrow{\partial_1} C_0 \xrightarrow{\partial_0} 0.$$

Here, the linear transformation ∂_k sends a k-cell of X (a basis element of C_k) to the abstract sum of its oriented $(k-1)$-dimensional faces (a sum of basis elements in C_{k-1}). For $\mathbb{F} = \mathbb{F}_2$ the field of integers modulo 2, this is a simple sum of the faces, each with coefficient 1. For other fields $\mathbb{F}$, one must assign an orientation to all basis cells and compute the image of ∂ with coefficients ± 1, depending on orientation. The reader for whom homology is unfamiliar may want to work with $\mathbb{F}_2$ coefficients at first, for which $-1 = +1$: *cf.* the treatment in [29]. Coefficients in $\mathbb{R}$ are motivated by, *e.g.*, currents in electrical networks. The most generally informative coefficients are $\mathbb{Z}$, prompting the use of free abelian chain groups C_k and homomorphisms ∂_k. For simplicity of exposition, we will use fields and linear-algebraic constructs where possible.

The boundary of a boundary is null: $\partial_{k-1} \circ \partial_k = 0$ for all k. As such, for all k, im ∂_{k+1} is a subspace of ker ∂_k. The HOMOLOGY of $\mathcal{C}$, $H_\bullet(\mathcal{C})$, is a sequence of $\mathbb{F}$-vector spaces built from the following features of ∂. A CYCLE of $\mathcal{C}$ is a chain with empty boundary, *i.e.*, an element of ker ∂. Homology is an equivalence relation on cycles of $\mathcal{C}$. Two cycles in $Z_k = \ker \partial_k$ are said to be HOMOLOGOUS if they differ by something in $B_k = \operatorname{im} \partial_{k+1}$. The homology of X is the sequence of quotients $H_k(X)$, for $k \in \mathbb{N}$, given by:

$$(5.2) \quad H_k(X) = Z_k/B_k = \ker \partial_k / \operatorname{im} \partial_{k+1} = \text{cycles}/\text{boundaries}.$$

To repeat: a HOMOLOGY CLASS $[\alpha] \in H_k(X)$ is an equivalence class of cycles, two cycles being declared homologous if their difference is a boundary. Homology $H_k(X)$ inherits the sequential structure or GRADING $(k = 0, 1, 2, \ldots)$ of $\mathcal{C}$ and will be denoted $H_\bullet(X)$ when no particular grading is intended.

It takes some effort to get an intuition for homology, and several perspectives and examples are useful to this end. Homology is:

(1) **Multifarious:** The homology of a space X can be defined in numerous ways, each of which counts some feature and cancels according to a boundary-like accounting. Homology theories which count cells [cellular], patches in a covering [Čech], critical points of a smooth map $f: X \to \mathbb{R}$ [Morse], maps of cells into X [singular], and more, are, under the right 'tameness' assumptions, isomorphic.

(2) **Functorial:** Homology applies not only to spaces, but to mappings between spaces. For $f: X \to Y$, there is an induced homomorphism (or linear transformation) $H(f): H_\bullet(X) \to H_\bullet(Y)$ that respects grading, identities, and composition of mappings. This $H(f)$ indicates how f transforms cycles of X over to cycles of Y.

(3) **Invariant:** The homology of X is invariant not only under changes in cell decompositions, but also under homotopy equivalences: for f: $X \xrightarrow{\simeq} Y$, H(f) is an isomorphism.
(4) **Excisive:** Given $A \subset X$ a subcomplex, there is a RELATIVE HOMOLOGY $H_\bullet(X, A)$ given by taking the quotients $C_k(X, A) = C_k(X)/C_k(A)$ and using the induced boundary maps ∂_k. This relative homology has the effect of collapsing the subcomplex A to an abstract point: $H_\bullet(X, A) \cong H_\bullet(X/A, \{A\})$.

6. Cohomology

An algebraic mirror image of homology will prove salient to defining the Euler characteristic on tame but non-compact spaces. A COCHAIN COMPLEX is a sequence $\mathcal{C} = (C^\bullet, d)$ of $\mathbb{F}$-vector spaces C^k (or free abelian groups) and linear transformations (homomorphisms) $d^k : C^k \to C^{k+1}$ with the property that $d^{k+1} \circ d^k = 0$ for all k. The arrows are reversed:

$$0 \longrightarrow C^0 \xrightarrow{d^0} C^1 \xrightarrow{d^1} \cdots \xrightarrow{d^{k-1}} C^k \xrightarrow{d^k} C^{k+1} \xrightarrow{d+1} \cdots .$$

The COHOMOLOGY of a cochain complex is,

(6.1) $$H^k(\mathcal{C}) = \ker d^k / \operatorname{im} d^{k-1}.$$

Cohomology classes are equivalence classes of COCYCLES in ker d. Two cocylces are COHOMOLOGOUS if they differ by a COBOUNDARY in im d.

The simplest means of constructing cochain complexes is to dualize a chain complex $(C_\bullet, \partial)$. Given such a complex (with vector spaces over $\mathbb{F}$), define $C^k = C_k^\vee$, the dual space of linear functionals $C_k \to \mathbb{F}$ (or, in the group setting, the group Hom of homomorphisms to the coefficient group). The coboundary d is the adjoint of the boundary ∂, so that

$$d \circ d = \partial^\vee \circ \partial^\vee = (\partial \circ \partial)^\vee = 0^\vee = 0.$$

The coboundary operator d can be presented explicitly: $(df)(\tau) = f(\partial \tau)$. For τ a k-cell, d implicates the COFACES — those (k+1)-cells having τ as a face.

All of the constructs of homology — exact sequences, functoriality, excision — pass naturally to cohomology by means of dualization. One further construct is necessary for use in Euler calculus. Given a cochain complex $\mathcal{C}^\bullet$ on a space X, consider the subcomplex $\mathcal{C}_c^\bullet$ of cochains which are compactly supported. The coboundary map restricts to $d : C_c^k \to C_c^{k+1}$ with $d^2 = 0$, yielding a well-defined COHOMOLOGY WITH COMPACT SUPPORTS, $H_c^\bullet(X)$. This cohomology satisfies the following:
(1) $H_c^k(\mathbb{R}^n) = 0$ for all k except $k = n$, in which case it is of rank 1.
(2) $H_c^\bullet$ is not a homotopy invariant, but is a proper homotopy (and hence a homeomorphism) invariant.
(3) $H_c^\bullet(X) \cong H^\bullet(X)$ for X compact.

Example 6.1. Homology and cohomology are closely related by a classical duality that springs from Poincaré's original conceptions of homology. For a simple example, consider a compact surface with a polyhedral cell structure, and let $\mathcal{C}$ be the cellular chain complex with $\mathbb{F}_2$ coefficients. There is a dual polyhedral cell structure, yielding a chain complex $\overline{\mathcal{C}}$, where the dual cell structure places a vertex in the center of each original 2-cell, has 1-cells transverse to each original 1-cell, and, necessarily, has as its 2-cells neighborhoods of the original vertices. Each dual 2-cell is an n-gon, where n is the degree of the original 0-cell. Note that these cell decompositions are truly dual and have the effect of reversing the dimensions of cells: k-cells generating C_k are in bijective correspondence with $(2-k)$-cells (on a surface) generating a modified cellular chain group $\overline{C}_{2-k}$. The dual complex $\overline{\mathcal{C}}^{\bullet}$ consisting of $\overline{C}^k = \overline{C}_k^{\vee}$ and $\overline{d} = \overline{\partial}^{\vee}$ entwines with $\mathcal{C}_{\bullet}$ in a diagram:

$$\begin{array}{ccccccccc} 0 & \longrightarrow & C_2 & \xrightarrow{\partial} & C_1 & \xrightarrow{\partial} & C_0 & \longrightarrow & 0 \\ & & \downarrow{\cong} & & \downarrow{\cong} & & \downarrow{\cong} & & \\ 0 & \longrightarrow & \overline{C}^0 & \xrightarrow{\overline{d}} & \overline{C}^1 & \xrightarrow{\overline{d}} & \overline{C}^2 & \longrightarrow & 0 \end{array}$$

The vertical maps are isomorphisms and, crucially, the diagram is commutative. The equivalence of singular and cellular (co)homology implies that, for a compact surface with $\mathbb{F}_2$ coefficients, $H_k \cong H^{2-k}$. Such a result fails for a non-compact surface (cf., $\mathbb{R}^2$), unless one switches to $H_c^{\bullet}$. With this, and using a similar proof as in the 2-d case, one obtains:

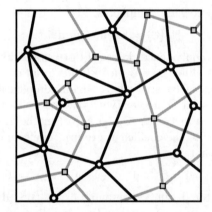

FIGURE 2. A polyhedral cell structure on a surface and its dual mirror homological-cohomological duality.

Theorem 6.2 (Poincaré duality). *For M an n-manifold, there is a natural isomorphism* $PD: H_c^k(M; \mathbb{F}_2) \to H_{n-k}(M; \mathbb{F}_2)$.

It follows that for M a compact n-manifold, $H_k(M; \mathbb{F}_2) \cong H_{n-k}(M; \mathbb{F}_2)$. The coefficients may be modified at the expense of worrying about orientability of the manifold and the torsional elements: see, *e.g.*, [50] for details. Poincaré duality does not hold in general for non-manifolds, but generalizations abound. In §11 and §13, a very broad and powerful extension will be used in defining Euler integrals.

7. Homotopy invariance of χ

Invariance of Euler characteristic can be pursued from many angles; *e.g.*, invariance under a refinement of cell structure can be ascertained through simple combinatorics. To get the full invariance under homotopy equivalence requires some basic homological algebra. This comes from lifting the notion of Euler characteristic and homology from a cell complex to an arbitrary (finite) chain complex

$\mathcal{C} = (C_\bullet, \partial)$, a sequence of free abelian groups C_k and homomorphisms ∂_k satisfying $\partial_{k-1}\partial_k = 0$ for all k. For such a sequence, the homology $H_\bullet(\mathcal{C})$ is, as before, ker ∂/im ∂ and the Euler characteristic is given via:

$$\tag{7.1} \chi(\mathcal{C}) = \sum_k (-1)^k \dim C_k.$$

Note that this is independent of the maps ∂_k and thus is sensible for any sequence of vector spaces. The reason for the alternating sum is to take advantage of cancelations that permit the following.

Lemma 7.1. *The Euler characteristic of a chain complex and its homology are identical, when both are defined.*

PROOF. Since $H_k = Z_k/B_k$, one has

$$\dim Z_k = \dim H_k + \dim B_k.$$

Since $\partial^2 = 0$, one has $C_k/Z_k \cong B_{k-1}$, so that

$$\dim C_k = \dim Z_k + \dim B_{k-1},$$

from which it follows that

$$\dim C_k = \dim H_k + \dim B_k + \dim B_{k-1}.$$

Multiply this equation by $(-1)^k$; the sum over k telescopes. □

Applying this to the chain complex for cellular homology and invoking the homotopy invariance of homology yields:

Corollary 7.2. *For X a finite compact cell complex,*

$$\tag{7.2} \chi(X) = \sum_k (-1)^k \dim H_k(X; \mathbb{R}).$$

Corollary 7.3. *χ is a homotopy invariant among finite compact cell complexes.*

This is helpful is computing integrals with respect to Euler characteristic: one need merely count holes as opposed to counting cells.

Example 7.4. The homological formulation of χ not only gives invariance — it also provides a potentially simple means of computing Euler integrals. Consider the integrand displayed in Figure 3. Constructing the appropriate triangulation and computing Euler characteristic may be involved. By combining Equation (4.5) with the homological definition, it is an easy matter (in this example at least) to compute the Euler integral.

8. Sequences

There is more to homology than simply counting cells or holes. The algebraic constructs built to support homology mirror the topological spaces implicated [35]. If a chain complex $\mathcal{C} = (C_\bullet, \partial)$ is the analogue of a topological space or cell complex, then the analogue of a continuous map is a CHAIN MAP — a map $\varphi_\bullet : C_\bullet \to C'_\bullet$ between chain complexes that is a homomorphism on chain groups respecting the

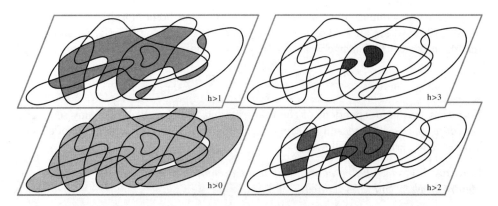

FIGURE 3. A simple example of Euler integration. Instead of triangulating, computing homology ranks of upper excursion sets is a simple method to determine the Euler integral.

grading and commuting with the boundary maps. This is best expressed in the form of a COMMUTATIVE DIAGRAM:

$$(8.1) \quad \begin{array}{ccccccc} \cdots & \longrightarrow & C_{n+1} & \xrightarrow{\partial} & C_n & \xrightarrow{\partial} & C_{n-1} & \xrightarrow{\partial} & \cdots \\ & & \downarrow{\varphi_\bullet} & & \downarrow{\varphi_\bullet} & & \downarrow{\varphi_\bullet} & & \\ \cdots & \longrightarrow & C'_{n+1} & \xrightarrow{\partial'} & C'_n & \xrightarrow{\partial'} & C'_{n-1} & \xrightarrow{\partial'} & \cdots \end{array}$$

Commutativity means that homomorphisms are path-independent in the diagram: $\varphi_\bullet \circ \partial = \partial' \circ \varphi_\bullet$. Since chain maps send neighbors to neighbors, the appropriate generalization of a homeomorphism to chain complexes is therefore an invertible chain map — one which is an isomorphism for all $C_k \to C'_k$. Clearly, a homeomorphism $f: X \to Y$ induces a chain map $f_\bullet: \mathcal{C}_X \to \mathcal{C}_Y$ from a cellular chain complex on X to the complex of the induced cell structure on Y: in this case, $f_\bullet$ is an isomorphism and, clearly, $H_\bullet(X) \cong H_\bullet(Y)$. Furthermore, if $f \simeq g: X \to Y$ are homotopic maps, then the induced homomorphisms are also isomorphisms: $H(f) \cong H(g)$.

Of equal importance is the algebraic analogue of a nullhomologous space. A chain complex $\mathcal{C} = (C_\bullet, \partial)$ is EXACT when its homology vanishes: $\ker \partial_n = \operatorname{im} \partial_{n+1}$ for all n. The most important examples of exact sequences are those relating homologies of various spaces and subspaces. The critical technical tool for the generation of such uses the technique of weaving an exact thread through a loom of chain complexes.

Theorem 8.1 (Snake Lemma). *Any short exact sequence of chain complexes*

$$0 \longrightarrow \mathcal{A}_\bullet \xrightarrow{i_\bullet} \mathcal{B}_\bullet \xrightarrow{j_\bullet} \mathcal{C}_\bullet \longrightarrow 0,$$

induces the LONG EXACT SEQUENCE:

$$(8.2) \quad \longrightarrow H_n(\mathcal{A}_\bullet) \xrightarrow{H(i)} H_n(\mathcal{B}_\bullet) \xrightarrow{H(j)} H_n(\mathcal{C}_\bullet) \xrightarrow{\delta} H_{n-1}(\mathcal{A}_\bullet) \xrightarrow{H(i)} ,$$

where δ is the induced CONNECTING MAP. Moreover, the long exact sequence is functorial: a commutative diagram of short exact sequences and chain maps

$$\begin{array}{ccccccccc} 0 & \longrightarrow & \mathcal{A}_\bullet & \longrightarrow & \mathcal{B}_\bullet & \longrightarrow & \mathcal{C}_\bullet & \longrightarrow & 0 \\ & & \downarrow f_\bullet & & \downarrow g_\bullet & & \downarrow h_\bullet & & \\ 0 & \longrightarrow & \tilde{\mathcal{A}}_\bullet & \longrightarrow & \tilde{\mathcal{B}}_\bullet & \longrightarrow & \tilde{\mathcal{C}}_\bullet & \longrightarrow & 0 \end{array}$$

induces a commutative diagram of long exact sequences

(8.3)
$$\begin{array}{ccccccccc} \longrightarrow & H_n(\mathcal{A}_\bullet) & \longrightarrow & H_n(\mathcal{B}_\bullet) & \longrightarrow & H_n(\mathcal{C}_\bullet) & \xrightarrow{\delta} & H_{n-1}(\mathcal{A}_\bullet) & \longrightarrow \\ & \downarrow H(f) & & \downarrow H(g) & & \downarrow H(h) & & \downarrow H(f) & \\ \longrightarrow & H_n(\tilde{\mathcal{A}}_\bullet) & \longrightarrow & H_n(\tilde{\mathcal{B}}_\bullet) & \longrightarrow & H_n(\tilde{\mathcal{C}}_\bullet) & \xrightarrow{\delta} & H_{n-1}(\tilde{\mathcal{A}}_\bullet) & \longrightarrow \end{array}$$

The exact sequence of chain complexes means that there is a short exact sequence for each dimension, and these short exact sequences fit into a commutative diagram with respect to the boundary operators. For details on the definition of δ, see [50, 35, 59] or any standard reference.

Example 8.2 (LES of pair). Given $A \subset X$ a subcomplex, the following short sequence is exact:

$$0 \longrightarrow C_\bullet(A) \xrightarrow{i_\bullet} C_\bullet(X) \xrightarrow{j_\bullet} C_\bullet(X, A) \longrightarrow 0 ,$$

where $i : A \hookrightarrow X$ is an inclusion and $j : (X, \varnothing) \hookrightarrow (X, A)$ is an inclusion of pairs. This yields the LONG EXACT SEQUENCE OF THE PAIR (X, A):

(8.4) $\quad \longrightarrow H_n(A) \xrightarrow{H(i)} H_n(X) \xrightarrow{H(j)} H_n(X, A) \xrightarrow{\delta} H_{n-1}(A) \longrightarrow ,$

The connecting map δ takes a relative homology class $[\alpha] \in H_n(X, A)$ to the homology class $[\partial \alpha] \in H_{n-1}(A)$.

Example 8.3 (Mayer-Vietoris). Another important sequence is derived from a decomposition of X into subcomplexes A and B. Consider the short exact sequence

(8.5) $\quad 0 \longrightarrow C_\bullet(A \cap B) \xrightarrow{\phi_\bullet} C_\bullet(A) \oplus C_\bullet(B) \xrightarrow{\psi_\bullet} C_\bullet(A + B) \longrightarrow 0,$

with chain maps $\phi_\bullet : c \mapsto (c, -c)$, and $\psi_\bullet : (a, b) \mapsto a + b$. The term on the right, $C_\bullet(A+B)$, consists of those chains which can be expressed as a sum of chains on A and chains on B. In cellular homology with A, B subcomplexes, $C_\bullet(A+B) \cong C_\bullet(X)$, resulting in the MAYER-VIETORIS SEQUENCE:
(8.6)
$$\longrightarrow H_n(A \cap B) \xrightarrow{H(\phi)} H_n(A) \oplus H_n(B) \xrightarrow{H(\psi)} H_n(X) \xrightarrow{\delta} H_{n-1}(A \cap B) \longrightarrow$$

These exact sequences allow one to re-derive the combinatorial properties of the Euler characteristic. The following come from careful use of Lemmas 7.1, exactness, and the two exact sequences above:

Proposition 8.4. *For $A, B \subset X$ compact and tame,*

(8.7) $\quad\quad\quad\quad \chi(A \cup B) = \chi(A) + \chi(B) - \chi(A \cap B)$

(8.8) $\quad\quad\quad\quad \chi(X - A) = \chi(X) - \chi(A)$

The combinatorial Euler characteristic χ is equal to χ_c, the Euler characteristic defined via cohomology with compact supports:

Lemma 8.5. *For X tame and locally compact,*

(8.9) $$\chi(X) := \sum_\sigma (-1)^{\dim \sigma} = \chi_c := \sum_{k=0}^\infty (-1)^k \dim H_c^k(X; \mathbb{R})$$

PROOF. If U is an open subset of a locally compact X we have the following long exact sequence in cohomology:

(8.10) $$\longrightarrow H_c^n(U) \longrightarrow H_c^n(X) \longrightarrow H_c^n(X - U) \longrightarrow ,$$

whence, from Lemma 7.1, $\chi_c(X) = \chi_c(U) + \chi_c(X - U)$. Invoking the Triangulation Theorem, fix a triangulation of X with U a cell of maximal dimension k in X. This is necessarily open and so applying the above result allows us to peel off a sum of $(-1)^k$. Repeat this for all (finitely many!) k-dimensional cells, and note that $X - (U_1 \cup \cdots \cup U_{n(k)})$ is a tame set of dimension $k - 1$. Repeating the argument inductively with respect to dimension concludes the proof. □

This is the impetus for using a cohomological version of the Euler characteristic — it allows one to manage non-compact spaces with a cell structure. Cohomological formulations quickly become complicated if local compactness is not assumed (*e.g.*, a compact triangle with one open face removed) [12]. A better method still is via categorifying a constructible set to a richer data structure: a constructible SHEAF.

The Sheaf-Theoretic Formulation

Tame sets possess triangulations of such virtue as to permit a combinatorial Euler characteristic. This simple counting procedure foreshadows the richer (co)homological formulation from algebraic topology, which, in turn, returns the principle of independence of triangulation. At yet higher elevations, this principle transmutes into invariance of the Euler integral by relating it to an intrinsic quantity — the Euler characteristic of a constructible sheaf. Far from being useless abstraction, this lifting provides both tools and insights for the applications to come. We proceed, therefore, via a brief introduction to sheaves.

It is of general interest to aggregate local observations into globally coherent ones. The connection between local and global properties has been captured in many famous mathematical theorems attached to equally famous names: Gauss-Bonnet, Poincaré-Hopf, Riemann-Roch, Atiyah-Singer and so on. In many of these results local quantities are integrated to produce interesting invariants. These invariants, traditionally algebraic in nature, can, for instance, provide obstructions to extending local constructions globally.

Sheaf theory provides a systematic means of describing and deriving many local-to-global results. This intricate machinery, often coupled with and cast in the language of categories, has garnered a reputation for being abstruse and abstract. The authors are of the opinion that sheaves are extremely useful as data management tools and will gradually shed their intimidating appearance as incarnations of sheaves relevant to Applied Mathematics are found. One such incarnation is given by Euler integration and constructible functions. Although Euler calculus can be appreciated separately from sheaves, the connection between the two, expressed most clearly in the work of Schapira [**71**], serves to both deepen the foundations of Euler calculus and elucidate sheaves.

9. Presheaves

The most basic assignment of local data is captured in a structure called a PRESHEAF. A presheaf is a map $\mathcal{P}$ from open sets of X to (say) abelian groups in a manner that respects *restriction to subsets*. Specifically, $\mathcal{P}$ is a contravariant functor from open sets under inclusion to a category (of, in this paper, abelian groups). For a less abstract definition, the following will suffice: given $V \subset U$ open in X, there is an induced homomorphism $\mathcal{P}(V) \xleftarrow{r} \mathcal{P}(U)$ that respects composition. Namely, the map $\mathcal{P}(U) \xleftarrow{r} \mathcal{P}(U)$ is the identity transformation, and for $W \subset V \subset U$ one has

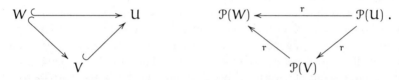

Note that, as in the case of cohomology and other CONTRAVARIANT theories, the algebraic maps reverse the direction of the topological maps. One calls $\mathcal{P}(U)$ the group of SECTIONS of $\mathcal{P}$ over U; an element $\gamma \in \mathcal{P}(U)$ is called a section. The presheaf condition implies that one can restrict sections uniquely. Although we define presheaves valued in abelian groups, one is free to use any algebraic data one might like to use: vector spaces, rings, modules, algebras, etc. Even more general data may be assigned: sets, spaces, spectra and even categories.[2]

The use of open sets for the assignees of data has many advantages over assigning data directly to points. However, pointwise data assignment can be recovered by a limiting process. For $x \in X$ and $\{U_i\}$ a nested sequence of open neighborhoods converging to x (that is, any neighborhood of x contains U_i for all i sufficiently large), one has a sequence of groups of sections and homomorphisms

$$\cdots \xrightarrow{r} \mathcal{P}(U_{i-1}) \xrightarrow{r} \mathcal{P}(U_i) \xrightarrow{r} \mathcal{P}(U_{i+1}) \xrightarrow{r} \cdots$$

The STALK of $\mathcal{P}$ at x, $\mathcal{P}_x$, is the group of equivalence classes of sequences $(\gamma_i)_i$ with $\gamma_{i+1} = r(\gamma_i)$ for $\mathcal{P}(U_i) \xrightarrow{r} \mathcal{P}(U_{i+1})$ and all i; and where two such sequences are equivalent if they eventually agree: $[(\gamma_i)] = [(\gamma'_i)]$ if and only if $\gamma_i = \gamma'_i$ for $i > N$, for some N. By using a more implicit definition — the stalk is the colimit over $\mathcal{P}(U_x)$ for all open sets U_x containing x — it follows that the stalk is a group and that $\mathcal{P}_x$ is independent of the system of neighborhoods chosen to limit to x.

10. Sheaves

A SHEAF is a presheaf $\mathcal{F}$ that respects *gluings* as well as restrictions. Consider two open subsets $U, V \subset X$. A presheaf is a sheaf if and only if for all U and V open in X, and sections γ_U, γ_V of U and V which agree on the overlap $U \cap V$, there exists a *unique* section of $U \cup V$ which agrees with γ_U and γ_V on the components.

---

[2]Presheaves of categories are often called PRESTACKS and have functors as restriction maps.

More generally, we require that this property hold for a family of open sets $\{U_i\}$ for i in some potentially large indexing set.

The canonical example of a non-sheaf presheaf is the presheaf that assigns to an open set U the group of constant functions $f : U \to \mathbb{Z}$, with restriction maps defined to be the identity. To see this is not a sheaf, consider two disjoint open sets U and V and the functions $f_U(x) = 5$ for $x \in U$ and $f_V(x) = 7$ for $x \in V$. Since the sets have empty intersection they vacuously agree there, despite the absence of a constant function $f : U \cup V \to \mathbb{Z}$ such that $f|_V = f_V$ and $f|_U = f_U$. One can mitigate the problem by instead assigning to any open set U the group of *locally constant functions*: this defines a sheaf, sometimes called $\widetilde{\mathbb{Z}}$. To an open subset it assigns continuous functions $f : U \to \mathbb{Z}$, which, by the discrete topology on $\mathbb{Z}$, are constant on connected components. Consequently the group of everywhere defined sections $\widetilde{\mathbb{Z}}(X)$ are exactly functions that are constant on connected components of X. The rank of this free abelian group calculates the number of connected components and is the most basic topological invariant of a space. This sheaf has even more information about X and has cohomological data exactly equal to the familiar cohomology of X with $\mathbb{Z}$ coefficients. This indicates another interpretation of sheaves — a sheaf is a generalized coefficient system for computing cohomology.

Sheaves, like simplicial complexes, are defined abstractly but have a geometric realization reminiscent of covering spaces. To any sheaf $\mathcal{F}$ is associated a topological space (also denoted $\mathcal{F}$) and a projection $\pi \colon \mathcal{F} \to X$. This ÉTALE SPACE is the disjoint union of all stalks $\mathcal{F}_x$, $x \in X$, outfitted with a topology so that:

(1) Fibers $\pi^{-1}(x)$ are discrete spaces with the structure of an abelian group.
(2) Algebraic operations in the fibers (addition/multiplication) yield continuous maps of the space $\mathcal{F}$.
(3) The projection π is a local homeomorphism: for each small neighborhood $U \subset X$, $\pi^{-1}(U)$ is a disjoint union of homeomorphic copies of U in $\mathcal{F}$.

A SECTION of $\mathcal{F}$ over an open set $U \subset X$ is a map $\gamma : U \to \mathcal{F}$ with $\pi \circ \gamma = \mathsf{Id}$. Denote by $\Gamma(U, \mathcal{F})$ the group of sections over U, defined in the obvious manner (via pointwise operations). The GLOBAL SECTIONS of $\mathcal{F}$ are denoted $\Gamma(X, \mathcal{F})$. The utility of the étale space perspective is that, as a space, the sheaf possesses global features whose detection and computation reveal important structure.

Note that the étale space of a sheaf provides an alternate path to defining sheaves. More precisely, let us temporarily term the pair $(\mathcal{F}, \pi)$ consisting of an étale space with its projection map an ÉTALE SHEAF. To such an étale sheaf we could associate a presheaf $\Gamma(\mathcal{F})$ that assigns to an open subset U in X the group of sections of the étale sheaf $\Gamma(U, \mathcal{F})$. It is a theorem (see [69] Thm 5.68) that $\Gamma(\mathcal{F})$ is actually a sheaf. Conversely, as noted above, starting with an abstract sheaf $\mathcal{F}$ and associating to it an étale sheaf will produce an isomorphic sheaf after applying $\Gamma(-)$, i.e., $\mathcal{F}(U) \cong \Gamma(U, \mathcal{F})$ for all U. Consequently one may use these two perspectives interchangeably. This also explains why in the literature one may see the terms *sheaf of sections* and *sheaf of groups* to refer to the same sheaf. The former term emphasizes a model of sheaves as étale sheaves whereas the latter term emphasizes the abstract assignment model we first described.

11. Sheaf operations

In all the examples of sheaves considered above, certain common features resolve into canonical constructions. These are the beginnings of sheaf theory — a means of working with data over spaces in a platform-independent manner. For example, every presheaf can be turned into a sheaf through a process called SHEAFIFICATION [16]. Conversely, every sheaf gives rise to a presheaf simply by forgetting the extra structure of sheaf, and these operations are, to a reasonable degree, inverses. Other operations in sheaf theory involve pushing forward and pulling back sheaves based on maps of the base spaces, defining the cohomology of a sheaf, and various constructions related to duality and the distinction between compactly and non-compactly supported sections. Instead of detailing these operations in full (as in, *e.g.*, [39, 16]), a few brief highlights with applications are given.

Morphisms: A MORPHISM $F\colon \mathcal{F} \to \mathcal{F}'$ of (pre)sheaves over X is a collection of maps $F(U)\colon \mathcal{F}(U) \to \mathcal{F}'(U)$ that are compatible with the restrictions internal to both $\mathcal{F}$ and $\mathcal{F}'$. One can pass to the stalk at x and get a well-defined map $F_x\colon \mathcal{F}_x \to \mathcal{F}'_x$. This allows one to address the notions of subsheaves, quotient sheaves, and exact sequences of sheaves on a stalk-by-stalk basis: the machinery is built to integrate based on stalk data. This becomes particularly relevant when defining cohomology for sheaves (§12), but is also requisite for understanding most other sheaf operations.

Direct image: Assume a continuous map $F\colon X \to Y$ of spaces. For $\mathcal{F}_X$ a sheaf over X, the DIRECT IMAGE or PUSHFORWARD of F is the sheaf $F_*\mathcal{F}_X$ on Y defined by $F_*\mathcal{F}_X(V) := \mathcal{F}_X(F^{-1}(V))$, for $V \subset Y$ open. A nice connection between the direct image and the group of global sections should be noted: If $p\colon X \to \star$ is the constant map then the pushforward of $\mathcal{F}$ from X to $\star$ returns a sheaf which is a single group $p_*\mathcal{F}_X = \Gamma(X, \mathcal{F})$. This example will be fundamental in understanding the theoretical connection between sheaf theory and the Euler calculus when X is compact. To handle the situation of non-compact X we use instead the direct image with compact supports

$$F_!\mathcal{F}_X(V) := \{s \in \mathcal{F}_X(F^{-1}(V)) \,;\, F|_{\mathsf{supp}(s)} \text{ proper}\}$$

Note that $\mathsf{supp}(s)$ is just the set of points x such that $s_x \neq 0$ — the image of s in the stalk at x. When F is already a proper continuous map we have $F_! = F_*$. If we again take p the constant map $p_!\mathcal{F} = \Gamma_c(X, \mathcal{F})$ is the group of global sections with compact support.

Inverse image: With $F\colon X \to Y$ as before, one can pull back a sheaf from the codomain to the domain. The INVERSE IMAGE or PULLBACK of a sheaf $\mathcal{F}_Y$ on Y is the sheaf on X defined (using the étale interpretation) as

$$F^*\mathcal{F}_Y := \{(x, y) \in X \times \mathcal{F}_Y : F(x) = \pi_Y(y)\},$$

where π is the canonical projection to Y. It is more awkward to define F^* in terms of presheaves, since one cannot guarantee that the forward image $F(U)$ of an open set $U \subset X$ is open in Y. This can be remedied by defining

$$F^*\mathcal{F}_Y(U) := \varinjlim_{F(U) \subset V} \mathcal{F}(V)$$

and this will be a presheaf. One can then sheafify this to define a pullback of sheaves without the étale perspective. One should note that the inverse image of $\mathcal{F}$ by F is sometimes written $F^{-1}\mathcal{F}$ as in [55]: our notation is closer to that of [52].

Duality: The canonical operations in sheaf theory are entwined. The morphisms between a direct and inverse image of a sheaf are related by a categorical ADJUNCTION — a form of duality. Let $\mathcal{F}$ and $\mathcal{G}$ be arbitrary sheaves on X and Y respectively and consider $F: X \to Y$ a continuous map; then

$$(11.1) \qquad \mathsf{Hom}(F^*\mathcal{G}, \mathcal{F}) \cong \mathsf{Hom}(\mathcal{G}, F_*\mathcal{F}),$$

where Hom denotes the group of sheaf morphisms. The motivation for calling this a form of duality comes from a related pair of operations: direct and inverse image with compact supports. As we will see later an adjunction between these two operations provides a vast generalization of Poincaré duality called VERDIER DUALITY [74, 27].

12. Sheaf Cohomology

There are two approaches to defining the cohomology of sheaves, each with particular advantages. One, defined using ČECH COHOMOLOGY, is computationally the most tractable and will be familiar to topologists. The other, which uses INJECTIVE RESOLUTIONS, is very useful for proving theorems and will be familiar to the reader comfortable with homological algebra.

Čech cohomology: Fix X a base space, $\{U_i\}$ an open cover of X, and a sheaf $\mathcal{F}$ over X. Now choose $k+1$ sets from $\{U_i\}$, say $U_{j_0}, \ldots, U_{j_k}$. Denote their intersection by $U_{j_0, \ldots, j_k}$, and let $\gamma_{j_0, \ldots, j_k}$ be an element of $\mathcal{F}(U_{j_0, \ldots, j_k})$. An assignment that takes every ordered tuple of $k+1$ sets from $\{U_i\}$ and specifies a section over the intersection is called a k-COCHAIN. The group of all such assignments,

$$C^k(\{U_i\}, \mathcal{F}) = \prod_{j_0 < \ldots < j_k} \mathcal{F}(U_{j_0, \ldots, j_k}),$$

is the kth cochain group of $\mathcal{F}$. For each $k \geq 0$ there is such a group and successive groups are connected via a differential described as follows: to define $d\alpha$ given $\alpha \in C^{k-1}(\{U_i\}, \mathcal{F})$, one must specify a value on every $k+1$-fold intersection. For each nonempty intersection of $k+1$ sets, one forgets each factor, say U_{j_i}, in the intersection, yielding a k-fold intersection with value $\alpha(U_{j_0, \ldots, \hat{j}_i, \ldots, j_k})$. Repeating this for $0 \leq i \leq k$, one obtains $k+1$ sections that can be combined after restriction:

$$d\alpha(U_{j_0, \ldots, j_k}) = \sum_{i=0}^{k} (-1)^i \alpha(U_{j_0, \ldots, \hat{j}_i, \ldots, j_k})|_{U_{j_0, \ldots, j_k}}.$$

Notice that the restriction is necessary, as the $\alpha(U_{j_0, \ldots, \hat{j}_i, \ldots, j_k})$ live in different abelian groups and so an alternating sum would not be well-defined, but the ability to restrict to a common abelian group is part of the definition of a sheaf. The situation is perhaps more easily visualized diagrammatically.

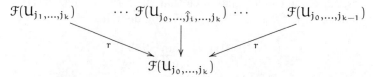

It is left to the reader to verify that $d^2 = 0$ and conclude that the cohomology $H^{\bullet}(C^{\bullet}(\{U_i\}, \mathcal{F}), d)$ is well-defined. For suitably nice covers $\{U_i\}$ this cohomology computes the cohomology $H^{\bullet}(X; \mathcal{F})$ of the sheaf $\mathcal{F}$. For example, to compute the cohomology of the constant sheaf $\widetilde{\mathbb{Z}}$ one may simply employ a GOOD COVER, *i.e.*, one whose intersections are contractible[3], and Čech cohomology will compute $H^\bullet(X;\mathbb{Z})$.

Injective resolutions: The Čech approach has the advantage of being computable, but, like many constructions, one would like to know if the output of said computation is independent of choices: the choice of cover, the choice of ordering, and so on. In order to put sheaf cohomology on intrinsic ground one appeals to injective resolutions. The basic operation one uses when dealing with sheaf cohomology is the global section functor:

$$\Gamma(X, -) : \mathsf{Sh}(X) \to \mathsf{Ab}$$

This is the map that takes a sheaf $\mathcal{F}$ on X and sends it to the abelian group $\mathcal{F}(X)$. The fact that this map is functorial comes from the observation that a map of sheaves includes maps on the level of global sections. More is true: if $F \colon \mathcal{F} \to \mathcal{G}$ is a map of sheaves that is injective on stalks, then $F_X \colon \mathcal{F}(X) \to \mathcal{G}(X)$ is injective ([**52**] II.2.2). The same statement is not true for surjections.

A canonical example of this asymmetry is given by the exponential sequence

$$0 \to \widetilde{\mathbb{Z}} \to \mathcal{O}_X \to \mathcal{O}_X^* \to 0$$

where $X = \mathbb{C} - 0$ is the punctured complex plane, $\mathcal{O}_X$ is the sheaf of holomorphic functions, and $\mathcal{O}_X^*$ nonvanishing holomorphic functions. The first map simply includes continuous integer valued functions into holomorphic ones. The second map takes a function $f \colon X \to \mathbb{C}$ and sends it to $g = e^{2\pi i f}$, which is manifestly nonzero. It is surjective on stalks because locally a non-zero function has a well defined logarithm, but this is not true globally. In particular, $g = z$ is a global section of $\mathcal{O}_X^*$ not hit by the exponential map.

This sequence is exact despite the fact that it is *not* an exact sequence of groups for every open set $\mathcal{U}$. Saying that a sequence of sheaves is exact is a local statement. One could take the following theorem as a definition of exactness:

Theorem 12.1 ([**52**] II.2.6). *A sequence of sheaves and sheaf morphisms*

$$\cdots \longrightarrow \mathcal{F}^{i-1} \longrightarrow \mathcal{F}^i \longrightarrow \mathcal{F}^{i+1} \longrightarrow \cdots$$

is exact if and only if the corresponding sequence of groups and group homomorphisms is exact for all $x \in X$

$$\cdots \longrightarrow \mathcal{F}^{i-1}_x \longrightarrow \mathcal{F}^i_x \longrightarrow \mathcal{F}^{i+1}_x \longrightarrow \cdots$$

The key feature of taking global sections is that it preserves exactness at the left endpoint. Namely, if

$$0 \longrightarrow \mathcal{F} \longrightarrow \mathcal{I}^0 \longrightarrow \mathcal{I}^1 \longrightarrow \cdots$$

is an exact sequence of sheaves then

$$0 \longrightarrow \mathcal{F}(X) \longrightarrow \mathcal{I}^0(X) \longrightarrow \mathcal{I}^1(X) \longrightarrow \cdots$$

is no longer an exact sequence of abelian groups, but $\mathcal{F}(X)$ still injects into $\mathcal{I}^0(X)$. Since $\mathcal{I}^\bullet(X)$ is no longer exact it has potentially interesting cohomology. When

---

[3]For non-good covers, Čech cohomology computes the E_2 page in a spectral sequence that converges to the full sheaf cohomology.

each of the $\mathfrak{I}^i$ are injective (see [52] for a good introduction) the cohomology of this complex is taken as the *definition* of sheaf cohomology of $\mathcal{F}$:

$$H^i(X, \mathcal{F}) := H^i(\mathfrak{I}^\bullet(X))$$

Injective sheaves can be quite mysterious. One may take on faith that an injective sheaf has the property that $r: \mathfrak{I}(X) \to \mathfrak{I}(U)$ is surjective for every open set U. Sheaves with this property are called FLABBY. In a flabby sheaf, one can always extend a section by zero to obtain a global section.

Since cohomology can also be calculated locally this data can be again assembled into a sheaf - called the ith COHOMOLOGY SHEAF. This is the sheaf $\mathcal{H}^i\mathcal{F}$ associated to the presheaf

$$\mathcal{H}^i\mathcal{F}: U \mapsto H^i(U, \mathcal{F}) = H^i(\mathfrak{I}^\bullet(U)).$$

Higher direct images: For a conceptual cartoon one may compare an injective resolution of a sheaf with a Taylor expansion of a function. Each term in the resolution has nicer algebraic (analytic) properties than the sheaf (function), and with infinitely many terms the resolution can replace the sheaf (function). The idea that a sheaf can be replaced by its injective resolution was instrumental to the development of the now standard technology of DERIVED CATEGORIES and functors. The sweeping generalization which Grothendieck introduced [44] allowed this construction of sheaf cohomology to be imitated with other operations — direct image, restriction and so on — to produce the HIGHER DIRECT IMAGE and other generalizations.

The higher direct image or right-derived direct image is the one most important to the sheaf-theoretic definition of the Euler integral. Take $F: X \to Y$ continuous and $\mathcal{F}$ a sheaf on X. Then the higher direct image is a sequence of sheaves $R^iF_*\mathcal{F}$ (one for each i) associated to the presheaves on Y

$$R^iF_*\mathcal{F}: V \mapsto H^i(F^{-1}(V), \mathcal{F})$$

One can define the higher direct image more intrinsically by replacing $\mathcal{F}$ with its injective resolution $\mathfrak{I}^\bullet$ and applying F_* level-wise to $\mathfrak{I}^\bullet$ then

$$RF_*\mathcal{F} := F_*\mathfrak{I}^\bullet \quad \text{and} \quad R^iF_*\mathcal{F} := \mathcal{H}^i(F_*\mathfrak{I}^\bullet).$$

This $RF_*\mathcal{F}$ is sometimes called the total right-derived pushforward and should be thought of as a complex of sheaves such that when you take the ith cohomology sheaf you get the ith right-derived pushforward $R^iF_*\mathcal{F}$. In the special case of a constant map $p: X \to \star$,

$$R^ip_*\mathcal{F} = H^i(X, \mathcal{F}),$$

the ith cohomology group of $\mathcal{F}$ on X. Similarly $RF_!$ is what you would expect — replace $\mathcal{F}$ with an injective resolution and apply $F_!$ level-wise to that. With $F = p: X \to \star$, we have

$$R^ip_!\mathcal{F} = H^i_c(X, \mathcal{F}),$$

the ith cohomology group with compact supports.

Since the combinatorial Euler characteristic only agrees with the intrinsic cohomological Euler characteristic when compact supports are used, an analogous construction is needed in the sheaf world. This construction is a form of duality.

Verdier Duality: One of the appealing features of sheaf theory is the access to the higher operations of the derived category. In this section we ask that our sheaves have some extra structure: for each open set U, $\mathcal{F}(U)$ is a $\Bbbk$-module for a fixed commutative ring $\Bbbk$ and the restriction maps respect this structure — they are $\Bbbk$-linear. In the language of Iversen [52], these are $\Bbbk$-sheaves.

For a simple example, consider an arbitrary $\Bbbk$-module D. This can be made into a sheaf $\widetilde{D}$ on X by taking the inverse image p^*D where $p \colon X \to \star$ is the constant map. Alternatively, this is the sheafification of the constant presheaf that assigns D to every open set.

Lemma 12.2.
$$\mathrm{Hom}(\widetilde{D}, \mathcal{F}) \cong \mathrm{Hom}(D, \Gamma(X, \mathcal{F}))$$

PROOF. This is immediate from the first adjunction recorded
$$\mathrm{Hom}(p^*D, \mathcal{F}) \cong \mathrm{Hom}(D, p_*\mathcal{F})$$
□

As a special case of this consider when $D = \Bbbk$. The isomorphism becomes much more useful to understand
$$\mathrm{Hom}(\widetilde{\Bbbk}, \mathcal{F}) \cong \mathrm{Hom}(\Bbbk, \Gamma(X, \mathcal{F})) \cong \Gamma(X, \mathcal{F})$$
where the last isomorphism comes from noting that any homomorphism is determined uniquely by where it sends $1 \in \Bbbk$. This is one instance of how adjunctions reveal information about sections.

As hinted earlier, adjunctions between other functors provide useful insight into sheaves and their sections.

Theorem 12.3 (Global Verdier Duality). *Suppose* $F \colon X \to Y$ *is a continuous map of locally compact spaces, and* $\mathcal{F}$ *and* $\mathcal{G}$ *are sheaves on* X *and* Y *then there exists an operation* $F^!$ *such that*
$$\mathrm{Hom}(RF_!\mathcal{F}, \mathcal{G}) \cong \mathrm{Hom}(\mathcal{F}, F^!\mathcal{G})$$
where the left and right hand sides live in the derived category of sheaves on Y *and* X *respectively.*

Example 12.4 (Classical duality). Consider $F = p \colon X \to \star$ the constant map, $\mathcal{F} = \widetilde{\Bbbk}$, $\mathcal{G} = \Bbbk$. Verdier duality then says
$$\mathrm{Hom}(Rp_!\widetilde{\Bbbk}, \Bbbk) \cong \mathrm{Hom}(\widetilde{\Bbbk}, p^!\Bbbk).$$
The second term is isomorphic to $\Gamma(X, p^!\Bbbk)$ as noted above. The first term is understood as usual: we replace $\widetilde{\Bbbk}$ with its injective resolution $\mathcal{I}^\bullet$ and by applying $Rp_!$ we are just taking compact supports. Our reinterpreted isomorphism then becomes
$$\mathrm{Hom}(\Gamma_c(X, \mathcal{I}^\bullet), \Bbbk) \cong \Gamma(X, p^!\Bbbk)$$
This may still seem discouraging. The Hom-term is still mysterious. To clarify, assume that $\Bbbk$ is a field and so everything in sight is a vector space and complexes thereof. Consequently Hom just indicates linear maps to the ground field $\Bbbk$, i.e. linear functionals. So when the complex of vector spaces,
$$\longrightarrow \Gamma_c(X, I^{k-1}) \longrightarrow \Gamma_c(X, I^k) \longrightarrow \Gamma_c(X, I^{k+1}) \longrightarrow$$
is dualized we must take the transposes of the connecting linear maps, thereby reversing them
$$\longleftarrow \Gamma_c(X, I^{k-1})^\vee \longleftarrow \Gamma_c(X, I^k)^\vee \longleftarrow \Gamma_c(X, I^{k+1})^\vee \longleftarrow$$

Notice that the grading in the dual complex decreases by one, as in homology, but we nevertheless want to take the cohomology of both sides. Fortunately, there is a general notational device that allows one to think of homology as cohomology, merely by allowing grading in negative degrees. We therefore place the dual of the kth vector space in degree $-k$, the dual of $(k-1)^{th}$ vector space in degree $1-k$ and so on. Notice that the dual map now augments $-k$ to $1-k$, so this agrees with cohomological conventions.[4]

In the case where X is an oriented n-manifold $p^!\Bbbk = \widetilde{\Bbbk}[n]$ the constant sheaf concentrated in degree $-n$. If we replace $\widetilde{\Bbbk}[n]$ with its injective resolution, $\mathcal{I}^\bullet[n]$, the isomorphism above is then interpreted as a chain complex isomorphism relating

$$\Gamma_c(X,\mathcal{I}^\bullet)^\vee \quad \text{and} \quad \Gamma(X,\mathcal{I}[n]^{-\bullet}) = \Gamma(X,\mathcal{I}^{n-\bullet}),$$

which, after taking cohomology, yields,

$$H^k_c(X,\widetilde{\Bbbk})^\vee = H^{n-k}(X,\widetilde{\Bbbk})$$

the classical Poincaré duality.

The operation $p^!$ is curious in that it associates to any complex of vector spaces a sheaf on X. This is called the DUALIZING COMPLEX $\omega_X := p^!\Bbbk$ and it relies only on the topological properties of X. The dualizing complex allows one to associate to any sheaf of real vector spaces $\mathcal{F}$ a dual sheaf $\mathbb{D}\mathcal{F}$ that comes from treating sheaf morphisms from $\mathcal{F}$ to ω_X as a sheaf — it is the sheaf that assigns to an open subsets U the group of sheaf morphisms from $\mathcal{F}|_U$ to $\omega_X|_U$.

The above example can be generalized as follows:

Theorem 12.5 ([27] Thm 3.3.10).

(12.1) $$H^k_c(X,\mathcal{F})^\vee \cong H^{-k}(X,\mathbb{D}\mathcal{F}).$$

and thus

(12.2) $$\chi_c(X,\mathcal{F}) = \chi(X,\mathbb{D}\mathcal{F}).$$

This will be useful below because it implies that the switch between compactly supported and regular cohomology is a switch between a sheaf and its dual.

13. Constructible Functions and Sheaves: A Dictionary

Persistent readers shall now be rewarded for their efforts. Euler calculus lifts to sheaf theory, whence its operations emanate. Recall that a constructible function is an integer-valued function $f: X \to \mathbb{Z}$ such that the level sets form a tame partition of X. The definition of a CONSTRUCTIBLE SHEAF is similar: X has a tame partition $\{X_\alpha\}$ such that $\mathcal{F}|_{X_\alpha}$ is locally constant. A sheaf $\mathcal{F}$ on a topological space X is locally constant if there is an open cover $\{U_i\}$ of X such that $\mathcal{F}|_{U_i}$ is isomorphic to the constant sheaf. By *constructible sheaf*, we will mean sheaves valued in real vector spaces of finite dimension: more general data values are possible.

The connection between constructible functions and sheaves is given by Euler characteristic, but the connection is subtle. Suppose one is given a constructible sheaf $\mathcal{F}$, then one could define a constructible function h as follows

$$h(x) := \chi(\mathcal{F})(x) = \dim(\mathcal{F}_x).$$

---

[4]The interested reader is encouraged to consult [35] or any other good text on homological algebra for notational conventions.

For example, suppose that $A \subset X$ is a tame set. Then consider the constant sheaf $\widetilde{\mathbb{R}}_A$ supported on A, i.e., it is the constant sheaf with stalk $\mathbb{R}$ for points on A and zero elsewhere. In this case
$$\chi(\widetilde{\mathbb{R}}_A)(x) = 1_A(x)$$
is the indicator function on A.

This definition will have trouble producing constructible functions that take values in the negative integers. To fix this we consider a bounded complex of constructible sheaves with zero maps connecting each sheaf:
$$\mathcal{F}^\bullet = \cdots \xrightarrow{0} \mathcal{F}^{i-1} \xrightarrow{0} \mathcal{F}^i \xrightarrow{0} \mathcal{F}^{i+1} \xrightarrow{0} \cdots$$
A wider class of constructible functions can be achieved simply by setting
$$h(x) := \chi(\mathcal{F}^\bullet)(x) = \sum_i (-1)^i \dim(\mathcal{F}^i_x).$$

In the case where the maps in $\mathcal{F}^\bullet$ are not identically zero one must instead use the LOCAL EULER-POINCARÉ INDEX
$$\chi(\mathcal{F}^\bullet)(x) = \sum_i (-1)^i \dim(\mathcal{H}^i \mathcal{F}^\bullet_x)$$
which generalizes the zero-map case because there $\mathcal{H}^i \mathcal{F}^\bullet = \mathcal{F}^i$.

To see that every constructible function is obtained via this lifted perspective, start with $h \in \mathrm{CF}(X)$ and consider the partition of X via level sets $h^{-1}(n)$, $n \in \mathbb{Z}$. Consider the following two-term complex of constructible sheaves $\widetilde{\mathbb{R}}_h$ (we omit the grading for simplicity):

(13.1) $$0 \xrightarrow{0} \bigoplus_{n<0} \widetilde{\mathbb{R}}^{-n}_{X_n} \xrightarrow{0} \bigoplus_{n>0} \widetilde{\mathbb{R}}^n_{X_n} \xrightarrow{0} 0.$$

The second term occupies the zeroth slot in the complex (such an assignment is necessary for the alternating sum to make sense) and $\widetilde{\mathbb{R}}^n$ indicates the constant sheaf of $\mathbb{R}^n$; thus $\chi(\mathcal{F}) = h(x)$.

One can now define Euler integration via sheaf theory: let $h \in \mathrm{CF}(X)$, for X compact, and $\mathcal{F}$ be a complex of sheaves on X such that $h(x) = \chi(\mathcal{F})(x)$. Let $p \colon X \to \star$ be the constant map. The EULER INTEGRAL of h is

(13.2) $$\int_X h \, d\chi := \chi(Rp_* \mathcal{F})$$

When X is not compact then we use the right-derived pushforward with compact supports:[5]

(13.3) $$\int_X h \, d\chi := \chi(Rp_! \mathcal{F})$$

The higher push forward of $\mathcal{F}$ to a point takes an injective resolution of $\mathcal{F} \to \mathcal{I}^\bullet$ and applies the direct image level-wise to $\mathcal{I}^\bullet$. From the definition of direct image in §11,
$$p_* \mathcal{I} = \mathcal{I}(X),$$
is the group (or vector space) of global sections. Thus we have a complex of vector spaces
$$\mathcal{I}^0(X) \longrightarrow \mathcal{I}^1(X) \longrightarrow \mathcal{I}^2(X) \longrightarrow \cdots$$

---

[5] The reader may note with pleasure that here, as in calculus class, integration over a non-compact domain can be ill-defined if one is not careful.

If one calculates the local Euler-Poincaré index over the one and only point $\star$, one gets

$$\begin{aligned}\chi(\mathrm{R}p_*\mathcal{F})(\star) &= \sum_i (-1)^i \dim(\mathcal{H}^i \mathcal{J}^\bullet(X)) \\ &= \sum_i (-1)^i \dim(H^i \mathcal{J}^\bullet(X)) \\ &= \sum_i (-1)^i \dim(H^i(X, \mathcal{F})) \\ &=: \chi(X, \mathcal{F})\end{aligned}$$

in the compact case. In the general case the Euler integral is actually computing $\chi_c(X, \mathcal{F})$. The moral of this section is this: *the Euler integral of a constructible function is the Euler characteristic of its associated constructible sheaf.*

The right hand side of the definition may now seem less mysterious, but the connection to the original definition of the Euler integral may still be elusive. The connection is given by o-minimal geometry and properties of constant sheaves like $\widetilde{\mathbb{R}}_A$. As noted for $\widetilde{\mathbb{Z}}$, the cohomology of the constant sheaf coincides with normal cohomology of the space. Specifically

$$H^i(X, \widetilde{\mathbb{R}}_A) = H^i(A; \mathbb{R})$$

and consequently

$$\chi(X, \widetilde{\mathbb{R}}_A) = \sum_k (-1)^k \dim H^k(A; \mathbb{R}).$$

Note that this is *not* the same combinatorial Euler characteristic χ used in Euler integration, because compact supports are not used for $H^\bullet$ above.

Euler characteristic and its properties are adaptable to any situation with complexes involved. In particular for two sheaves (or complexes of sheaves) $\mathcal{F}, \mathcal{G}$ over X

$$\begin{aligned}\chi(X, \mathcal{F} \oplus \mathcal{G}) &= \chi(X, \mathcal{F}) + \chi(X, \mathcal{G}) \\ \chi(X, \mathcal{F} \otimes \mathcal{G}) &= \chi(X, \mathcal{F}) \cdot \chi(X, \mathcal{G})\end{aligned}$$

The same equations hold for χ_c, the compactly-supported cohomological Euler characteristic of §6. As a particular instance of the above one has

$$\chi(X, \widetilde{\mathbb{R}}_A^n) = n\chi(X, \widetilde{\mathbb{R}}_A) = n\chi(A)$$

To see this result and more we prove the following lemma:

Lemma 13.1 ([27], 2.5.4 pg. 49). *For $\mathcal{F}$ a locally constant sheaf on X valued in $\mathbb{R}$-vector spaces,*

$$\chi(X, \mathcal{F}) = n\chi(X) \quad \text{and} \quad \chi_c(X, \mathcal{F}) = n\chi_c(X)$$

where n is the (locally constant) stalk dimension.

PROOF. By hypothesis there is a cover of $\{U_i\}$ of X so that $\mathcal{F}|_{U_i} \cong \widetilde{\mathbb{R}}_{U_i}^n$. Let us first assume that the cover $\{U_i\}$ consists of a single open set U. In this case

$$\mathcal{F} \cong \widetilde{\mathbb{R}}^n \cong \bigoplus_{i=1}^n \widetilde{\mathbb{R}}.$$

Consequently an injective resolution of $\widetilde{\mathbb{R}}^n$ can be constructed as follows: Take a resolution of $\widetilde{\mathbb{R}}$, say $\mathcal{I}^\bullet$. Now define an injective sheaf $\mathcal{J}^i = \oplus_{k=1}^n \mathcal{I}^i$. The map from $\mathcal{J}^i \to \mathcal{J}^{i+1}$ is the n-fold direct sum (block matrix) of the map $\mathcal{I}^i \to \mathcal{I}^{i+1}$. This shows that for each i

$$H^i(X, \widetilde{\mathbb{R}}^n) = \oplus H^i(X, \widetilde{\mathbb{R}})$$

and thus

$$\chi(X, \widetilde{\mathbb{R}}^n) = n\chi(X, \widetilde{\mathbb{R}}).$$

Now assume that $X = U \cup V$ where $\mathcal{F}|_U$ and $\mathcal{F}|_V$ are isomorphic to the constant sheaf. To handle this case, we look to the MAYER-VIETORIS SEQUENCE FOR SHEAVES for inspiration. For U, V open there is a long exact sequence in sheaf cohomology:
(13.4)
$$\longrightarrow H^n(U \cup V, \mathcal{F}) \longrightarrow H^n(U, \mathcal{F}) \oplus H^n(V, \mathcal{F}) \longrightarrow H^n(U \cap V, \mathcal{F}) \longrightarrow .$$

The proof of this comes from the fact that if we replace $\mathcal{F}$ with its injective (flabby) resolution $\mathcal{J}^\bullet$, then the following complex of vector spaces (or groups) is exact:

$$0 \longrightarrow \Gamma(U \cup V, \mathcal{J}^\bullet) \longrightarrow \Gamma(U, \mathcal{J}^\bullet) \oplus \Gamma(V, \mathcal{J}^\bullet) \longrightarrow \Gamma(U \cap V, \mathcal{J}^\bullet) \longrightarrow 0.$$

As a consequence of this short exact sequence of complexes we get for free that

$$\chi(U, \mathcal{F}) + \chi(V, \mathcal{F}) = \chi(U \cup V, \mathcal{F}) + \chi(U \cap V, \mathcal{F}).$$

Since we have assumed that $\mathcal{F}$ is isomorphic to $\widetilde{\mathbb{R}}^n$ on the open sets $U, V, U \cap V$ the Euler characteristic formula above yields for $U \cup V = X$

$$\chi(X, \mathcal{F}) = n\chi(U) + n\chi(V) - n\chi(U \cap V) = n\chi(X).$$

Applying this argument inductively to a countable cover yields the result. To repeat the proof for χ_c, note that Mayer-Vietoris for compact supports is also valid, except for the reversal of arrows, as per the general covariant behavior of compact supports. □

It follows that, for a constructible — that is, piecewise-constant — sheaf, the Euler characteristic of the sheaf is simple to compute on constant patches. One more argument is necessary to show that the Euler characteristic of a constructible sheaf is an Euler integral.

Theorem 13.2 ([27], Theorem 4.1.22; [74], Lemma 2.0.2). *For $\mathcal{F}$ a constructible sheaf and $h(x) = \chi(\mathcal{F})(x)$, the Euler characteristic of $\mathcal{F}$ equals the integral of h with respect to $d\chi$:*

(13.5)
$$\chi_c(X, \mathcal{F}) = \int_X h \, d\chi.$$

PROOF. Let X_i be the partition of X into tame sets so that $\mathcal{F}|_{X_i}$ is locally constant. Since each X_i is tame, consider a triangulation of each X_i into cells S_{ij}. Let U be the union of the (disjoint) top dimensional cells, which is necessarily open. In analogy with the compactly supported cohomology we have the following long exact sequence:

(13.6)
$$\longrightarrow H_c^n(U, \mathcal{F}) \longrightarrow H_c^n(X, \mathcal{F}) \longrightarrow H_c^n(X - U, \mathcal{F}) \longrightarrow$$

and so

$$\chi_c(X, \mathcal{F}) = \chi_c(U, \mathcal{F}) + \chi_c(X - U, \mathcal{F}) = \sum_i n_i \chi_c(U_i) + \chi_c(X - U, \mathcal{F})$$

Here the U_i are the open cells belonging to each of the X_i (there may be none), and $\mathcal{F}$ has potentially different rank on each of X_i. This is the inductive step. The theorem holds if X is zero dimensional. Induction allows us to excise the top dimensional cells and pass from dimension n to $n-1$. Thus,

$$\chi_c(X, \mathcal{F}) = \sum_{i,j} n_{ij} \chi_c(S_{ij}) = \sum_i n_i \chi_c(X_i) = \int h\, d\chi.$$

□

Now duality makes another nice appearance. Recall that to any sheaf we can associate a dual sheaf $\mathbb{D}\mathcal{F}$ such that $\chi_c(X, \mathcal{F}) = \chi(X, \mathbb{D}\mathcal{F})$, so the Euler integral is also computing the Euler characteristic of the dual sheaf. If one considers the associated dual function

$$\mathbb{D}h(x) = \chi(\mathbb{D}\mathcal{F})(x)$$

and integrates this instead, one obtains:

$$\int_X \mathbb{D}h\, d\chi = \chi_c(X, \mathbb{D}\mathcal{F}) = \chi(X, \mathbb{D}\mathbb{D}\mathcal{F})$$

As duality is involutive — $\mathbb{D}\mathbb{D}\mathcal{F} \cong \mathcal{F}$ — the Euler characteristic of a sheaf is computed via duality.

This lift of constructible functions to constructible sheaves fits nicely within the story (if not the rigorous definition) of categorification that inspired our use of homology in §5. Here one lifts from the group of constructible functions to the category of constructible sheaves and then "decategorifies" by taking Euler characteristic.

The sheaf perspective does more than simply give a definition of the Euler integral. If one pays special attention to the definition of $F_!$ and the limiting procedure of the stalk, one gets the BASE CHANGE theorem [74]:

$$(RF_!\mathcal{F})_y \cong R\Gamma_c(F^{-1}(y), \mathcal{F})$$

The integral over the fiber computes the compactly supported sheaf cohomology there, simply by applying χ to both sides. The correspondence between constructible sheaves and functions is totally functorial: one may operating on sheaves first and then decategorify, or decategorify first and then operate on functions.

Applications to Sensor & Network Data Aggregation

Although sheaf theory provides a systematic approach to Euler calculus, the real utility of sheaves lies in a conceptual broadening of the notion of a function. Instead of assigning a single number to a point, a vector, group or other type of datum can be substituted. Moreover, this assignment is local in nature — one is recording observations valid only in a neighborhood of the observer. Sheaf cohomology answers the question whether these local observations can be aggregated into globally coherent ones and provides the obstruction to consistency.

The core insight then is the following: there is a rigorous framework for organizing observations valued in arbitrarily complicated data types, that is local for whatever topology desired, and a global calculus for inference-making. This global calculus *is* sheaf theory. Euler calculus is just a small part of this larger theory that exploits the following: constructible sheaves can be decategorified into constructible functions via local Euler-Poincaré index, so general data types are packaged into integer data. Integer data is computationally easier to manipulate, but the interpretation of these quantities in the context of an applied problem only makes sense on the sheaf level. We thus turn from the theory of Euler calculus to its recent applications in management of data in engineering systems and sensor networks, beginning with an example having no sensible distinction between sheaf and function — counting anonymous targets locally.

14. Target enumeration

Consider a finite collection of TARGETS, represented as discrete points in a domain W. There is a large collection of sensors, each of which observes some subset of W and counts the number of targets therein. The sensors will be assumed to be distributed so densely as to be approximated by a topological space X (typically a manifold, in the continuum limit). Note that various factors important in sensing (geometry of the domain, obstacles, time-dependence, moving targets, etc.) may be accounted for by enlarging the domains X and/or W. For example, the case of moving distinct targets may require setting W to be a configuration space of points.

There are many modes and means of sensing: infrared, acoustic, optical, magnetometric, and more are common. To best abstract the idea of sensing away from the engineering details, it is proper to give a topological definition of sensing. In a particular system of sensors in X and targets in W, let the SENSING RELATION be the relation $\mathcal{S} \subset W \times X$ where $(w, x) \in \mathcal{S}$ iff a sensor at $x \in X$ detects a target at $w \in W$. The horizontal and vertical fibers (inverse images of the projections of $\mathcal{S}$ to X and W respectively) have simple interpretations. The vertical fibers — TARGET SUPPORTS — are those sets of sensors which detect a given target in W. The horizontal fibers — SENSOR SUPPORTS — are those targets observable to a given sensor in X.

Assume that the sensors are additive but anonymous: each sensor at $x \in X$ counts the number of targets in W detectable and returns a local count $h(x)$, but the identities of the sensed targets are unknown. This counting function $h \colon X \to \mathbb{Z}$ is, under the assumption of tameness, constructible. Given h and some minimal information about the sensing relation $\mathcal{S}$, what is the total number of targets? This is in essence a problem of computing a global section (number) from a collection of local sections (target counts): it is no surprise that the tools of sheaf theory are applicable.

Theorem 14.1 ([5]). *If $h \in CF(X)$ is a counting function of target supports U_α with $\chi(U_\alpha) = N \neq 0$ for all α, then $\#\alpha = \frac{1}{N} \int_X h \, d\chi$.*

The proof is trivial given that $\int d\chi$ is an integration operator with measure $d\chi$.

$$(14.1) \qquad \int_X h \, d\chi = \int_X \left(\sum_\alpha \mathbf{1}_{U_\alpha} \right) d\chi = \sum_\alpha \int_X \mathbf{1}_{U_\alpha} \, d\chi = \sum_\alpha \chi(U_\alpha) = N \#\alpha.$$

This is really a problem in aggregation of redundant data, since many nearby sensors with the same reading are detecting the same targets; in the absence of target identification (an expensive signal processing task, in practice), it seems very difficult. Notice that the restriction $N \neq 0$ is nontrivial. If $h \in CF(\mathbb{R}^2)$ is a finite sum of characteristic functions over annuli, it is not merely inconvenient that

$\int_{\mathbb{R}^2} h\, d\chi = 0$, it is a fundamental obstruction to disambiguating sets. Some sums of annuli may be expressed as a union of different numbers of embedded annuli.

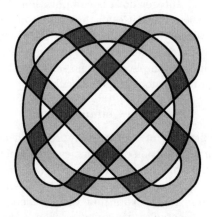

FIGURE 4. An example for which the integral with respect to Euler characteristic vanishes. Note that the integrand can be represented as a sum of indicator functions of embedded annuli, the number of which is indeterminate.

15. Enumeration and Fubini

The Fubini theorem is very significant in applications of the Euler calculus, both classical (*e.g.*, the Riemann-Hurwitz Theorem [**78, 45**]) and contemporary.

15.1. Enumerating vehicles. One application of the Fubini Theorem is to time-dependent targets. Consider a collection of vehicles, each of which moves along a smooth curve $\gamma_i : [0, T] \to \mathbb{R}^2$ in a plane filled with sensors that count the passage of a vehicle and increment an internal counter. Specifically, assume that each vehicle possesses a *footprint* — a support $U_i(t) \subset \mathbb{R}^2$ which is a compact contractible neighborhood of $\gamma_i(t)$ for each i, varying tamely in t. At the moment when a sensor $x \in \mathbb{R}^2$ detects when a vehicle comes within proximity range — when x crosses into $U_i(t)$ — that sensor increments its internal counter. Over the time interval $[0, T]$, the sensor field records a counting function $h \in CF(\mathbb{R}^2)$, where $h(x)$ is the number of times x has entered a support. As before, the sensors do not identify vehicles; nor, in this case, do they record times, directions of approach, or any ancillary data. It is helpful to think of the TRACE of a vehicle: the union over t of $U_i(t)$ in $\mathbb{R}^2$. Note the possibility that the trace can be a non-contractible set. On an overlap, however, the counting function reads a value higher than 1. This is the key to resolving the true vehicle count.

Proposition 15.1 ([**5**]). *Assuming the above, the number of vehicles is equal to* $\int_{\mathbb{R}^2} h\, d\chi$.

PROOF. Consider the projection map $p \colon \mathbb{R}^2 \times [0, T] \to \mathbb{R}^2$. Each target traces out a compact tube in $\mathbb{R}^2 \times [0, T]$ given by the union of slices $(U_i(t), t)$ for $t \in [0, T]$. Each such tube has $\chi = 1$. The integral over $\mathbb{R}^2 \times [0, T]$ of the sum of the characteristic functions over all N tubes is, by Theorem 14.1, N, the number of targets. By the Fubini Theorem, this also equals $\int_{\mathbb{R}^2} h\, d\chi$, where $h(x)$ is the value of the integral of the aforementioned sum of tubes over the fiber $p^{-1}(x)$. Since $p^{-1}(x)$ is $\{x\} \times [0, T]$, the integral over $p^{-1}(x)$ records the number of (necessarily compact) connected intervals in the intersection of $p^{-1}(x)$ with the tubes in $\mathbb{R}^2 \times [0, T]$. This number is precisely the sensor count (the number of times a sensor detects a vehicle coming into range). □

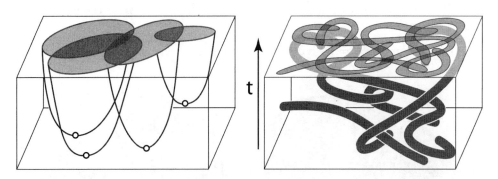

FIGURE 5. Time-dependent phenomena (waves [left] or tracks [right]) can be enumerated by sensors without clocks.

15.2. Enumerating wavefronts. Consider a finite collection of points $\mathcal{O}_\alpha$ in $W \subset \mathbb{R}^n$. Each $\mathcal{O}_\alpha$ represents an event which occurs at some time and which triggers a wavefront that propagates for a finite extent. Assume that each sensor has the ability to record the presence of a wavefront which passes through its vicinity. Each node has a simple counter memory which allows it to store the number of wavefronts that have passed. Under the continuum assumption, this yields a counting function $h : W \to \mathbb{N}$ that returns wavefront detection counts *a posteriori*. The problem is to determine the number of source events $\mathcal{O}_\alpha$.

In this setting, there is no temporal data associated to the sensors. With a particular assumption on the sensing modality, this problem is solved as a corollary of Proposition 15.1, using the same Fubini argument. Assume that the 'wavefront' associated to each event $\mathcal{O}_\alpha$ induces a continuous definable map F_α from a compact ball D^n to W whose restriction to rays from the origin are geodesic rays in W based at $\mathcal{O}_\alpha$. It is not enough to model sensors which count wavefronts by recording the number of 'fronts' that have passed, as one must account for singularities. To that end, the cleanest assumption for the counting sensors is the following:

Corollary 15.2. *In the context of this subsection, assume that each sensor at $w \in W$ increments its internal counter by $\chi(F_\alpha^{-1}(w))$ as the wavefront of $\mathcal{O}_\alpha$ passes over. Under this assumption, the number of triggering events is $\#\alpha = \int_W h \, d\chi$.*

PROOF. Apply the Fubini theorem to $h = \sum_\alpha (F_\alpha)_* \mathbf{1}_{D^n}$, where F_α is the mapping of the wavefront into W. □

The assumption is academic, but not outrageous, since we have suppressed the time variable. Since $n = \dim(W) = \dim(D^n)$, the inverse image $F_\alpha^{-1}(w)$ is generically discrete, and the assumption on the sensor modality boils down to a count of the number of passing wavefronts over the entire time interval. Of course, certain complications can arise in practice. For example, very coarse binary sensors may not be able to distinguish between one wavefront and several wavefronts passing over simultaneously: this can lead to positive-codimension defects in the counting function h. A similar loss of upper semi-continuity occurs when there is reflection of wavefronts along the boundary ∂W. For a compact domain $W \subset \mathbb{R}^n$ with smooth boundary ∂W, consider a wavefront-counting integrand $h = \sum_\alpha (F_\alpha)_* 1_{D^n}$ whose projection maps F_α may have fold singularities (reflections) along ∂W. Let $h^+ : W \to \mathbb{N}$ be the upper semi-continuous extension of h. Then, as the reader may show:

$$(15.1) \qquad \#\alpha = \int_W h \, d\chi = \int_W h^+ \, d\chi - \frac{1}{2} \int_{\partial W} h^+ \, d\chi.$$

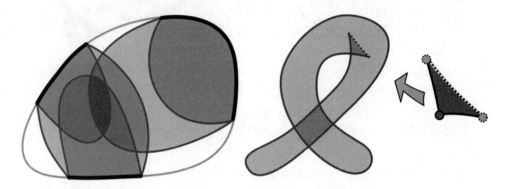

FIGURE 6. Singularities in a system of wavefronts can be resolved by numerical approximation in the case of fold (reflection) singularities [left], but not in the case of cusps [right].

This allows for numerical approximation of integrals involving fold singularities (in the context of signal reflection). We are unaware of how to meaningfully compute integrals when wavefronts have cusp singularities. There are several other challenges associated to measuring signals in the context of waves and obstacles: besides reflections and caustics, diffractions around corners are nontrivial and seemingly difficult to excise.

15.3. Enumerating via beams. Fix a Euclidean target space in $\mathbb{R}^n$ and consider a variant of the counting problem in which a sensor at each $x \in \mathbb{R}^n$ senses targets via a "beam" that is a round compact k-dimensional ball in $\mathbb{R}^n$ centered at x (the term beam evoking the case $k = 1$). Each target $\mathcal{O}_\alpha$ has spatial extent equal to a closed convex domain V_α in $\mathbb{R}^n$. Each sensor at $x \in \mathbb{R}^n$ performs a sweep of its k-ball beam over all possible bearings. At each such location-bearing pair, the sensor counts the number of target shapes V_α which intersect the beam. The sensor field is parameterized by the affine Grassmannian $\mathsf{AGr}_k^n = \mathbb{R}^n \times \mathsf{Gr}_k^n$, where Gr_k^n is the Grassmannian of k-planes in $\mathbb{R}^n$. (For example, the projective space $\mathbb{P}^{n-1}$ is Gr_1^n.) Thus, the sensor field returns a counting function $h \colon \mathsf{AGr}_k^n \to \mathbb{Z}$.

Theorem 15.3. *Under the above assumptions and the additional assumption that if n if even then so is k, the number of targets is equal to*

$$(15.2) \qquad \#\alpha = \binom{\lfloor \frac{n}{2} \rfloor}{\lfloor \frac{k}{2} \rfloor} \int_{\mathsf{AGr}_k^n} h \, d\chi,$$

where $\lfloor \cdot \rfloor$ denotes the floor function.

PROOF. Target supports $U_\alpha \subset AGr_k^n$ are computed as follows. Fix an α and fix a *bearing* k-plane in the Grassmannian Gr_k^n. The set of nodes in $\mathbb{R}^n$ at which this k-ball intersects V_α is star-convex with respect to (the centroid of) V_α. Thus, the target support U_α is homeomorphic to $V_\alpha \times Gr_k^n$ and thus to Gr_k^n. The Euler characteristic of Gr_k^n is:

$$\chi(Gr_k^n) = \begin{cases} 0 & : n \text{ even}, k \text{ odd} \\ \binom{\lfloor \frac{n}{2} \rfloor}{\lfloor \frac{k}{2} \rfloor} & : \text{else} \end{cases}$$

The result follows from Theorem 14.1. □

Of course, such sensors ("k-dimensional planes in n-space") at continuum-level densities are fanciful at best. Worse still, in the most physically plausible setting — the case of the plane ($n = 2$) with linear beams ($k = 1$) — the theorem fails to be useful, since target supports have the homotopy type of $\mathbb{S}^1$, a measure zero set in Euler calculus.

16. Numerical approximation

What if (as is true in practice) sensor counting data is not given over a continuous space of sensors but rather a discrete set of points? In expressing the answer to the aggregation problem as a formal integral, one is led to the notion of numerical integration. The task of approximating an integral based on a discrete sampling has a long and fruitful history. One must be cautious, however; this integration theory is not at all like that of Riemann or Lebesgue. An annulus in the plane is a set of Euler measure zero; a single point is a set of Euler measure one. Thus, $d\chi$ gives tremendous flexibility, as one can integrate target supports with vastly dissimilar geometry; however, noise, in the form of errors at individual sensors, can be fatal to the estimation of the integral. The development of numerical integration theory for $d\chi$ is in its infancy.

16.1. Euler integration based on a triangulation. Let $h: \mathbb{R}^2 \to \mathbb{N}$ be constructible and assume that h is known only over the vertex set of a triangulation of $\mathbb{R}^2$ and that $\tilde{h}$ is the constructible function whose value on an open k-simplex equals the minimum of all boundary vertex values. The most straightforward approach to the approximation of the integral is (since $\tilde{h} \geq 0$ and upper semicontinuous) to use the following formula:

$$(16.1) \qquad \sum_{s=0}^{\infty} \#V\{\tilde{h} > s\} - \#E\{\tilde{h} > s\} + \#F\{\tilde{h} > s\},$$

where $\#V$, $\#E$, and $\#F$ refer to the number of vertices, edges, and faces, respectively. This formula can, however, fail in several ways. The sampling can be too sparse relative to the features of h, as in Figure 7. This is not too surprising, since a too-sparse sampling impedes the approximation of Riemann integrals likewise.

Unfortunately, such discretization errors can persist, even with triangulations of arbitrarily high density. A generic codimension two singularity in $h \in CF^+(\mathbb{R}^2)$ is locally equivalent (up to a constant) to the sum of Heaviside step functions $h(x, y) := H_0(x) + H_0(y) \in CF^+(\mathbb{R}^2)$. A random triangulation will be with positive probability locally equivalent to that in Figure 8. The level sets of h with respect

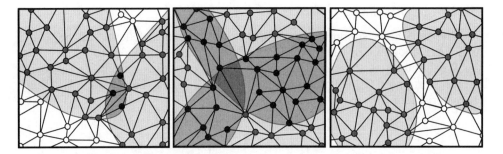

FIGURE 7. These types of triangulations leads to incorrect Euler integrals due to mis-sampling of connected components of upper or lower excursion sets of the integrand: the relevant features are too small for the triangulation to accurately sample the Euler characteristics.

to the triangulation produce a spurious path component, adding 1, wrongly, to the Euler integral. This example is scale-invariant and not affected by increased density of sampling. In a random placement of N discs in a bounded planar domain, the expected number of such intersection points is quadratic in N, leading to potentially large errors in numerical approximations to $\int \cdot d\chi$, regardless of sampling density.

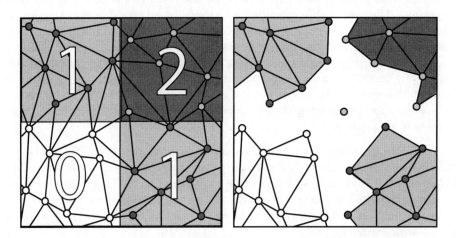

FIGURE 8. At a typical discontinuity in an integrand $h \in \mathsf{CF}(\mathbb{R}^2)$, a triangulation [left] can lead to an error in the topology of the level sets [right], causing an error in the integral computation. This is a scale-independent phenomenon.

16.2. Exploiting numerical errors in Fubini. Sometimes, numerical errors are serendipitous. Consider h a sum of characteristic functions over compact annuli A in $\mathbb{R}^2$ that are *convex* in the sense that each annulus is a compact convex disc minus a smaller open convex sub-disc. As noted $\int h \, d\chi = 0$. However, assume (1) that h is sampled over a rectilinear lattice; (2) that the geometric "feature size" of h is sufficiently large relative to the sampling density (meaning that the aspect ratios, size, and distances between boundary curves are suitably bounded). Then, to put it in a less-than-technical form,

$$(16.2) \qquad \#\text{annuli} = \frac{1}{2} \int_{\text{the other}} \int_{\text{one way}} \tilde{h} \, d\chi \, d\chi,$$

where the integration is performed numerically (using Equation (16.1) of §16.1). The reasoning is as follows. By additivity and general position, it suffices to demonstrate in the case of a single annulus. Performing the first integral means counting the number of connected intersections of lines in the rectilinear lattice with the annulus. The result is a function which increases from zero to two, then back to zero. Assuming transversality, the (numerical!) integral of the resulting function on $\mathbb{R}$ is equal to two, *not zero*, since codimension-1 values of the integrand are ignored by the numerical sampling. The curious reader should, armed with Fubini, contrast the case of a field of sensors over $\mathbb{R} \times \mathbb{R}$ with the discretized setting.

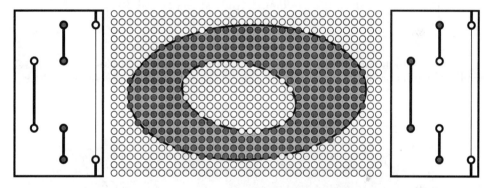

FIGURE 9. In a dense lattice [center], integrating exactly along (horizontal) directions [left] yields an integrand with Euler integral zero. However, performing numerical integration introduces errors [right] which allows one to count fat convex annuli in general position.

17. Planar *ad hoc* networks

A highly effective numerical method for performing integration with respect to Euler characteristic over a planar network uses the homological properties of the Euler characteristic and is an excellent argument for the utility of that approach to χ. We consider an integrand $h \in \mathsf{CF}(\mathbb{R}^2)$ sampled over a network $\mathcal{G}$ in which values of h are recorded at the vertices of $\mathcal{G}$ and edges of $\mathcal{G}$ correlate (roughly) to distance in $\mathbb{R}^2$. However, no coordinate data are assumed — the embedding of $\mathcal{G}$ into $\mathbb{R}^2$ is unknown. This is a realistic model of an *ad hoc* coordinate-free network, such as might occur when simple sensors with no GPS set up a wireless communications network. This lack of coordinates or distances makes the use of a triangulation problematic, though not impossible. Equation (4.5) suggests that the estimation of the Euler characteristics of the upper excursion sets is an effective approach. However, if the sampling occurs over a network with communication links, then it is potentially difficult to approximate those Euler characteristics, since (inevitable) undersampling leads to holes that ruin an Euler characteristic approximation. Duality is the key to mitigating this phenomenon.

Theorem 17.1. *For* $h\colon \mathbb{R}^2 \to \mathbb{N}$ *constructible and upper semi-continuous,*

$$(17.1) \qquad \int_{\mathbb{R}^2} h\, d\chi = \sum_{s=0}^{\infty} \left(\beta_0\{h > s\} - \beta_0\{h \leq s\} + 1\right),$$

where $\beta_0 = \dim H_0$, *the number of connected components of the set.*

PROOF. Let A be a compact nonempty subset of $\mathbb{R}^2$. From (1) the homological definition of the Euler characteristic and (2) Alexander duality, we note:

$$\chi(A) = \dim H_0(A) - \dim H_1(A) = \dim H_0(A) + \dim H_0(\mathbb{R}^2 - A) - 1.$$

Since h is upper semi-continuous, the set $A = \{h > s\}$ is compact. Noting that $\mathbb{R}^2 - A = \{h \leq s\}$, one has:

$$\int h\, d\chi = \sum_{s=0}^{\infty} \chi\{h > s\} = \sum_{s=0}^{\infty} \dim H_0(\{h > s\}) - (\dim H_0(\{h \leq s\}) - 1).$$

□

This result gives both a criterion and an algorithm for correctly computing Euler integrals based on nothing more that an *ad hoc* network of sampled values.

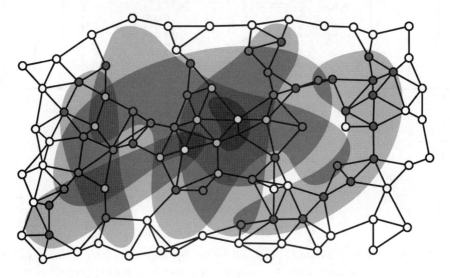

FIGURE 10. A sparse sampling over an *ad hoc* network retains enough connectivity data to evaluate the integral exactly.

Corollary 17.2. *The degree of sampling required to ensure exact approximation of* $\int h\, d\chi$ *over a planar network* $\mathcal{G}$ *is that* $\mathcal{G}$ *correctly samples the connectivity of all the upper and lower excursion sets of* h.

In other words, if, given h on the vertices of $\mathcal{G}$, one extends to an upper semi-continuous $\tilde{h} \in CF(\mathcal{G})$ in the usual manner, then the criterion is that the upper and lower excursion sets of $\tilde{h}$ on $\mathcal{G}$ have the same number of connected components as those of h on $\mathbb{R}^2$.

18. Software implementation

The formula in Theorem 17.1 has been implemented in Java as a general *Eu*ler *ch*aracteristic *i*ntegration *s*oftware, *Eucharis* [30]. The determination of the number of connected components of the upper and lower excursion sets is a simple clustering problem, computable in logspace with respect to the number of network nodes. The software implementation has the following features:

(1) Lattice or *ad hoc* networks of arbitrary size and communication distance can be generated.
(2) Targets with supports of predetermined shape (circular, polygonal, etc.) can be placed at will; targets with support a neighborhood of a drawn path are also admissible.
(3) The value of the counting function h is represented by colors on the nodes.
(4) Euler integrals are estimated via Equation (17.1) and compared to true values.
(5) Various advanced transforms are available, including addition of noise, smoothing via convolution with Gaussians , and other integral transforms (using ideas from §24-28).

Screenshots are included as Figures 11.

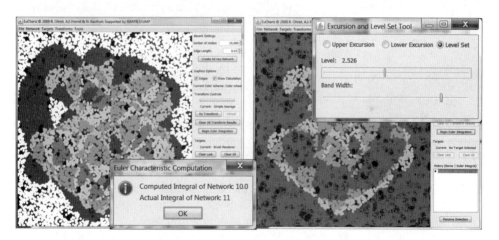

FIGURE 11. Screenshots of *Eucharis*, including an Euler integral approximation [left] and a level-set explorer [right]. The true integral (11) is approximated (10) numerically, but imperfectly so, due to the existence of small, poorly-sampled chambers.

Of course, the criterion of Corollary 17.2 is not always verifiable: to know the connected components of the excursion sets of the true integrand is, under realistic circumstances, a luxury. Approximating integrals of Riemann integrals based of discrete sampling suffers from similar problems, as mass can concentrated in a small region. Unlike that situation, we cannot easily bound derivatives and work in that manner. There is a wealth of open problems relating to how one properly does numerical estimation in Euler calculus. It is remarkable that, though the Euler calculus has existed in its fullest form for more than thirty years, only a few works (*e.g.*, [57]) have addressed the problem of estimating Euler characteristics of sets based on discrete data. The broader problem of numerical computation of the Euler integral based on discrete sampling seems to be in its infancy.

Integral Transforms & Signal Processing

The previous chapter presented simple applications of basic Euler integrals to data aggregation problems, using little more than the definition and well-definedness of the integral, along with the corresponding Fubini Theorem. Euler *Calculus* is more comprehensive than the integral alone: Euler integration admits a variety of operations which mimic analytic constructs. The operations surveyed in this chapter are the beginnings of a rich calculus blending analytic, combinatorial, and topological perspectives into a package of particular relevance to signal processing, imaging, and inverse problems.

19. Convolution and duality

On a finite-dimensional real vector space V, a CONVOLUTION operation with respect to Euler characteristic is straightforward. Given $f, g \in CF(V)$, one defines

$$(19.1) \qquad (f * g)(x) = \int_V f(t) g(x - t) \, d\chi(t).$$

Lemma 19.1.

$$(19.2) \qquad \int f * g \, d\chi = \int f \, d\chi \int g \, d\chi$$

PROOF. Fubini. □

This convolution operator is of interest to problems of computational geometry, given the close relationship to the MINKOWSKI SUM: for A and B convex, $1_A * 1_B = 1_{A+B}$, where $A + B$ is the set of all vectors expressible as a sum of a vector in A and a vector in B [78, 71, 9, 43]. The work of Guibas and collaborators [47, 46, 65] contains a wealth of results on the computational complexity of using Euler convolution instead of the usual Minkowski sums in computational geometry, with applications in [65] to robot motion planning and obstacle avoidance.

Where there is convolution, a deconvolution operator lurks, if as nothing more than desideratum. In this case, the appropriate avenue to deconvolution is via a formal (Verdier) duality operator on sheaves [71]. In the context of CF, this vast generalization of Poincaré duality takes on a simple and concrete form. Define the DUAL of $h \in CF(X)$ to be:

$$(19.3) \qquad Dh(x) = \lim_{\epsilon \to 0^+} \int_X h 1_{B(x,\epsilon)} \, d\chi,$$

where $B(x, \epsilon)$ denotes an *open* ball of radius ϵ about x. This limit is well-defined thanks to Theorem 3.2 applied to $h: X \to \mathbb{Z}$. Duality provides a de-convolution operation — a way to undo a Minkowski sum.

Lemma 19.2 ([72]). *For any convex closed $A \subset V$ with non-empty interior, the convolution inverse of 1_A is $D1_{-A}$, where $-A$ denotes the reflection of A through the origin. Specifically, $1_A * D1_{-A} = \delta_0$, where δ is the indicator function of the origin.*

The related LINK transform Λ is defined as in **D** but by integrating over the boundary sphere:

$$\Lambda h(x) = \lim_{\epsilon \to 0^+} \int_X h \mathbf{1}_{\partial B(x,\epsilon)} d\chi. \tag{19.4}$$

The link transform acts, in a sense, like a derivative in the Euler calculus. For $h \in \mathsf{CF}(\mathbb{R}^{2n})$, Λh, like a derivative, vanishes everywhere except at the discontinuities of h. On an n-manifold, Λ acts as multiplication by $1-(-1)^n$ on open regions where h is constant. By now, the reader is used to the particulars of this dimensional parity. The relationship to **D** is made precise in the following:

Lemma 19.3. $\Lambda = \mathsf{Id} - \mathbf{D}$.

20. Radon transforms and inversion

One of the most flexible integral transforms is the RADON TRANSFORM of Schapira [72, 18, 23]. This Fredholm-type transform integrates over a kernel via pushforward and pullback operations on CF. Recall from §4 the pushforward $F_*: \mathsf{CF}(X) \to \mathsf{CF}(Y)$ associated to a definable $F: X \to Y$ via fibrewise Euler integration. The PULLBACK $F^*: \mathsf{CF}(Y) \to \mathsf{CF}(X)$ is the obvious composition: given $g \in \mathsf{CF}(Y)$ constructible, $(F^*g) := g \circ F$. Functoriality of these operations is expressed in the PROJECTION FORMULA: for $g \in \mathsf{CF}(Y)$ and $h \in \mathsf{CF}(X)$,

$$F_*(h(F^*g)) = F_*(h)g, \tag{20.1}$$

Given a locally closed definable set $S \subset W \times X$, let P_W and P_X denote the projection maps of $W \times X$ to their factors. The RADON TRANSFORM is the map $\mathbf{R}_S: \mathsf{CF}(W) \to \mathsf{CF}(X)$ given by

$$\mathbf{R}_S h = (P_X)_*((P_W^* h)\mathbf{1}_S). \tag{20.2}$$

The push/pull language is apt. One pulls back $h \in \mathsf{CF}(W)$ via projection to the support S (along vertical fibers) then pushes this down (via integration along horizontal fibers) to $\mathbf{R}_S h \in \mathsf{CF}(X)$. This is reminiscent of a Fredholm transform with kernel $\mathbf{1}_S$.

Example 20.1. Duality on $\mathsf{CF}(X)$ is the Radon transform associated to the relation $S \subset X \times X$ where S is a sufficiently small open tubular neighborhood of the diagonal $\Delta = \{(x,x) : x \in X\}$.

In the context of sensor networks, the Radon transform is entirely natural. As in §14 let W denote the target space and X denote the sensor space. The sensing relation is $\mathcal{S} = \{(w,x) : \text{the sensor at } x \text{ senses a target at } w\}$. This lies in the product space $W \times X$ as a relation whose vertical fibers $\mathcal{S}_w = P_X(P_W^{-1}(w) \cap \mathcal{S})$ are target supports and whose horizontal fibers $\mathcal{S}_x = P_W(P_X^{-1}(x) \cap \mathcal{S})$ are sensor supports. Consider a finite set of targets $T \subset W$ as defining an atomic function $\mathbf{1}_T \in \mathsf{CF}(W)$. Observe that the counting function $h \in \mathsf{CF}(X)$ which the sensor field on X returns is precisely the Radon transform $\mathbf{R}_\mathcal{S} \mathbf{1}_T$. In this language, Theorem 14.1 is implied by the following:

Lemma 20.2. *Assume that* $\mathcal{S} \subset W \times X$ *has vertical fibers* $P_W^{-1}(w) \cap \mathcal{S}$ *with constant Euler characteristic* N. *Then,* $\mathbf{R}_\mathcal{S}: \mathsf{CF}(W) \to \mathsf{CF}(X)$ *scales Euler integration by a factor of* N:

$$\int_X d\chi \circ \mathbf{R}_\mathcal{S} = N \int_W d\chi.$$

PROOF. Consider a compact contractible stratum $V \subset W$. Then $(P_W)^*1_W = 1_{S'}$ for some $S' \subset S$. Knowing that $(P_X)_*1_{S'} = R_S 1_V$ and that the vertical fibers of S' have $\chi = N$ yields

$$\int_X R_S 1_{S'} d\chi = \int_{W \times X} 1_{S'} d\chi = \int_W \int_{F^{-1}(w)} 1_{S'} d\chi(s) \, d\chi(w) = N \int_W 1_V d\chi = N,$$

thanks to the Fubini theorem. Linearity of the integral extends the result from 1_V to all of $CF(W)$. □

20.1. Schapira's inversion formula.
A similar regularity in the Euler characteristics of fibers allows a general inversion formula for the Radon transform [72].

Theorem 20.3 (Schapira). *Assume that $S \subset W \times X$ and $S' \subset X \times W$ have fibers S_w and S'_w in X satisfying (1) $\chi(S_w \cap S'_w) = \mu$ for all $w \in W$; and (2) $\chi(S_w \cap S'_{w'}) = \lambda$ for all $w' \neq w \in W$. Then for all $h \in CF(W)$,*

$$(20.3) \qquad (R_{S'} \circ R_S)h = (\mu - \lambda)h + \lambda \left(\int_W h \, d\chi \right) 1_W.$$

PROOF. We prove a generalization as from [7]. It is a simple matter to generalize from Radon transforms to Fredholm transforms with arbitrary constructible kernels having an Euler regularity in the fibers. Given a kernel $K \in CF(W \times X)$, one defines the weighted Radon transform as $R_K h = (P_X)_*((P_W^* h)K)$. Inversion requires an inverse kernel $K' \in CF(X \times W)$ such that there exist constants μ and λ with $\int_X K(w,x)K'(x,w')d\chi = (\mu - \lambda)\delta_{w-w'} + \lambda$ for all $w, w' \in W$. Then, for all $h \in CF(W)$, we claim that:

$$(20.4) \qquad (R_{K'} \circ R_K)h = (\mu - \lambda)h + \lambda \left(\int_W h \, d\chi \right) 1_W.$$

To show this, note that for any $w' \in W$,

$$\begin{aligned}
(R_{K'} \circ R_K h)(w') &= \int_X \left[\int_W h(w)K(w,x) \, d\chi \right] K'(x,w') \, d\chi \\
&= \int_W h(w) \left[\int_X K(w,x)K'(x,w')d\chi \right] d\chi \\
&= \int_W [(\mu - \lambda)h(w)\delta_{w-w'} + \lambda h(w)] \, d\chi \\
&= (\mu - \lambda)h(w') + \lambda \int_W h \, d\chi,
\end{aligned}$$

where the Fubini theorem is used in the second line. □

If the conditions of Theorem 20.3 are met and $\lambda \neq \mu$, then the inverse Radon transform $R_{S'} h = R_{S'} R_S 1_T$ is equal to a multiple of 1_T plus a multiple of 1_W. Thus, one can localize and identify a collection of targets — and determine the exact shape of T — by performing the inverse transform.

A more general proof still (by D. Lipsky [7]) uses the following COCYCLE CONDITION. Let X_1, X_2, and X_3 be spaces. Supposing the constructible kernels $K_i \in CF\left(\prod_{j \neq i} X_j\right)$ and the projection maps $P_i : \prod_i X_i \to \prod_{j \neq i} X_j$ satisfy the cocycle condition

$$(20.5) \qquad K_3 = (P_3)_*(P_1^* K_1 \cdot P_2^* K_2).$$

Then, consider the diagram

(20.6)

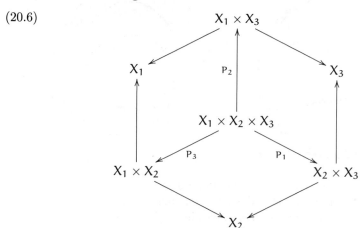

It follows from commutativity and Equation (20.1) that $\mathbf{R}_{K_2} \circ \mathbf{R}_{K_1} = \mathbf{R}_{K_3}$.

20.2. Examples of Radon inversion.

Example 20.4 (Hyperplanes). Schapira's paper outlines a potential application to imaging. Assume that $W = \mathbb{R}^3$ and one scans a compact subset $T \subset W$ by slicing $\mathbb{R}^3$ along all flat hyperplanes, recording simply the Euler characteristics of the slices of T. Since a compact subset of a plane has Euler characteristic the number of connected components minus the number of holes (which, in turn, equals the number of bounded connected components of the complement), it may be feasible to compute an accurate Euler characteristic, even in the context of noisy readings (using convolution to smooth the data).

This yields a constructible function on the sensor space $X = \mathsf{AGr}_2^3$ (the affine Grassmannian manifold of 2-planes in $\mathbb{R}^3$) equal to the Radon transform of $\mathbf{1}_T$. Using the same sensor relation to define the inverse transform is effective. Since $\mathcal{S}_w \cong \mathbb{P}^2$ and $\mathcal{S}_w \cap \mathcal{S}_{w'} \cong \mathbb{P}^1$, one has $\mu = \chi(\mathbb{P}^2) = 1$, $\lambda = \chi(\mathbb{P}^1) = 0$, and the inverse Radon transform, by (20.3), yields $\mathbf{1}_T$ exactly: one can recover the shape of T based solely on connectivity data of black and white regions of slices.

Example 20.5 (Sensor beams). Let the targets T be a finite disjoint collection of points in $W = \mathsf{D}^n$, the open unit disc in $\mathbb{R}^n$. Assume that the boundary ∂W is lined with sensors, each of which sweeps a ray over W and counts, as a function of bearing, the number of targets intersected by the beam. The sensor space X is homeomorphic to $\mathsf{T}_* \mathbb{S}^{n-1}$, the tangent bundle of ∂W. (To see this, note that the bearing of a ray at a point $p \in \partial W$ lies in the open hemisphere of the unit tangent bundle to W at p. This open hemisphere projects to the open unit disc in $\mathsf{T}_p \partial W$.)

Any point in W is seen by any sensor in ∂W along a unique bearing angle. Thus, the sensor relation S has vertical fibers (target supports) which are sections of $\mathsf{T}_* \mathbb{S}^{n-1}$ and hence spheres of Euler measure $\mu = \chi(\mathbb{S}^n) = 1 + (-1)^n$. Any two distinct vertical fibers project to X and intersect along the subset of rays from ∂W that pass through both points in W: this is a discrete set of cardinality (hence χ) $\lambda = 2$. For n even, one has $\lambda = 2$, $\mu = 0$, and Equation (20.3) implies that the inverse Radon transform gives $-2\mathbf{1}_T + 2(\#T)\mathbf{1}_W$. Thus, on even-dimensional spaces, targets may be localized with boundary beam sensors. One may likewise assume

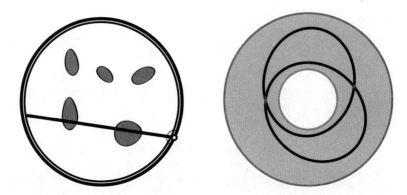

FIGURE 12. Convex targets occupying a disc $W = D^2$ are sensed by a beam emitted from the edge [left]; any pair of distinct vertical fibers correspond to intersecting circles in X [right].

that the target set T is not a collection of points but rather a disjoint collection of compact sets in W; target shapes will then be recovered, assuming the ability of beams to measure Euler characteristic of slices exactly. This is perhaps reasonable for convex-shaped targets.

Example 20.6 (distance-based sensing). Non-self-dual examples of Radon inversion can be generated easily with complementary supports. For example, let $X = W = \mathbb{R}^n$ with $\mathcal{S}_w$ a closed ball about w and $\mathcal{S}'_w$ the closure of the complement in $\mathbb{R}^n$. Physically, this means that the sensor relation detects proximity-within-range, and the inverse sensor relation counts targets out-of-range. The inversion formula applies, because of the singular nature of the "eclipse" that occurs when targets coalesce. Specifically: (1) $\chi(\mathcal{S}_w \cap \mathcal{S}'_{w' \neq w}) = 1$; and (2) $\chi(\mathcal{S}_w \cap \mathcal{S}'_w) = \chi(\mathbb{S}^{n-1}) \neq 1$.

There are technicalities in applying the inversion formula in settings where the supports are non-compact, since we have defined CF in terms of compactly supported functions. This does not interfere with inversion in this example: all non-compact integrands are still tame. In addition, this example may be easily modified so that the supports (and the corresponding complements) change smoothly from point-to-point within the domain, so long as $\mathcal{S}_w$ is, say, always compact and contractible and varies "slowly" enough so that $\chi(\mathcal{S}_w \cap \mathcal{S}'_{w' \neq w}) = 1$.

Example 20.7 (discrete distance). Using discrete distances can lead to inversions. Let $W = \mathbb{R}^{2n} = X$ and define $\mathcal{S}_w$ as the unit sphere about w: sensors count targets at a fixed distance. The inverse kernel $\mathcal{S}'$ has fibers equal to concentric spheres about the basepoint of radius $r = 1, 3, 5, \ldots$. Distinct fibers of $\mathcal{S}$ and $\mathcal{S}'$ always intersect in a set homeomorphic either to $\mathbb{S}^{2n-2}$ or, if the points line up at integer distances, $\mathbb{S}^0$: either way, $\chi(\mathcal{S}_w \cap \mathcal{S}'_{w' \neq w}) = 2$. In the instance when the basepoints coincide, the intersection is precisely the unit sphere $\mathbb{S}^{2n-1}$ with $\chi = 0$: full invertibility follows.

Example 20.8 (weighted kernels). Consider the case of Example 20.7, modified so that $W = X = \mathbb{R}^{2n+1}$. On odd-dimensional spaces, the spheres of intersection have dimension $2n-1$ or 0, depending on position. Thus λ is not a constant, preventing invertibility. However, if we weight the inverse kernel $\mathcal{S}'$ so that concentric spheres of radius $r = 1, 3, 5, \ldots$ have weight $(-1)^{r+1}$, then $\chi(\mathcal{S}_w \cap \mathcal{S}'_{w' \neq w}) = 0$, independent

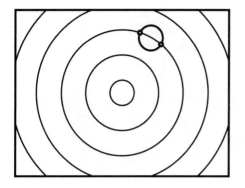 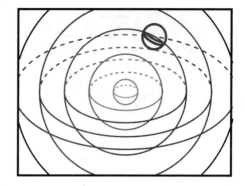

FIGURE 13. Using supports for a Radon transform based on discrete distances works in even dimensions [left] but requires a signed weighting scheme in odd dimensions [right].

of position. Full inversion is therefore possible using the generalization of Theorem 20.3 given in the proof.

21. The Microlocal Fourier transform

One of the deep tools of sheaf theory is a Fourier (or Fourier-Sato) transform [55] that is perhaps best described as *microlocal*. In the context of $\mathbf{CF}(\mathbb{R}^n)$, this transform reveals itself to be extremely concrete and important in applications. Fix $\mathbb{R}^n$ with the standard Euclidean structure. The MICROLOCAL FOURIER TRANSFORM of $h \in \mathbf{CF}(\mathbb{R}^n)$ at x takes as its argument a unit vector $\xi \in \mathbb{S}^{n-1}$ and returns:

$$(21.1) \qquad (_\mu F h)_x(\xi) = \lim_{\epsilon \to 0^+} \int_{B_\epsilon(x)} \mathbf{1}_{\xi \cdot (y-x) \geq 0} h \, d\chi(y),$$

where, as before B_ϵ denotes the *open* ball of radius ϵ. Like the classical Fourier transform, $_\mu F$ takes a frequency vector ξ and integrates over isospectral sets defined by dot product. The microlocal character of the transform stems from its dependence on the location x as well as the direction ξ; in §22.1, we will outline a more global Fourier transform.

The microlocal Fourier transform is strongly connected to the (differential) geometry of the integrand. For Y a compact tame set, $_\mu F \mathbf{1}_Y(\xi)$ is the constructible function on Y that records the EULER-POINCARÉ INDEX of the (generically Morse) function pointing in the direction ξ (*cf.* the results from §25). This function, when averaged over ξ with respect to Haar measure on $\mathbb{S}^{n-1}$ yields the curvature measure $d\kappa_Y$ on Y implicit in the Gauss-Bonnet Theorem (as shown by Bröcker and Kuppe [19]): for any open Borel set $U \subset \mathbb{R}^n$,

$$\int_U d\kappa_Y = \frac{1}{\text{vol } \mathbb{S}^{n-1}} \int_{\mathbb{S}^{n-1}} \int_U {}_\mu F \mathbf{1}_Y \, d\chi \, d\xi.$$

For Y a smooth surface in $\mathbb{R}^3$, $d\kappa$ is (up to 2π) Gauss curvature times the area form; on the boundary of such a surface, $d\kappa$ is geodesic curvature times the length form; and on a polyhedral domain, $d\kappa$ is supported on the vertex set and returns the exterior angles. The Gauss-Bonnet Theorem, properly interpreted in this language,

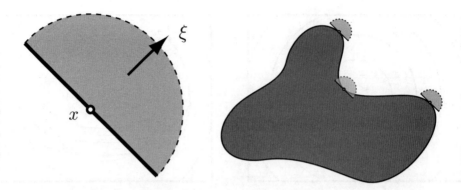

FIGURE 14. The microlocal Fourier transform at x has support on a half-open half-disc in the direction ξ [left]. This directional component captures critical-point features of the integrand with respect to the Morse function pointing in the direction ξ [right].

says that

$$(21.2) \qquad \frac{1}{\text{vol } \mathbb{S}^{n-1}} \int_{\mathbb{S}^{n-1}} \int_Y {}_\mu F 1_Y \, d\chi \, d\xi = \int_Y d\chi = \chi(Y)$$

It is apparent that the Gauss-Bonnet theorem extends via linearity of ${}_\mu F$ to all of $CF(Y)$, and one may sensibly speak of the curvature form $d\kappa_h$ of $h \in CF(Y)$:

$$(21.3) \qquad \int_U d\kappa_h = \frac{1}{\text{vol } \mathbb{S}^{n-1}} \int_{\mathbb{S}^{n-1}} \int_U {}_\mu F h \, d\chi \, d\xi.$$

This is yet another concrete and elegant example in the style of categorification that the sheaf-theoretic Euler calculus permits: the Gauss-Bonnet Theorem applies not just to sets but also to (constructible) data over sets.

22. Hybrid Euler-Lebesgue integral transforms

Blending Euler characteristic with Lebesgue integration has a long history in integral geometry and its applications [1, 28, 34, 80]. There are a number of interesting integral transforms based on a mixture of $d\chi$ and standard Lebesgue measure. We introduce (from [38]) two Euler integral transforms on real vector spaces for use in signal processing problems.

22.1. Fourier. The following integral transform is best thought of as a global version of the microlocal Fourier transform ${}_\mu F$ on a finite-dimensional real vector space V with inner product. The Fourier transform takes as its argument a covector $\xi \in V^\vee$. For $h \in CF(V)$ define the FOURIER TRANSFORM of h in the direction $\xi \in V^\vee$ as

$$(22.1) \qquad Fh(\xi) = \int_0^\infty \int_{\xi^{-1}(r)} h \, d\chi \, dr.$$

The integration domain $\xi^{-1}(r)$ is the ISOSPECTRAL SET of the transform; in this case, as in the classical Fourier transform, a hyperplane normal to ξ. One may reasonably rescale ξ to a unit covector: this has the effect of rescaling the normal dr measure.

Example 22.1. For A a compact convex subset of $\mathbb{R}^n$ and $\|\xi\| = 1$, $(\mathbf{F1}_A)(\xi)$ equals the projected width of A along the ξ-axis.

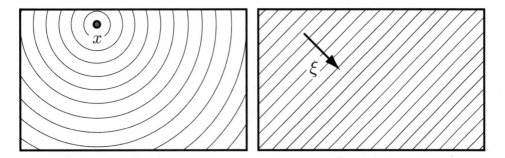

FIGURE 15. Isospectral sets for Bessel [left] and Fourier [right] transforms. In these transforms, one integrates with respect to $d\chi$ along isospectral sets, then across the isospectral sets with respect to Lebesgue measure.

22.2. Bessel. For the Eulerian generalization of a classical Bessel (or Hankel) transform, let V denote a finite-dimensional real vector space with norm $\|\cdot\|$, and let $D_r(x)$ denote the compact disc of points $\{y : \|y - x\| \leq r\}$. Recall that CF denotes compactly-supported definable integer-valued functions. For $h \in \mathsf{CF}(V)$ define the BESSEL TRANSFORM of h via

$$(22.2) \qquad \mathbf{B}h(x) = \int_0^\infty \int_{\partial D_r(x)} h \, d\chi \, dr.$$

This transform Euler-integrates h over the concentric spheres at x of radius r, and Lebesgue-integrates these spherical Euler integrals with respect to r. For the Euclidean norm, these isospectral sets are round spheres. Given our convention that $\mathsf{CF}(V)$ consists of compactly supported functions, $\mathbf{B} \colon \mathsf{CF}(V) \to \mathsf{Def}(V)$ is well-defined using Theorem 3.2 (Hardt Theorem).

The Bessel transform can be seen as a Fourier transform of the log-blowup. This perspective leads to results like the following.

Proposition 22.2. *The Bessel transform along an asymptotic ray is the Fourier transform along the ray's direction: for $h \in \mathsf{CF}(V)$ and $x \neq 0 \in V$,*

$$(22.3) \qquad \lim_{\lambda \to \infty} (\mathbf{B}h)(\lambda x) = (\mathbf{F}h)\left(\frac{x^\vee}{\|x^\vee\|}\right).$$

where $x^\vee$ is the dual covector.

PROOF. The isospectral sets restricted to the (compact) support of h converge in the limit and the scalings are identical. □

While the Fourier transform obviously measures a *width* associated to a constructible function, the geometric interpretation of the Bessel transform is more involved and related to image processing.

Example 22.3. Consider $h \in \mathsf{CF}(\mathbb{R}^2)$ equal to $h = \sum_\alpha \mathbf{1}_{D_\alpha}$ for D_α a closed disc of unknown location and radius. For each α, the Bessel transform of $\mathbf{1}_{D_\alpha}$ takes on a minimum of 0 precisely at the center of the disc (since $\chi(\mathbb{S}^1) = 0$) and has maximum equal to the diameter of D_α (equal to $\mathsf{F}(\mathbf{1}_{D_\alpha})$) asymptotically via Proposition 22.2.

Section 27 will use index theory to give computational formulae and interpretations for both **F** and **B**, and will apply these observations to target localization and classification problems.

23. Wavelet transforms

Wavelet transforms have found numerous applications in recent years for representing multiscale signals [76, 15, 79]. Though first appearing in the early 20th century [48], the idea lay dormant until the 1970s, when it began to find application in signal processing, culminating in a substantial literature [26, 42, 51]. Underlying all wavelet theory is the idea that information can be extracted by examining the inner product between a signal of interest and a collection of localized reference signals. Typical classes of localized reference signals are the Haar basis [48], the Mexican hat [17], and the Daubechies wavelet [26], though there are many others.

This section gives some preliminary observations about possible wavelet transforms in Euler calculus. The obvious choice for an Euler inner product on $\mathsf{CF}(X)$ is:

$$(23.1) \qquad (f, g)_\chi = \int_X f\, g\, d\chi.$$

This product is bilinear but lacks definiteness as for $f = \mathbf{1}_{(0,1)}$ one has $(f, f)_\chi = -1$. It is unreasonable to expect the typical Fourier series decomposition of functions to hold in Euler calculus. Nevertheless, let S be a set of mutually (χ-)orthogonal constructible functions. What can be learned about $f \in \mathsf{CF}(X)$ by its Euler products with the elements of S?

As a first step, we will define two families of wavelets that together define an Euler calculus version of the Haar wavelet transform. We work over $\mathbb{R}^n$ via induction on n. Define the first family

$$(23.2) \qquad H^{(0)}_{s,t}(x) = \begin{cases} 1 & \text{if } 2^s x = t \\ 0 & \text{otherwise} \end{cases}$$

consisting of atomic functions on dyadic points of the real line, where s and t are integers. The integers s and t are called *scale* and *translation* respectively. We note that there is some harmless duplication of functions with this indexing, for instance, all of the functions with $t = 0$ are identical. The second family consists of step functions of the following form:

$$(23.3) \qquad H^{(1)}_{s,t}(x) = \begin{cases} 1 & \text{if } 0 < 2^s x - t < 1/2 \\ -1 & \text{if } 1/2 < 2^s x - t < 1 \\ 0 & \text{otherwise} \end{cases}$$

for $s, t \in \mathbb{Z}$.

It should be immediately clear that distinct elements of $\{H^{(0)}_{s,t}\}$ are orthogonal, and that they are orthogonal to each element of $\{H^{(1)}_{\sigma,\tau}\}$ for which $\sigma \le s$. Similarly, the set $\{H^{(1)}_{s,t}\}$ is orthogonal but not orthonormal since $(H^{(1)}_{s,t}, H^{(1)}_{s,t})_\chi = -2$.

We extend this definition to functions on $\mathbb{R}^n$ as follows. Let $p \in \{0,1\}^n$ be a binary n-tuple, and $s, t \in \mathbb{Z}^n$ be integer n-tuples. We define the Haar wavelet functions on $x = (x_1, \ldots, x_n)$ multiplicatively as

$$(23.4) \qquad H_{s,t}^{(p)}(x) = H_{s_1,t_1}^{(p_1)}(x_1) \ldots H_{s_n,t_n}^{(p_n)}(x_n),$$

where we again have some harmless duplication of functions, due to the indexing of $H^{(0)}$.

We define the *Euler-Haar wavelet transform* of a function $f \in \mathsf{CF}(\mathbb{R}^n)$ to be

$$(23.5) \qquad (\mathcal{W}f)(p, s, t) = (f, H_{s,t}^{(p)})_\chi,$$

for $p \in \{0,1\}^n$ and $s, t \in \mathbb{Z}^n$.

Theorem 23.1. *The Euler-Haar wavelet transform is injective on* $\mathsf{CF}(\mathbb{R}^n)$.

PROOF. The result follows trivially from showing injectivity in the case $n = 1$. Suppose that $f \in \mathsf{CF}(\mathbb{R})$. Discern two cases

(1) If some nonzero level set of f contains a dyadic point $t/2^s$, then $\int f H_{s,t}^{(0)} d\chi \neq 0$.
(2) Otherwise, $f \neq 0$ for only a finite set of points, since f is constructible. Therefore, there are integers s, t so that the support of $H_{s,t}^{(1)}$ contains exactly one point on which $f \neq 0$. Hence $\int f H_{s,t}^{(1)} d\chi \neq 0$.

These cases are non-overlapping, so both types of Haar wavelet functions, the $H^{(0)}_{s,t}$ and the $H^{(1)}_{s,t}$, are necessary for injectivity to hold. □

We note from the proof that there is a function orthogonal to each $H_{s,t}^{(p)}$, so removing any element of the $\{H_{s,t}^{(p)}\}$ destroys the injectivity of $\mathcal{W}$ on the real line.

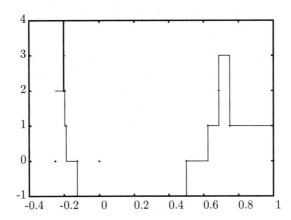

FIGURE 16. This is the result of using the Fourier summation formula with $\mathcal{W}$ coefficients. The original function was the indicator function on the compact interval $[-0.2, 0.7]$.

We reiterate that $\mathcal{W}$ cannot be used to directly reconstruct functions by the usual Fourier series formula

$$\sum_{p,s,t} (f, H_{s,t}^{(p)})_\chi H_{s,t}^{(p)},$$

by lack of mutual orthogonality. A rather spectacular example of this failure is shown in Figure 16 which is supposed to be the reconstruction of an indicator function on an interval.

Example 23.2. As an example of how $\mathcal{W}$ operates on $\mathsf{CF}(\mathbb{R}^2)$, consider the indicator function shown in Figure 17[left], with the $\mathcal{W}$ transform (with $\mathsf{p} = (1,1)$, all scales) shown at right, with all relevant scales shown. The image is broken up into rows and columns divided by heavy lines. Each column corresponds to a particular choice of horizontal scale (starting at zero at the left and increasing to the right), and each row corresponds to a particular vertical scale (starting at zero at the top and increasing as one proceeds downward). Within each subframe are the coefficients corresponding to various translates of the $\mathsf{H}^{(1)}$.

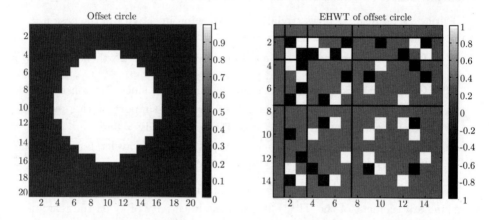

FIGURE 17. An indicator function on a disc (left) and its $\mathcal{W}$ transform (right) with $\mathsf{p} = (1,1)$.

From the definition of the $\mathsf{H}^{(1)}$ functions, it is clear that the transform is sensitive to jumps in the original function. In particular, in the lower-right corner of the transformed function in Figure 17 (corresponding to the smallest scale) is a rough copy of the gradient of the original function. However, since the original function is not aligned with any dyadic points in x or y, there is not perfect symmetry in the transform. It is also immediately apparent that the transform is rather sparse, though this is more a reflection of of the sparsity of the original function.

Another example of a more complicated constructible function and its $\mathcal{W}$ transform is shown in Figure 18. Together, these two examples indicate that the transform's coefficients depend on the function in a fairly complicated and sensitive manner and is currently not well understood. Indeed, $\mathcal{W}$ is *not* translation invariant, not even in the weak sense of the Fourier transform in which translation results in a multiplication by a phase. Although this sensitivity to translation might be perceived as a difficulty, it is shared (to a lesser extent) with more traditional wavelet transforms. Both of these kinds of wavelet transforms are *dyadic* translation invariant, so that $\tau\mathcal{W} = \mathcal{W}\tau$ for a translation τ of a power-of-2 amount in a cardinal direction. Its increased sensitivity to translation makes $\mathcal{W}$ potentially useful for detecting when signals fail to be aligned. We suspect (but do not here argue) that $\mathcal{W}$ may prove useful in prefiltering mechanisms for correlation processes.

These find extensive application in synchronization contexts, the most prominent being in localization such as GPS signals and radar.

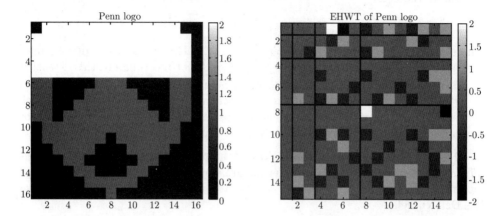

FIGURE 18. A two dimensional image (left) and its $\mathcal{W}$ transform (right) with $p = (1,1)$.

Toward a $\mathbb{R}$-valued Euler Calculus

24. Real-valued integrands

The Euler calculus, being integer-valued, has a delimited purview: it would be beneficial to have a calculus that manages other types of data than integral. Extending the Euler integral from $h \colon X \to \mathbb{Z}$ to (tame) functions with range in a discrete subgroup of a ring R is a simple exercise for the reader. For purposes of constructing an honest numerical analysis for the constructible setting of Euler integration, it would be beneficial to have an extension to $\mathbb{R}$-valued continuous functions. Any such extension will require the appropriate notion of tameness for integrands. Fortunately, this is easy in the o-minimal setting. Given X and Y tame spaces in an o-minimal structure, let $\mathsf{Def}(X,Y)$ denote the definable functions $f \colon X \to Y$ — those whose graphs in $X \times Y$ are tame. Recall from §3 that such maps are only piecewise smooth/continuous in general; thus, denote by $\mathsf{Def}^k(X,Y)$ those definable maps which are C^k smooth in the usual sense. This section focuses on extensions of Euler integration to $\mathsf{Def}(X) := \mathsf{Def}(X, \mathbb{R})$. We will, as per the remainder of this article, carry an implicit assumption of compactly supported integrands; this assumption can be removed with care and additional work if needed.

24.1. The Rota-Chen definition. One such extension was proposed by Rota [**68**] and used and enriched by Chen [**22**]. For $h \in \mathsf{Def}(\mathbb{R})$, define the integral to be:

$$(24.1) \quad \int_{\mathbb{R}} h \, d\chi := \int_{\mathbb{R}} \mathcal{J}h \, d\chi \quad : \quad (\mathcal{J}h)(x) := h(x) - \frac{1}{2}\left(\lim_{y \to x^-} h(y) + \lim_{y \to x^+} h(y)\right).$$

The operator $J\colon \mathsf{Def}(\mathbb{R}) \to \mathsf{CF}(\mathbb{R})$ records the (finite) jumps of h, so that the integral of Jh becomes a finite sum. For $h \in \mathsf{Def}(\mathbb{R}^n)$, the integral is defined so as to satisfy the Fubini theorem:
$$\int_{\mathbb{R}^n} h\,d\chi := \int\!\!\int \cdots \int h\,d\chi \cdots d\chi\,d\chi,$$
where each integral is over a coordinate axis of $\mathbb{R}^n$. This yields a well-defined integral operator which agrees with the usual $\int d\chi$ on CF. Unfortunately, this extension of $\int d\chi$ to $\mathsf{Def}(\mathbb{R}^n)$ has the definable continuous functions Def^0 in its kernel. This is singularly unhelpful as a tool for doing numerical analysis in Euler calculus in which we want to, say, take a piecewise-linear approximation to a sampled integrand, or perhaps smooth out a noisy integrand via convolution with a Gaussian.

24.2. A Riemann-sum definition.
A more simple-minded definition yields an entirely different extension of Euler integrals over $\mathsf{Def}(X)$. Following [4], extend integration to $\mathsf{Def}(X)$ via upper- or lower- step function approximations. Given $h \in \mathsf{Def}(X)$, define:

$$(24.2) \qquad \int_X h\,\lfloor d\chi\rfloor = \lim_{n\to\infty} \frac{1}{n}\int_X \lfloor nh\rfloor\,d\chi \quad : \quad \int_X h\,\lceil d\chi\rceil = \lim_{n\to\infty}\frac{1}{n}\int_X \lceil nh\rceil\,d\chi,$$

where $\lfloor\cdot\rfloor$ is the floor function and $\lceil\cdot\rceil$ is the ceiling function (returning the nearest integer below/above respectively). The notations $\lfloor d\chi\rfloor$ and $\lceil d\chi\rceil$ are chosen to denote which step-function approximation (lower/upper) is used in the limit. These limits exist and are well-defined, though not equal.

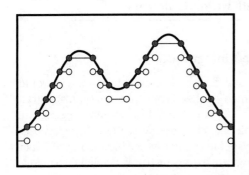 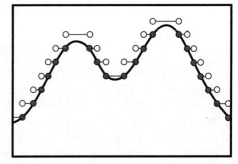

FIGURE 19. Lower and upper step function approximations lead to different limits due to, in this case, endpoints.

Lemma 24.1. *Given an affine function $h \in \mathsf{Def}(\sigma)$ on an open k-simplex σ,*

$$(24.3) \qquad \int_\sigma h\,\lfloor d\chi\rfloor = (-1)^k \inf_\sigma h \quad : \quad \int_\sigma h\,\lceil d\chi\rceil = (-1)^k \sup_\sigma h.$$

PROOF. For h affine on σ, $\chi\{\lfloor nh\rfloor > s\} = (-1)^k$ for all $s < \lfloor n\inf_\sigma h\rfloor$, and 0 otherwise. One computes
$$\lim_{n\to\infty}\frac{1}{n}\int_\sigma \lfloor nh\rfloor\,d\chi = \lim_{n\to\infty}\frac{1}{n}\sum_{s=0}^{\infty}\chi\{\lfloor nh\rfloor > s\} = (-1)^k \inf_\sigma h.$$

The analogous computation holds with $\chi\{\lceil nh\rceil > s\} = (-1)^k$ for all $s < \lceil n\sup_\sigma h\rceil$, and 0 otherwise. $\square$

This integration theory is robust to changes in coordinates.

Lemma 24.2. *Integration on* $\mathsf{Def}(X)$ *with respect to* $\lfloor d\chi \rfloor$ *and* $\lceil d\chi \rceil$ *is invariant under the right action of definable bijections of* X.

PROOF. The claim is true for Euler integration on $\mathsf{CF}(X)$; thus, it holds for $\int_X \lfloor nh \rfloor \, d\chi$ and $\int_X \lceil nh \rceil \, d\chi$. □

Lemma 24.3. *The limits in Equation (24.2) are well-defined.*

PROOF. An extension of the Triangulation Theorem to $\mathsf{Def}(X)$ [77] states that to any $h \in \mathsf{Def}(X)$, there is a definable triangulation of X on which h is definably affine, meaning that there is a Euclidean simplicial complex K and a definable homeomorphism $\phi \colon K \to X$ for which $h \circ \phi$ is affine on each open simplex of K. The result now follows from Lemmas 24.1 and 24.2. □

24.3. Nonlinearity.
One is tempted to apply all the sheaf-theoretic perspectives of §13 to $\int_X \colon \mathsf{Def}(X) \to \mathbb{R}$. However, for this formulation of the integral on $\mathsf{Def}(X)$, functoriality, and indeed, linearity, fails.

Lemma 24.4. $\int_X \colon \mathsf{Def}(X) \to \mathbb{R}$ *(via* $\lfloor d\chi \rfloor$ *or* $\lceil d\chi \rceil$*) is not a homomorphism for any* X *with* $\dim X > 0$.

PROOF. By explicit computation,

$$1 = \int_{[0,1]} 1 \lfloor d\chi \rfloor \neq \int_{[0,1]} x \lfloor d\chi \rfloor + \int_{[0,1]} (1-x) \lfloor d\chi \rfloor = 1 + 1 = 2.$$

□

Nonlinearity of these integral operators is due to the fact that the floor and ceiling functions used are nonlinear: $\lfloor f + g \rfloor$ agrees with $\lfloor f \rfloor + \lfloor g \rfloor$ up to a set of Lebesgue measure zero, but not Euler measure zero. In particular, the problems arise when increasing and decreasing functions compete. Though the change of variables formula (Lemma 24.2) holds, the more general Fubini theorem does not:

Corollary 24.5. *The Fubini theorem fails for* $\int_X \colon \mathsf{Def}(X) \to \mathbb{R}$ *(via* $\lfloor d\chi \rfloor$ *or* $\lceil d\chi \rceil$*) in general.*

PROOF. Let $F \colon X = Y \sqcup Y \to Y$ be the projection map with fibers $\{p\} \sqcup \{p\}$. Any $h \in \mathsf{Def}(X)$ is decomposable as $h = f \sqcup g$ for $f, g \in \mathsf{Def}(Y)$. The Fubini theorem applied to F is equivalent to the statement

$$\int_Y f + \int_Y g = \int_X h = \int_Y F_* h = \int_Y f + g$$

(where the integration is with respect to $\lfloor d\chi \rfloor$ or $\lceil d\chi \rceil$ as desired). Lemma 24.4 completes the proof. □

The Fubini theorem *does* hold when the map respects fibers.

Theorem 24.6. *For* $h \in \mathsf{Def}(X)$, *let* $F \colon X \to Y$ *be definable and* h-*preserving (*h *is constant on fibers of* F*). Then* $\int_Y F_* h \lfloor d\chi \rfloor = \int_X h \lfloor d\chi \rfloor$, *and* $\int_Y F_* h \lceil d\chi \rceil = \int_X h \lceil d\chi \rceil$.

PROOF. By Theorem 3.2, Y has a definable partition into tame sets Y_α such that $F^{-1}(Y_\alpha)$ is definably homeomorphic to $U_\alpha \times Y_\alpha$ for U_α tame, and that $F : U \times Y_\alpha \to Y_\alpha$ acts via projection. Since h is constant on fibers of F, one computes

$$\int_{Y_\alpha} F_*h \lfloor d\chi \rfloor = \int_{Y_\alpha} h\chi(U_\alpha) \lfloor d\chi \rfloor = \int_{U_\alpha \times Y_\alpha} h \lfloor d\chi \rfloor.$$

The same holds for $\int \lceil d\chi \rceil$. □

Corollary 24.7. *For $h \in \text{Def}(X)$, $\int_X h = \int_\mathbb{R} h_* h$. In other words,*

(24.4) $$\int_X h \lfloor d\chi \rfloor = \int_\mathbb{R} s\chi\{h = s\} \lfloor d\chi \rfloor,$$

and likewise for $\lceil d\chi \rceil$.

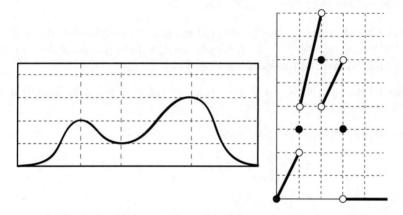

FIGURE 20. The application of Corollary 24.7 to a smooth univariate function [left] gives a piecewise-linear integrand [right] with the same Euler integral.

24.4. Computation. There are several definable analogues of computational formulae for integration over CF. The following results parallel those of §4.

Proposition 24.8. *For $h \in \text{Def}(X)$,*

(24.5) $$\int_X h \lfloor d\chi \rfloor = \int_{s=0}^\infty \chi\{h \geq s\} - \chi\{h < -s\} \, ds$$

(24.6) $$\int_X h \lceil d\chi \rceil = \int_{s=0}^\infty \chi\{h > s\} - \chi\{h \leq -s\} \, ds.$$

PROOF. For $h \geq 0$ affine on an open k-simplex σ,

$$\int_\sigma h \lfloor d\chi \rfloor = (-1)^k \inf_\sigma h = \int_0^\infty \chi(\sigma \cap \{h \geq s\}) ds,$$

and for $h \leq 0$, the equation holds with $-\chi(\sigma \cap \{h < -s\})$. The result for $\int \lceil d\chi \rceil$ follows from an identical computation. □

It is tempting to guess that $\int_X h \lfloor d\chi \rfloor = \int_0^\infty s \chi\{h = s\} \, ds$: the reader will easily verify that this is false. However, Equation (4.3) does have a definable analogue as follows.

Proposition 24.9. *For* $h \in \text{Def}(X)$,

$$\text{(24.7)} \quad \int_X h \lfloor d\chi \rfloor = \lim_{\epsilon \to 0^+} \frac{1}{\epsilon} \int_{\mathbb{R}} s\chi\{s \leq h < s + \epsilon\}\, ds$$

$$\text{(24.8)} \quad \int_X h \lceil d\chi \rceil = \lim_{\epsilon \to 0^+} \frac{1}{\epsilon} \int_{\mathbb{R}} s\chi\{s < h \leq s + \epsilon\}\, ds.$$

PROOF. For h affine on an open k-simplex σ, and $0 < \epsilon$ sufficiently small, $\int_{\mathbb{R}} s\chi\{s \leq h < s+\epsilon\}\,ds = \epsilon(-1)^k\left(-\frac{\epsilon}{2} + \inf_\sigma h\right)$ and $\int_{\mathbb{R}} s\chi\{s < h \leq s+\epsilon\}\,ds = \epsilon(-1)^k\left(-\frac{\epsilon}{2} + \sup_\sigma h\right)$. □

24.5. Interpretation. The measures $\lfloor d\chi \rfloor$ and $\lceil d\chi \rceil$, though not equal, are clearly related. A measured perusal of the formulae derived in this section suggest a type of *duality* between them (sup versus inf; $\geq$ versus $<$). One general result reinforcing that duality is the following: these measures are related by conjugation with -1.

Lemma 24.10.

$$\text{(24.9)} \quad \int h \lceil d\chi \rceil = -\int -h \lfloor d\chi \rfloor.$$

PROOF. Triangulate. Note that $\sup_\sigma h = -\inf_\sigma -h$. □

As to the question of what these integrals measure, the following direct computation gives a clue: both integrals are generalizations of total variation.

Corollary 24.11. *If* M *is a 1-dimensional manifold and* $h \in \text{Def}^0(M)$, *then*

$$\text{(24.10)} \quad \int_M h \lfloor d\chi \rfloor = -\int_M h \lceil d\chi \rceil = \frac{1}{2}\text{totvar}(h).$$

PROOF. Choose an affine triangulation of h which triangulates M with the maxima $\{p_i\}$ and minima $\{q_j\}$ as 0-simplices and the intervals between them as 1-simplices. To each minimum q_j is associated two open 1-simplicies, since M is a 1-manifold. Thus, via Lemma 24.1:

$$\int_M h \lfloor d\chi \rfloor = \sum_i h(p_i) + \sum_j h(q_j) - 2\sum_j h(q_j) = \frac{1}{2}\text{totvar}(h).$$

This equals $-\int_M h \lceil d\chi \rceil$ by either an analogous computation or by Lemma 24.10. □

Equation (24.10) *concentrates* the measures $\lfloor d\chi \rfloor$ and $\lceil d\chi \rceil$ on the critical points of the (univariate) integrand. This is the key to understanding what happens in the higher-dimensional generalization of total variation.

25. Morse theory

Morse theory is a means of relating the global features of (in the classical setting) a Riemannian manifold M with the local features of critical points of smooth $\mathbb{R}$-valued functions on M. Recall that $h : M \to \mathbb{R}$ is MORSE if all critical points of h are nondegenerate, in the sense of having a nondegenerate Hessian matrix of second partial derivatives. Denote by $\text{Cr}(h)$ the set of critical points of h. For each $p \in \text{Cr}(h)$, the MORSE INDEX of p, $\mu(p)$, is defined as the number of negative eigenvalues of the Hessian at p, or, equivalently, the dimension of the unstable manifold $W^u(p)$ of the gradient vector field $-\nabla h$ at p.

We will show, as a consequence of a more general approach, the following Morse-theoretic interpretation of $\int \cdot \lfloor d\chi \rfloor$:

Theorem 25.1. *If h is a Morse function on a closed n-manifold M, then:*

$$(25.1) \qquad \int_M h \lfloor d\chi \rfloor = \sum_{p \in Cr(h)} (-1)^{n-\mu(p)} h(p);$$

$$(25.2) \qquad \int_M h \lceil d\chi \rceil = \sum_{p \in Cr(h)} (-1)^{\mu(p)} h(p).$$

PROOF. The Euler characteristic of upper/lower excursion sets is piecewise-constant, changing only at critical values. Assume distinct critical values of h (using localization for the more general case). For $p \in Cr(h)$, $s = h(p)$, and $\epsilon \ll 1$, classical Morse theory [62] says that $\{h \geq s + \epsilon\}$ differs from $\{h \geq s - \epsilon\}$ by the addition of a product of discs $D^{\mu(p)} \times D^{n-\mu(p)}$ glued along $D^{\mu(p)} \times \partial D^{n-\mu(p)}$. The change in Euler characteristic resulting from this handle addition is $(-1)^{n-\mu(p)}$. Equation (24.5) and duality complete the proof. □

A more general result arises from a more general Morse theory. Stratified Morse theory [41] is a powerful extension to singular spaces, including definable sets with respect to an o-minimal structure [19, 74]. Let X be a tame subset of $\mathbb{R}^n$ and $h \colon X \to \mathbb{R}$ a smooth map which extends to $\tilde{h} \colon \mathbb{R}^n \to \mathbb{R}$ smooth. The LOCAL MORSE DATA of h at a point $x \in X$ is not a number (like the Morse index μ) but rather (the homotopy type of) a pair of compact spaces:

$$\mathrm{LMD}(h, x) = \lim_{\epsilon' \ll \epsilon \to 0^+} \left(\overline{B_\epsilon(x)} \cap \{|h - h(x)| \leq \epsilon'\}, \overline{B_\epsilon(x)} \cap \{h - h(x) - \epsilon'\} \right),$$

where $\overline{B_\epsilon(x)}$ is the closed ball of radius ϵ about $x \in X$. The limit (in homotopy type) is well-defined (indeed, the homeomorphism type stabilizes) thanks to Theorem 3.2. The EULER-POINCARÉ INDEX of h at x is, equivalently,

$$\begin{aligned}
(\mathfrak{I}_* h)(x) &= \chi(\mathrm{LMD}(h, x)) \\
&= 1 - \lim_{\epsilon' \ll \epsilon \to 0^+} \chi\left(\overline{B_\epsilon(x)} \cap \{h = h(x) - \epsilon'\} \right) \\
&= \lim_{\epsilon' \ll \epsilon \to 0^+} \chi\left(\overline{B_\epsilon(x)} \cap \{h > h(x) - \epsilon'\} \right).
\end{aligned}$$

There is a dual CO-INDEX given by $\mathfrak{I}^* h = \mathfrak{I}_*(-h)$. Note that $\mathfrak{I}_*, \mathfrak{I}^* \colon \mathrm{Def}(X) \to \mathrm{CF}(\overline{X})$, and the restriction of these operators to $\mathrm{CF}(X)$ is the identity (every point of a constructible function is a critical point). The two types of integration on $\mathrm{Def}(X)$ correspond to the Euler-Poincaré (co)index.

Theorem 25.2. *For $h \in \mathrm{Def}^0(X)$,*

$$(25.3) \qquad \int_X h \lfloor d\chi \rfloor = \int_{\overline{X}} h \mathfrak{I}^* h \, d\chi \quad ; \quad \int_X h \lceil d\chi \rceil = \int_{\overline{X}} h \mathfrak{I}_* h \, d\chi.$$

PROOF. On an open k-simplex $\sigma \subset X \subset \mathbb{R}^n$ in an affine triangulation of h, the co-index $\mathfrak{I}^* h$ equals $(-1)^{\dim(\sigma)}$ times the characteristic function of the closed face of σ determined by $\inf_\sigma h$. Since h is continuous, $\int_{\overline{\sigma}} h \mathfrak{I}^* h \, d\chi = (-1)^{\dim(\sigma)} \inf_\sigma h$. Lemma 24.1 and additivity complete the proof; duality gives a proof for $\mathfrak{I}_*$ and $\lceil d\chi \rceil$. □

This yields a concise proof of Theorem 25.1: for $p \in \mathsf{Cr}(h)$ a nondegenerate critical point on an n-manifold, $\mathfrak{I}^*h(p) = (-1)^{n-\mu(p)}$ and $\mathfrak{I}_*h(p) = (-1)^{\mu(p)}$. Furthermore, one sees clearly that the relationship between $\lfloor d\chi \rfloor$ and $\lceil d\chi \rceil$ is regulated by Poincaré duality — the canonical *flip* $\mathfrak{I}^*h = \mathfrak{I}_*(-h)$ is in play. For example, on continuous definable integrands over an n-dimensional manifold M,

$$(25.4) \qquad \int_M h \lceil d\chi \rceil = (-1)^n \int_M h \lfloor d\chi \rfloor.$$

The generalization from continuous to general definable integrands requires weighting $\mathfrak{I}^*h$ by h directly. To compute $\int_X h \lfloor d\chi \rfloor$, one integrates the weighted co-index

$$(25.5) \qquad \lim_{\epsilon' \ll \epsilon \to 0^+} h(x + \epsilon')\chi\left(\overline{B_\epsilon(x)} \cap \{h < h(x) + \epsilon'\}\right)$$

with respect to $d\chi$. Details are left to the curious reader.

It is worth noting that although the integral operators $\int \lfloor d\chi \rfloor$ and $\int \lceil d\chi \rceil$ are not linear, not functorial, and have seemingly little relationship to the sheaf-theoretic tools of §9-13, it is nevertheless true that at heart these operators stem from stratified Morse theory, a subject born from and intrinsic to the techniques of constructible sheaves [41, 74].

26. Incomplete data and harmonic interpolation

Our first application of the definable integration theory concerns management of uncertain or incomplete data in estimation of an Euler integral over $\mathsf{CF}(\mathbb{R}^2)$. It is common in sensor networks to possess regions of undersampling (or *holes*) in the underlying domain, through incomplete coverage or node failures. In this case, one wants to estimate the number of targets relative to the missing information. This translates to the following relative problem: if one knows $h: X \to \mathbb{N}$ only on some subset $A \subset X$, how well can one estimate $\int_X h \, d\chi$ from the restriction $h|_A$?

26.1. Bounds. We give bounds for the planar case, following [6].

Theorem 26.1. *Assume $h: \mathbb{R}^2 \to \mathbb{N}$ is the sum of indicator functions over a collection of compact contractible sets in $\mathbb{R}^2$, none of which is contained entirely within D, a fixed open contractible disc. Then*

$$(26.1) \qquad \int_{\mathbb{R}^2} \hat{h} \, d\chi \leq \int_{\mathbb{R}^2} h \, d\chi \leq \int_{\mathbb{R}^2} \check{h} \, d\chi,$$

where

$$\hat{h}(y) = \begin{cases} \max_{\partial D} h & : y \in \overline{D} \\ h & : else \end{cases}$$

$$\check{h}(y) = \begin{cases} \min_{\partial D} h & : y \in D \\ h & : else \end{cases}$$

PROOF. Via additivity of χ over domains, Eqn. (26.1) follows from the corresponding inequalities over the compact domain $\overline{D}$. Explicitly, if $\overline{h} = h$ on $\mathbb{R}^2 - \overline{D}$, then

$$\int_{\mathbb{R}^2} \overline{h} \, d\chi = \int_{\mathbb{R}^2 - D} h \, d\chi - \int_{\partial D} h \, d\chi + \int_{\overline{D}} \overline{h} \, d\chi.$$

Denote by $\mathcal{V} = \{V_\beta\}$ the collection of nonempty connected components of intersections of all target supports U_α with $\overline{D}$. Since we work in $\mathbb{R}^2$, each V_β is a compact contractible set which intersects ∂D. By Theorem 14.1, $\int_{\overline{D}} h \, d\chi$ equals the number of components $|\mathcal{V}|$. There are at least $\max_{\partial D} h$ such pieces; hence

$$\int_{\overline{D}} \hat{h} \, d\chi \leq \int_{\overline{D}} h \, d\chi.$$

Consider $\min_{\partial D} h$ and remove from the collection $\mathcal{V}$ this number of elements, including all such V_β equal to $\overline{D}$ (which is possible since we remove at most $\min_{\partial D} h$ such elements). Each remaining $V_\beta \in \mathcal{V}$ is not equal to $\overline{D}$ and thus intersects ∂D in a set with strictly positive Euler characteristic. Thus,

$$\int_{\overline{D}} h \, d\chi \leq \int_{\overline{D}} \check{h} \, d\chi.$$

□

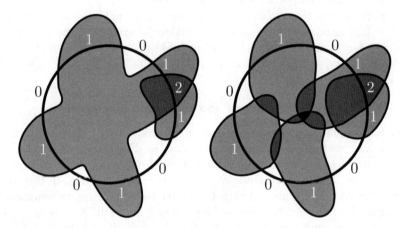

FIGURE 21. An example for which the upper and lower bounds of Equation (26.1) are sharp.

Example 26.2. Consider the example illustrated in Figure 21. The upper and lower estimates for the number of targets are 2 and 4 respectively. For this example, the estimates are sharp in that one can have collections of target supports over compact contractible sets which agree with h outside of $\overline{D}$ and realize the bounds.

Remark 26.3. The lower bound $\int \hat{h} \, d\chi$ may fail to be sharp in several ways. For example, a target support can intersect D in multiple components, causing $\hat{h}$ to not have a decomposition as a sum of characteristic functions over contractible sets (but rather with annuli). One can even find examples for which each target support intersects $\overline{D}$ in a contractible set but for which $\int \hat{h} \, d\chi$ is negative. The fact that the lower bound $\int \hat{h}$ can be so defective follows from the difficulty associated with annuli in the plane — these are 'large sets of measure zero' in $d\chi$.

The bounds of Theorem 26.1 allow one to conclude when a hole is inessential and no ambiguity about the integral exists.

Corollary 26.4. *Under the hypotheses of Theorem 26.1, the upper and lower bounds are equal when there is a unique connected local maximum of h on ∂D.*

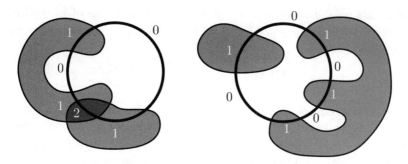

FIGURE 22. Two examples in which the lower bound $\int \hat{h}\, d\chi$ fails: [left] the lower bound is 1, but $\hat{h}$ cannot have a contractible support; [right] $\int \hat{h}\, d\chi$ is negative.

PROOF. In the case where h is constant on ∂D, $\check{h} = \hat{h}$ and the result is trivial. Otherwise, both $\check{h}$ and $\hat{h}$ have connected (and thus contractible) upper excursion sets; thus,

$$\int_{\mathbb{R}^2} \check{h}\, d\chi - \int_{\mathbb{R}^2} \hat{h}\, d\chi = \sum_s \chi\{\check{h} \geq s\} - \chi\{\hat{h} \geq s\}$$
$$= \sum_s 1 - 1 = 0.$$

$\square$

26.2. Harmonic extension and "expected" target counts. We continue the results of the previous subsection, considering the case of a planar domain with a contractible hole on which the integrand is unknown. As shown, upper and lower bounds are realized by extending the integrand across the hole via minimal and maximal values on the boundary of the hole. Inspired by the result that the PL-extension of a discretely sampled integrand yields correct integrals with respect to Euler characteristic, we consider extensions over holes via *continuous* functions.

The following result says that there is a principled interpolant between the upper and lower extensions. Roughly speaking, an extension to a harmonic function (discrete or continuous, solved over the hole with Dirichlet boundary conditions) provides an approximate integrand whose integral lies between the bounds given by upper and lower convex extensions. Analytically harmonic functions are unneeded: any form of weighted averaging will lead to an extension which respects the bounds. A specific criterion follows.

Theorem 26.5. *Given* $h: \mathbb{R}^2 - D \to \mathbb{N}$ *satisfying the assumptions of Theorem* 26.1, *let* $\overline{h}$ *be any definable extension of* h *which has no strict local maxima or minima on* D. *Then*

(26.2) $$\int_{\mathbb{R}^2} \hat{h}\, d\chi \leq \int_{\mathbb{R}^2} \overline{h}\, \lfloor d\chi \rfloor \leq \int_{\mathbb{R}^2} \check{h}\, d\chi,$$

PROOF. Consider an open neighborhood of $\overline{D}$ in $\mathbb{R}^2$ and modify $\overline{h}$ so that it preserves critical values, is Morse, and falls off to zero quickly outside of D. This perturbed function, denoted $\tilde{h}$, has isolated maxima on ∂D, isolated saddles in the interior of D (since there are no local extrema in D by hypothesis) and no other critical points outside of D. Since $\tilde{h}$ is a small perturbation of $\bar{h}$, the integral of $\bar{h}$ with respect to $\lfloor d\chi \rfloor$ is equal to $\int_{\overline{D}} \tilde{h} \lfloor d\chi \rfloor$. Via the Morse-theoretic formula of Equation (25.1),

$$\int \tilde{h} \lfloor d\chi \rfloor = \sum_{p \in Cr(\tilde{h})} (-1)^{2-\mu(p)} \tilde{h}(p).$$

The integral thus equals the sum of h over the maxima on ∂D minus the sum of $\bar{h}$ over the saddle points in the interior of D, since saddles have Morse index $\mu = 1$.

Denote by $\{p_i\}_1^M$ the maxima of $\tilde{h}$, ordered by their (increasing) $\tilde{h}$ values. Denote by $\{q_i\}_1^N$ the saddles of $\tilde{h}$, ordered by their (increasing) $\tilde{h}$ values. By the Poincaré index theorem,

$$1 = \chi(\overline{D}) = \#\mathrm{maxima}(\tilde{h}) - \#\mathrm{saddles}(\tilde{h}),$$

hence, $N = M - 1$. Note that, since there are no local minima, $\tilde{h}(q_i) < \tilde{h}(p_i)$ for all $i = 1 \ldots M - 1$. Thus,

$$\begin{aligned}
\int_{\overline{D}} \bar{h} \, d\chi &= \int_{\overline{D}} \tilde{h} \, d\chi \\
&= \tilde{h}(p_M) + \sum_{i=1}^{M-1} \tilde{h}(p_i) - \tilde{h}(q_i) \\
&\geq \tilde{h}(p_M) = \max_{\partial D} h = \int_{\overline{D}} \hat{h} \, d\chi.
\end{aligned}$$

For the other bound,

$$\begin{aligned}
\int_{\overline{D}} \bar{h} \, d\chi &= \tilde{h}(p_M) + \sum_{i=1}^{M-1} \tilde{h}(p_i) - \tilde{h}(q_i) \\
&\leq \sum_{i=1}^{M} \tilde{h}(p_i) = \int_{\overline{D}} \check{h} \, d\chi.
\end{aligned}$$

$\square$

A harmonic or harmonic-like function $\tilde{h}$ will often lead to an integral with non-integer value. One is tempted to interpret such an integral as an *expected* target count, though no formal notion of expected values for the Euler calculus as yet exists.

Example 26.6. Consider a hole D and a function h which is known only on ∂D and which has two maxima with value 1 and two minima with value 0. Without knowing more about the possible size and shape of the target supports which make up h, it is not clear whether this is more likely to come from one target support (which crosses the hole) or from two separate target supports. Computing a harmonic extension of this h over the interior of D yields a function $\tilde{h}$ with one saddle-type critical point in D. The value of the saddle is c and satisfies $0 < c < 1$, depending on the geometry of h on ∂D. This yields $\int \bar{h} \, d\chi = 2 - c$, reflecting the uncertainty of either one or two targets. In the perfectly symmetric case of Figure 23[left], $c = \frac{1}{2}$ and the integral is $\frac{3}{2}$, reflecting the balanced geometric ambiguity in target counts. In Figure 23[right], the harmonic extension has $c < \frac{1}{2}$, meaning that it is more likely that there are two target supports.

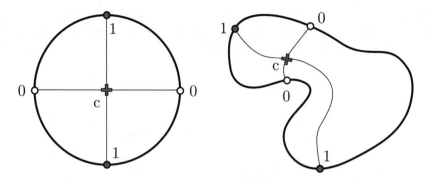

FIGURE 23. An integrand with a hole has two minima at height 0 and two maxima at height +1. Filling in by a harmonic function $\tilde{h}$ has an interior saddle at height $0 < c < 1$, depending on the geometry of h on ∂D: [left] $c = \frac{1}{2}$; [right] $c < \frac{1}{2}$.

In the network setting, holes often arise due to node failure or lack of sufficient node density. In these scenarios, one may reasonably employ any weighted local averaging scheme across dead nodes to recover a function which will respect the bounds of Theorem 26.1. Different weighting schemes may be more appropriate for different systems. For example, node readings can be assigned a *confidence measure*, which, when used as a weighting for the averaging over the dead zone, returns a value of the integral which reflects the fidelity of the data.

Example 26.7. We have assumed that targets are interrogated by a fixed network of stationary counting sensors: it is desirable to generalize to mobile sensors, especially in the context of robotics. Consider the following scenario: a collection of fixed target supports $\{U_\alpha\}$ lie in the plane. One or more mobile robots R_i can maneuver in the plane along chosen paths $x_i(t)$, returning sensed counting functions $h_i(t) = \#\{\alpha : x_i(t) \in U_\alpha\}$. How should the paths x_i be chosen so as to effectively determine the correct target count? If target supports are extremely convoluted, no guarantees are possible: therefore, assume that some additional structure is known (*e.g.*, an injectivity radius) giving a lower bound on how "thin" the target supports may be. Assume that the robots initially explore the planar domain along a rectilinear graph Γ that tiles the domain into rectangles. If desired, one can make these rectangles have either width or height small enough in order to guarantee that all the U_α intersect Γ. Consider the sensor function $h \colon \Gamma \to \mathbb{N}$. The integral $\int_\Gamma h\, d\chi$ is likely to give the wrong answer, even (especially!) for a dense Γ. Two means of getting a decent approximation are (1) use the network approximation formula of Equation (17.1); or (2) perform a harmonic extension over the holes of Γ.

Neither approach is guaranteed to give a good *a priori* approximation to the target count. How can one tell if Γ should be filled in more? The simplest criterion follows from Corollary 26.4. Consider a basic cycle $\Gamma' \subset \Gamma$ in the tiling induced by Γ. If there is a single connected local maximum on Γ', then (assuming that no small U_α lies entirely within the hole) the harmonic extension over Γ' gives an accurate contribution to the integral.

If, on the other hand, there are multiple maximal sets on Γ', then one must refine Γ into smaller cycles for which the criterion holds. The obvious approach is to guide the mobile sensors so as to try and connect disjoint maxima and/or disjoint minima. Figure 24 gives the sense of the technique. The crucial observation is that Corollary 26.4 provides a stopping criterion.

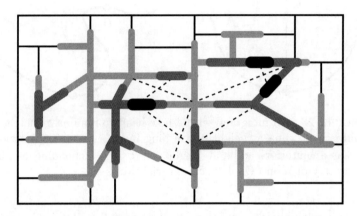

FIGURE 24. Mobile agents determine target counts over a network. Holes with multiple maxima require further refinement (dashed lines).

27. Index formulae for integral transforms

The definable Euler integration theory has important implications in the efficient computation of certain integral transforms from §22. Recall the Euler-Bessel transform $\mathbf{B}\colon CF(V) \to Def(V)$ on a finite-dimensional normed real vector space V, given by integrating along concentric spherical isospectral sets with respect to $d\chi$ and along the radius in Lebesgue measure. This transform, though acting on $CF(V)$, is interpretable in terms of definable Euler integrals, and, hence, Morse theory. The following is from [**38**].

Lemma 27.1. *For $A \subset V$ a compact tame codimension-0 ball, star-convex with respect to $x \in A$,*

$$(27.1) \qquad \mathbf{B}1_A(x) = \int_{\partial A} d_x \lfloor d\chi \rfloor,$$

where d_x is the distance-to-x function $d_x : V \to [0, \infty)$.

PROOF. By definition,

$$\mathbf{B}h(x) = \int_0^\infty \chi(A \cap \partial B_r(x)) dr.$$

For A a tame ball, star-convex with respect to x, $A \cap \partial B_r(x)$ is homeomorphic to $\partial A \cap \{d_x \geq r\}$. By Equation (4.5),

$$\mathbf{B}h(x) = \int_0^\infty \chi(\partial A \cap \{d_x \geq r\}) dr = \int_{\partial A} d_x \lfloor d\chi \rfloor.$$

□

This theorem is a manifestation of Stokes' Theorem: the integral of the distance over ∂A equals the integral of the 'derivative' of distance over A. For non-star-convex domains, it is necessary to break up the boundary into positively and negatively oriented pieces. These orientations implicate $\lfloor d\chi \rfloor$ and $\lceil d\chi \rceil$ respectively.

Theorem 27.2. *For $A \subset V$ a codimension-0 submanifold with corners in V and $x \in V$, decompose ∂A into $\partial A = \partial_x^+ A \cup \partial_x^- A$, where $\partial_x^\pm A$ are the (closure of) subsets of ∂A on which the outward-pointing halfspaces contain (for ∂_x^-) or, respectively, do not contain (for ∂_x^+) x. Then,*

$$(27.2) \qquad \mathbf{B1}_A(x) = \int_{\partial_x^+ A} d_x \lfloor d\chi \rfloor - \int_{\partial_x^- A} d_x \lceil d\chi \rceil$$

$$(27.3) \qquad\qquad\quad = \int_{Cr_x \cap \partial_x^+ A} d_x \, \mathcal{J}^* \, d\chi - \int_{Cr_x \cap \partial_x^- A} d_x \, \mathcal{J}_* \, d\chi.$$

where Cr_x denotes the critical points of $d_x \colon \partial A \to [0, \infty)$.

PROOF. Assume, for simplicity, that A is the closure of the difference of C_x^+, the cone at x over A_x^+, and C_x^-, the cone over A_x^-, with the case of multiple cones following by induction. These cones, being star-convex balls with respect to x, admit Lemma 27.1. The crucial observation is that, by additivity of χ,

$$\chi(\partial B_r(x) \cap A) = \chi(\partial C_x^+ \cap \{d_x \geq r\}) - \chi(\partial C_x^- \cap \{d_x > r\}).$$

Integrating both sides with respect to dr and applying Equation (4.5) gives

$$\mathbf{B1}_A(x) = \int_{\partial C_x^+} d_x \lfloor d\chi \rfloor - \int_{\partial C_x^-} d_x \lceil d\chi \rceil.$$

By Theorem 25.2, this reduces to an integral over the critical sets of the (stratified Morse) function d_x. The only critical point of d_x on $C_x^+ - \partial A$ or $C_x^- - \partial A$ is x itself, on which the integrand d_x takes the value 0 and does not contribute to the integral. Therefore the integrals over the cone boundaries may be restricted to $\partial^+ A$ and $\partial^- A$ respectively. The index-theoretic result follows from Theorem 25.2. □

In even dimensions, the $\lfloor d\chi \rfloor$-vs-$\lceil d\chi \rceil$ dichotomy dissolves:

Corollary 27.3. *For $\dim V$ even and $A \subset V$ a codimension-0 submanifold with corners,*

$$(27.4) \qquad \mathbf{B1}_A(x) = \int_{\partial A} d_x \lfloor d\chi \rfloor = \int_{Cr_x} d_x \, \mathcal{J}_* \, d\chi.$$

PROOF. In this setting, ∂A is an odd-dimensional definable manifold. Equation (25.4) implies that $\int \lceil d\chi \rceil = -\int \lfloor d\chi \rfloor$. Equation (27.2) completes the proof. □

Given the index theorem for the Euler-Bessel transform, that for the Euler-Fourier is a trivial modification that generalizes Example 22.1.

Theorem 27.4. *For $A \subset V$ a codimension-0 submanifold with corners in V and $\xi \in V^\vee - \{0\}$ a nonzero dual vector, decompose ∂A into $\partial A = \partial_\xi^+ A \cup \partial_\xi^- A$, where $\partial_\xi^\pm A$ are the (closure of) subsets of ∂A on which ξ points out of (∂^+) or into (∂^-)*

A. Then,

(27.5) $$\mathbf{F1}_A(\xi) = \int_{\partial_\xi^+ A} \xi \lfloor d\chi \rfloor - \int_{\partial_\xi^- A} \xi \lceil d\chi \rceil$$

(27.6) $$= \int_{\mathsf{Cr}_\xi \cap \partial_\xi^+ A} \xi \mathcal{J}^* \, d\chi - \int_{\mathsf{Cr}_\xi \cap \partial_\xi^- A} \xi \mathcal{J}_* \, d\chi.$$

where Cr_ξ denotes the critical points of $\xi \colon \partial A \to [0, \infty)$. For dim V even, this becomes:

(27.7) $$\mathbf{F1}_A(\xi) = \int_{\partial A} \xi \lfloor d\chi \rfloor = \int_{\mathsf{Cr}_\xi} \xi \mathcal{J}_* \, d\chi.$$

The proof follows that of Theorem 27.2 and is an exercise. Figure 25 gives a simple example of the Bessel and Fourier index theorems in $\mathbb{R}^2$.

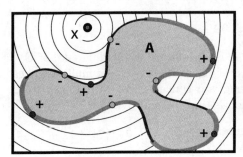 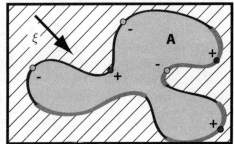

FIGURE 25. The index formula for **B** [left] and **F** [right] applied to $\mathbf{1}_A$ localizes the transform to (topological or smooth) tangencies of the isospectral sets with ∂A.

Example 27.5. By linearity of **B** and **F** over CF(V), one derives index formulae for integrands in CF(V) expressible as a linear combination of $\mathbf{1}_{A_i}$ for A_i the closure of definable bounded open sets. For a set A which is not of dimension dim V, it is still possible to apply the index formula by means of a limiting process on compact tubular neighborhoods of A. For example, let $A \subset \mathbb{R}^2$ be a compact straight line segment: see Figure 26. Let A_0 and A_1 denote the endpoints of A. The reader may compute directly that

$$\mathbf{B1}_A(x) = d(x, A_0) + d(x, A_1) - 2d(x, A),$$

where $d(x, A)$ denotes the distance from x to the segment A. When one of the endpoints minimizes this distance, one gets $\mathbf{B1}_A(x) = \max_{\partial A} d_x - \min_{\partial A} d_x$, exactly as Equation (27.2) would suggest. This correspondence seems to fail when there is a tangency between the interior of A and the isospectral circles, as in Figure 26[right]: why the factor of 2? The index interpretation is clear, however, upon taking a limit of neighborhoods, in which case a pair of negative-index tangencies are revealed.

These index formulae are useful in explaining the geometric content of the Fourier and Bessel transforms.

Corollary 27.6. *For A a compact round ball about* $p \in \mathbb{R}^{2n}$, *the Bessel transform* $\mathbf{B1}_A$ *is a nondecreasing function of the distance to* p, *having unique zero at* p.

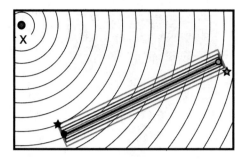 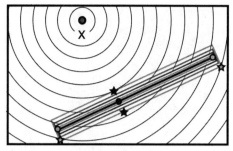

FIGURE 26. The Euler-Bessel Transform of a line segment in $\mathbb{R}^2$ has an index formula determined by d_x at the endpoints and at an interior tangency. This follows from Theorem 27.2 by limits of compact tubular neighborhoods.

PROOF. Convexity of balls and Corollary 27.3 implies that
$$\mathbf{B}\mathbf{1}_A(x) = \int_{\partial A} d_x \lfloor d\chi \rfloor = \max_{\partial A} d_x - \min_{\partial A} d_x,$$
which equals diam$A$ for $x \notin A$ and is monotone in distance-to-p within A. □

Note that Corollary 27.6 is vacuous in odd dimensions; the Bessel transform of a ball in $\mathbb{R}^{2n+1}$ is constant, and $\mathbf{B}$ obscures all information. However, for even dimensions, Corollary 27.6 provides a basis for target localization and shape detection. For targets with convex supports (regions detected by counting sensors), the local minima of the Euler-Bessel transform can reveal target locations: see Figure 27[left] for an example. Note that in this example, not all local minima are target centers: interference creates ghost minima. However, given $h \in \mathsf{CF}(V)$, the integral $\int_V h \, d\chi$ determines the number of targets. This provides a guide as to how many of the deepest local minima to interrogate. The deepest minima correspond to perfectly round targets: this gives a basis for performing shape discrimination based solely on enumerative data, as in Figure 27[right].

There are significant limitations to superposition by linearity for this application. When targets are nearby or overlapping, their individual transforms will have overlapping sidelobes, which results in uncertainty when the transform is being used for localization.

28. Integral transforms with definable kernels

The previous subsection examined integral transforms over CF by means of index theory. There are numerous open questions concerning integral transforms over $\mathsf{Def}(X)$.

28.1. Continuity. Though the integral operators with respect to $\lfloor d\chi \rfloor$ and $\lceil d\chi \rceil$ are not linear on $\mathsf{Def}(X)$, they nevertheless retain some nice properties. The following remarks are adapted from [4]. All properties below stated for $\int \lfloor d\chi \rfloor$ hold for $\int \lceil d\chi \rceil$ via duality.

Lemma 28.1. *The integral* $\int \lfloor d\chi \rfloor \colon \mathsf{Def}(X) \to \mathbb{R}$ *is positively homogeneous.*

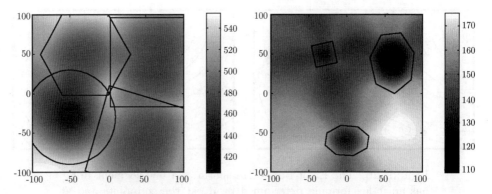

FIGURE 27. The Euler-Bessel transform of a collection of convex targets [left] has local minima at the target centers. The deepest minima in the Bessel-transform correspond to round targets, and may be useful for shape detection [right].

PROOF. For $f \in \text{Def}(X)$ and $\lambda \in \mathbb{R}^+$, the change of variables variables $s \mapsto \lambda s$ in Eqn. [24.5] gives $\int \lambda f \lfloor d\chi \rfloor = \lambda \int f \lfloor d\chi \rfloor$. □

Example 28.2. Integration is not continuous on $\text{Def}(X)$ with respect to the C^0 topology. An arbitrarily large change in $\int h \lfloor d\chi \rfloor$ may be effected by small changes to h. An example appears in [**81**]: consider $h(x, y) = 1 - 2\left|x - \frac{1}{2}\right| \in \text{Def}([0, 1]^2)$ as in Figure 28[left]. Since there is one maximal value at 1 along the compact interval $\{x = \frac{1}{2}\}$, the integral $\int h \lfloor d\chi \rfloor = 1$. However, sampling over a triangulation that does not sample the topology of the maximal set correctly yields a number of maxima and saddles. It may be arranged (as in Figure 28[right]) so that successively fine triangulations cause the net variation to blow up to infinity, even as the graphs converge pointwise. In some situations the "complexity" of the definable functions

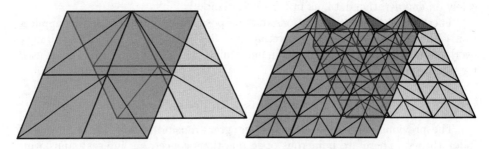

FIGURE 28. The integrand h has PL approximations that converge pointwise, but diverge in integration with respect to $\lfloor d\chi \rfloor$.

can be controlled in a way sufficient to ensure continuity [**4**]. The above example fails if sufficiently smoothed. It remains an interesting open question to explore topologies on $\text{Def}(X)$ and the relationship to continuity of Euler integration.

28.2. Duality.

The duality operator $\mathbf{D}\colon \mathsf{CF}(X) \to \mathsf{CF}(X)$ of §19 extends seamlessly to integrals on $\mathsf{Def}(X)$ in the obvious manner:

$$(28.1) \qquad \mathbf{D}h(x) = \lim_{\epsilon \to 0^+} \int_X h \mathbf{1}_{B_\epsilon(x)} \lfloor d\chi \rfloor = \lim_{\epsilon \to 0^+} \int_X h \mathbf{1}_{B_\epsilon(x)} \lceil d\chi \rceil,$$

where B_ϵ is an open metric ball of radius ϵ.

Lemma 28.3. *Duality* $\mathbf{D}\colon \mathsf{Def}(X) \to \mathsf{Def}(X)$ *is well-defined and independent of whether the integration in (28.1) is with respect to* $\lfloor d\chi \rfloor$ *or* $\lceil d\chi \rceil$.

PROOF. The limit is well-defined thanks to the Hardt Theorem. To show that it is independent of the upper- or lower-semicontinuous approximation, take $\epsilon > 0$ sufficiently small. Note that by triangulation, h can be assumed to be piecewise-affine on open simplices. Pick a point x in the support of h and let $\{\sigma_i\}$ be the set of open simplices whose closures contain x. Then for each i, the limit $h_i(x) := \lim_{y \to x} h(y)$ for $y \in \sigma_i$ exists. One computes

$$(28.2) \qquad \mathbf{D}h(x) = \sum_i (-1)^{\dim \sigma_i} h_i(x),$$

independent of the measure $\lfloor d\chi \rfloor$ or $\lceil d\chi \rceil$. $\square$

For a continuous definable function h on a manifold M, $\mathbf{D}h = (-1)^{\dim M} h$, as one can verify. This is commensurate with the result of Schapira [71] that $\mathbf{D}$ is an involution on $\mathsf{CF}(X)$.

Theorem 28.4. *Duality is involutive on* $\mathsf{Def}(X)$: $\mathbf{D} \circ \mathbf{D}h = h$.

PROOF. Given h, fix a triangulation on which h is affine on open simplices. Note that the dual of h at x is completely determined by the trivialization of h at x. Let $L_x h$ be the constructible function on $B_\epsilon(x)$ which takes on the value $h_i(x)$ on strata $\sigma_i \cap B_\epsilon(x)$. (Though this is not necessarily an integer-valued function, its range is discrete and therefore is constructible.) As $L_x h$ is close to h in $B_\epsilon(x)$ (this follows from the continuity of h on each of the strata), $\mathbf{D}h$ is close to $\mathbf{D}L_x h$ in $B_\epsilon(x)$: indeed, the total Betti number of intersections of strata with any ball $B_\epsilon(y)$ is bounded, and Euler integral of a function small in absolute value is small as well. Hence the definable function $\mathbf{D}^2 h$ is close to the constructible function $\mathbf{D}L_x h$ with ϵ small. As $\mathbf{D}^2 L_x h(x) = L_x h(x) = h(x)$, the result follows. $\square$

Duality may be used to define a link operator on definable functions as

$$(28.3) \qquad \Lambda h(x) = \lim_{\epsilon \to 0^+} \int_X h \mathbf{1}_{\partial B_\epsilon(x)} d\chi.$$

The link of a continuous function on an n-manifold M is multiplication by $1+(-1)^n$, as a simple computation shows. In general, $\Lambda = \mathsf{Id} - \mathbf{D}$, where Id is the identity operator.

28.3. Linearity.

The nonlinearity of the integration operator prevents most straightforward applications of Schapira's inversion formula. Fix a kernel $K \in \mathsf{Def}(X \times Y)$ and consider the general integral transform $\mathcal{T}_K \colon \mathsf{Def}(X) \to \mathsf{Def}(Y)$ of the form $(\mathcal{T}_K h)(y) = \int_X h(x) K(x,y) \lfloor d\chi \rfloor(x)$. In general, this operator is non-linear, via Lemma 24.4. However, some vestige of (positive) linearity survives within CF^+, the *positive* linear combinations of indicator functions over tame top-dimensional subsets of X.

Lemma 28.5. *The integral transform $\mathcal{T}_K$ is positive-linear over* $\mathsf{CF}^+(X)$.

PROOF. Any $h \in \mathsf{CF}^+(X)$ is of the form $h = \sum_k a_k \mathbf{1}_{U_k}$ for $a_k \in \mathbb{N}$ and $U_k \in \mathsf{Def}(X)$. For $h = \mathbf{1}_A$, $\mathcal{T}_K h = \int_A K \lfloor d\chi \rfloor$. Additivity of the integral in $\lfloor d\chi \rfloor$ combined with Lemma 28.1 completes the proof. □

This implies in particular that when one convolves a function $h \in \mathsf{CF}^+(\mathbb{R}^n)$ with a smoothing kernel (*e.g.*, a Gaussian) as a means of filtering noise or taking an average of neighboring data points, that convolution may be analyzed one step at a time (decomposing h). The restriction to positive-linearity is critical, since $\int -h \lfloor d\chi \rfloor \neq -\int h \lfloor d\chi \rfloor$. However, integral transforms which combine $\lfloor d\chi \rfloor$ and $\lceil d\chi \rceil$ compensate for this behavior. Define the measure $[d\chi]$ to be the average of $\lfloor d\chi \rfloor$ and $\lceil d\chi \rceil$:

$$(28.4) \quad \int_X h[d\chi] = \frac{1}{2}\left(\int_X h \lfloor d\chi \rfloor + \int_X h \lceil d\chi \rceil\right).$$

Theorem 28.6. *Any integral transform of the form*

$$(28.5) \quad (\mathcal{T}_K h)(y) = \int_X h(x) K(x,y) [d\chi](x)$$

for $K \in \mathsf{Def}(X \times Y)$ *is a linear operator* $\mathsf{CF}(X) \to \mathsf{Def}(Y)$.

PROOF. From Lemma 28.5, $\mathcal{T}$ is positive-linear over $\mathsf{CF}^+(X)$. Full linearity follows from the observation that $\int_X -h[d\chi] = -\int_X h[d\chi]$, which follows from Lemma 24.1 by triangulating h. □

Example 28.7. Consider the transform with $X = \mathbb{R}^n$, $Y = \mathbb{S}^{n-1} \subset X$, and kernel $K(x, \xi) = \langle x, \xi \rangle$. This transform with respect to $[d\chi]$ applied to $\mathbf{1}_A$ for A compact and convex returns the 'centroid' of A along the ξ-axis: the average of the maximal and minimal values of ξ on ∂A. Note how the dependence on critical values of the integrand on ∂A reflects the Morse-theoretic interpretation of the integral in this case. The interested reader may wish to derive a more general index-theoretic result over $\mathsf{CF}(X)$ using the ideas of §25.

Integration with respect to $[d\chi]$ seems suitable only for integral transforms over CF, since, on a continuous integrand, the integral with respect to $[d\chi]$ either returns zero (cf. the Rota-Chen definition of §24.1) or else the integral with respect to $\lfloor d\chi \rfloor$, depending on the parity of the dim X.

28.4. Convolution. Recall the convolution operator from §19: $(f * g)(x) = \int_V f(t) g(x-t) \, d\chi$. Convolution is well-defined on $\mathsf{Def}(V)$ by integrating with respect to $\lfloor d\chi \rfloor$ or $\lceil d\chi \rceil$. However, the product formula (Lemma 19.1) for $\int f * g$ fails in general, since one relies on the Fubini theorem to prove it in $\mathsf{CF}(V)$. By Lemma 28.5, convolution with a definable $g \in \mathsf{Def}(V)$ is positive-linear over $\mathsf{CF}^+(V)$. We indicate (following [4]) how to use this to smooth noise in an integrand $h \in \mathsf{CF}^+(V)$.

Integration over $\mathsf{CF}^+(X)$ is poorly behaved with respect to noise (Figure 29[left]), owing to the fact that points have full measure in $d\chi$. Assume a sampling of $f \in \mathsf{CF}^+(X)$ over a network, with an error of ± 1 on random nodes: specifically, $h = f + e$, where $e: \mathcal{N} \to \{-1, 0, 1\}$ is an error function that is nonzero on a sparse subset of nodes $\mathcal{N}' \subset \mathcal{N}$. For typical choices of $\mathcal{N}'$, $\left|\int h \, d\chi - \int f \, d\chi\right|$ will be a normal of variance $O(|\mathcal{N}'|^{\frac{1}{2}})$.

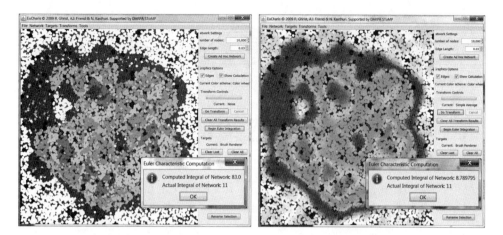

FIGURE 29. Screenshots of *Eucharis*, demonstrating the deleterious impact of noise [left] and the resulting mitigation via convolution with a bump function via neighbor-averaging [right]. The true integral (11) is poorly approximated (83) when the integrand has a 10% noise added, but is more reasonably approximated (8.8) after convolution.

One strategy for mitigating noise is to dissipate via convolution with a smoothing kernel. For discrete data, this is best accomplished via a weighted average of neighboring node values, with weights inverse to distance (hop metric for the network). Such an averaging, for a sufficiently dense network of randomly paced nodes, approximates a convolution of the constructible data with a bump function. This poses the problem of when such a convolution returns the true integral. Assume a convolution $h * K$ of $h \in CF^+(\mathbb{R}^n)$ with an appropriate kernel K, unimodal and of appropriately small support relative to h. To quantify the optimal size of K, we use a characteristic length (related to the WEAK FEATURE SIZE of [21]) which encodes the fragility of an integrand $h \in CF^+(\mathbb{R}^n)$ with respect to $\lfloor d\chi \rfloor$. Define the CONSTRUCTIBLE FEATURE SIZE of $h \in CF^+(\mathbb{R}^n)$ at $x \in \mathbb{R}^n$ to be $CFS_x(h)$, the supremum of all R such that, for any closure $\mathcal{C}$ of any connected component of upper or lower excursion sets of h, the convex hull of all outward-oriented normals to $\partial \mathcal{C} \cap B_R(x)$ contains 0: see Figure 30. Minimizing over x yields $CFS(h) = \inf_x CFS_x(h)$. Constructible feature size regulates the impact of a smoothing kernel, whether coming from data diffused at the hardware level, or purposefully smoothed to mitigate noise.

Theorem 28.8 ([4]). *For $h \in CF^+(\mathbb{R}^n)$ and $K \in Def^1(\mathbb{R}^n)$ a radially-symmetric kernel with support of radius $R < CFS(h)/2$,*

$$(28.6) \qquad \int_{\mathbb{R}^n} h * K \lfloor d\chi \rfloor = \int_{\mathbb{R}^n} h \, d\chi.$$

PROOF. Note that $h * K \in Def^1(\mathbb{R}^n)$. By Lemma 28.5, $h * K$ is the sum over α of $\mathbf{1}_{U_\alpha} * K$. Each such convolution is a smoothing of $\mathbf{1}_{U_\alpha}$ on an exterior tubular neighborhood of radius R of U_α. The integral of each $\mathbf{1}_{U_\alpha} * K$ is thus unchanged. Lack of linearity for $\int \lfloor d\chi \rfloor$ forbids concluding that the integral of the full $h * K$ is

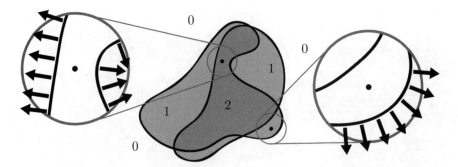

FIGURE 30. Constructible feature size CFS of $h \in \mathsf{CF}(\mathbb{R}^n)$ detects fragility in excursion sets' topology.

unchanged. It does suffice, however, to show that no changes in critical points arise in summing these R-neighborhood smoothed supports.

Consider x in the intersection of the exterior R-neighborhoods of (after re-indexing) a collection $\{U_\beta\}$. The gradient $\nabla_x(h * K)$ is a weighted vector sum over β of $\nabla_x(1_{U_\beta} * K)$. By radial symmetry of K, each such vector is parallel (and oppositely oriented) to the normal from x to U_β. By definition of CFS, this sum cannot be zero, and thus no new critical points arise in the convolution $h * K$. Invoking Theorem 25.2 and verifying that the indices of previously existing critical sets are unchanged by smoothing, one has that the integrals agree. □

Thus, integrating the (convolved) intensity data in $\lfloor d\chi \rfloor$ returns the $d\chi$ integral of the (true) constructible data, when feature size is sufficiently large. Stochastic geometry techniques reveal that typical feature size is small [4]. More recent results [14] give strong results on convolution formulae for Gaussian smoothing of noisy integrands.

Closing Thoughts

29. Open questions

29.1. Sampling and estimation. The success of Alexander duality in computing Euler integrals over *ad hoc* planar networks as in §17 stands in marked contrast to situation in dimensions three and higher. In dimension three, *ad hoc* networks are still a sensible way to deploy sensors to sample a function, motivating the need for numerically efficient and stable algorithms for computing Euler integrals. Alexander duality permits one to pair H_0 and H_2, thus avoiding the need to determine voids in the sampling by means of clustering; however, H_1 is self-dual, and this is a nontrivial difficulty. *Is there a fast method for estimating Euler integrals over a network-sampled integrand on $\mathbb{R}^k$ for $k > 2$?*

29.2. Numerical analysis. This survey leaves open the broader question of rigorous computation and estimation: *is there a sensible numerical analysis for the Euler calculus?* It seems that classical tools and techniques for approximating integrals fail to hold in the Euler calculus. The typical *sine qua non* — convergence of integration with arbitrarily dense discrete sampling — can fail, as shown in §16. Any sceptic would feel justified in giving up. Nevertheless, we believe that the potential for applications demands the careful investigation of numerical issues in Euler calculus. It is our belief (offered without proof!) that a well-defined and applicable numerical analysis for Euler calculus exists and is worthwhile.

29.3. Efficient integral transforms. The potential of Euler-based integral transforms seems very significant in image and radar signal processing. Besides the difficulties of determining inverse transforms, etc., the sheer numerical complexity of computation is an impediment: *are there fast algorithms for computing and inverting Euler integral transforms?* As described in §27, one way to dramatically improve the burden of computing certain Euler integral transforms is to specifically exploit the underlying index theory, focusing the transforms onto critical sets. As a small step in this direction, it is possible to construct explicit formulae for the integral transforms for geometric figures in, *e.g.*, the Euler-Bessel and Fourier transform. In computer graphics contexts, where a (usually semialgebraic) description of sets exist, it may be efficient to exploit critical point computations on surfaces of positive codimension. This would enable dramatically less computation than computing Euler integrals directly, and may also incur less numerical error.

29.4. DSP and Euler calculus. One of the major enablers in the signal processing revolution was the discovery of the Fast Fourier Transform (FFT). By recursively interleaving the domain, a discrete-time signal's Fourier transform can be computed in $O(N \log N)$ operations rather than $O(N^2)$. The crucial insight is that the transform can be quickly computed on an interval of length $N_1 N_2$ by first computing it on the much shorter intervals of length N_1 and N_2. *What is the Eulerian FFT?* It is presently unclear if such a reduction in computation can be made for the Euler integral on regularly-sampled domains: the underlying mechanism for combining the interleaved computations in the usual FFT is orthogonality, and the Euler product does not support the usual synthesis formula. As a result, the relationship of a function's Euler integral to subsampled copies is likely to be complicated. That there is any possibility of an Eulerian FFT at all comes from the scissors relation for definable sets: definable level sets of a function on a regular grid have a decomposition (via a definable homeomorphism) that may enable computational simplifications.

29.5. Euler wavelets. As shown in §23, the Euler-Haar wavelet transform is injective, and can be used to synthesize a function from its transform algorithmically. However, the primary use of the analysis-synthesis properties of transforms is to enable filtering. *How do particular classes of Euler wavelet-based filters affect a given signal?* The inversion algorithm we have exhibited is delicate: not every set of wavelet coefficients corresponds to a transform of a function, which could cause nontermination of the algorithm. Worse, since we do not have an explicit characterization of what the family of appropriate wavelet coefficients are, naive filtering may be destructive. In the definable category of functions, especially, there are likely many other wavelet families to be studied. As in the Lebesgue case, Haar wavelets represent the crudest (and not terribly effective) family. It is to be hoped that other Euler wavelet families that have substantially better properties may be found.

29.6. Data fusion. The Euler integral and its associated operations arise from a subtle connection to the sheaf of constructible functions. These functions take values in the integers, which provides much of the motivation for considering integer-valued data in this context. Substantially more complicated algebraic data can be stored into sheaves; appropriately generalized integration theories *may* follow. It remains an open problem: *how to fuse disparate data types over a space (or network) into a consistent global datum?* These data types may be as simple as $\mathbb{Z}$ versus $\mathbb{R}$ (some sensors count; others record intensities); more complex data types (local logical statements about targets) would be of phenomenal impact. In the latter case, one suspects a relationship with geometric logic and topoi.

29.7. Target classification. The Euler calculus may provide a promising way to address the problem of identifying characteristics of a particular family of target supports. *How can one perform target identification with only enumerative data?* For instance, the number of wheels on a vehicle may be an identifying characteristic that can be counted by the Euler integral if applied to field of pressure sensors or remote imaging data. More refined applications can be envisioned that use the integral transforms we have described, most notably the Bessel and Fourier transforms [38]. In particular, families of invariants might be extracted from transforms of functions (rather than the Euler integral alone) that would aid in robust identification even if errors are present.

29.8. Lifting topological invariants. The Euler integral is a lift of the classical Euler characteristic χ from an invariant of spaces to an invariant of *data* (constructible or definable) over spaces. The success of this lifting prompts an examination of other classical topological invariants. One prime candidate is the Lusternik-Schnirelmann category of a space [25], which has been lifted recently to an invariant of $\mathbb{R}$-valued distributions [8]. Others are more geometric: there is a valuation μ_k on tame subsets of $\mathbb{R}^n$, one for each $k = 0, \ldots, n$ that interpolate between Euler (μ_0) and Lebesgue (μ_n) measure. These INTRINSIC VOLUMES[6] have been lifted not only to $\mathsf{CF}(\mathbb{R}^n)$ but also to $\mathsf{Def}(\mathbb{R}^n)$ in dual pairs $\lfloor \mu_k \rfloor$, $\lceil \mu_k \rceil$ [81]. *Are there other algebraic invariants that can be lifted from spaces to data over spaces? Is it possible to lift algebraic-topological invariants of maps between spaces (e.g., Lefshetz index) to morphisms between data structures over spaces?*

29.9. Applied sheaf theory. Euler calculus is the vanguard of an emerging family of techniques in applied mathematics based on constructible sheaves. Any locally-defined algebraic data has a sheaf (or cosheaf) interpretation, and constructible sheaves are especially suited to applications due to their ease of manipulation and computation. Sheaf cohomology permits local-to-global inference and is incisive, even in simple examples. We suggest that many scientifically-motivated problems that exhibit locality of information can be addressed via the cohomology of an appropriate sheaf. For instance, in communications networks, nodes and links typically only retain information about messages from their immediate neighbors: information in the network is local. Work in progress [67, 66, 37] indicates a sheaf-theoretic interpretation of network information, with sheaf cohomology measuring features of messages passed through the network at large. Examples with higher-dimensional base spaces would permit the use of sheaf cohomology beyond

---

[6]Also known as Hadwiger measures, mixed volumes, quemasseintegrals, Lipshitz-Killing curvatures, etc.

H^0 and H^1. *What are other examples in which higher-dimensional constructible sheaf cohomology classes solve data aggregation problems?*

• ── •

Acknowledgements

The applications described in this survey would not have been possible without the guidance and hard work of two individuals. Yuliy Baryshnikov was the first to appreciate applications of Euler calculus to sensing and is directly or jointly responsible for most of the applications appearing in this survey. Benjamin Mann oversaw the DARPA program *SToMP: Sensor Topology & Minimalist Planning* under which the sensing applications of Euler calculus were developed.

References

[1] R. Adler, *The Geometry of Random Fields*, Wiley, London, 1981. Reprinted by SIAM, 2010.
[2] R. Adler, O. Bobrowski, M. Borman, E. Subag and S. Weinberger, "Persistent homology for random fields and complexes," in *Institute of Mathematical Statistics Collections* vol. 6, 2010, 124–143.
[3] R. Adler and J. Taylor, *Random Fields and Geometry*, Springer, 2007.
[4] Y. Baryshnikov and R. Ghrist, "Euler integration for definable functions," *Proc. National Acad. Sci.*, 107(21), May 25, 9525-9530, 2010.
[5] Y. Baryshnikov and R. Ghrist, "Target enumeration via Euler characteristic integration," *SIAM J. Appl. Math.*, 70(3), 2009, 825–844.
[6] Y. Baryshnikov and R. Ghrist, "Target enumeration via integration over planar sensor networks," in *Proc. Robotics: Science & Systems*, 2008.
[7] Y. Baryshnikov, R. Ghrist, and D. Lipsky "Inversion of Euler integral transforms with applications to sensor data," *Inverse Problems*, to appear. Preprint, 2010.
[8] Y. Baryshnikov and R. Ghrist, "Unimodal category and topological statistics," in Proceedings of *NOLTA*, 2011.
[9] J. Basch, L. Guibas, G. Ramkumar, and L. Ramshaw, "Polyhedral tracings and their convolutions," in *Algorithms for Robotic Motion and Manipulation, J.-P. Laumond and M. Overmars, eds.*, A. K. Peters, 1996.
[10] S. Basu, "On bounding the Betti numbers and computing the Euler characteristic of semialgebraic sets." *Discrete Comput. Geom.*, 22(1), 1999, 1–18.
[11] S. Basu and T. Zell, "Polynomial hierarchy, Betti numbers and a real analogue of Toda's theorem," *Foundations of Computational Mathematics* 10, 2010, 429–454.
[12] T. Beke, "Topological invariance of the combinatorial euler characteristic of tame spaces," preprint, 2011.
[13] W. Blaschke, *Vorlesungen über Integralgeometrie*, Vol. 1, 2nd edition. Leipzig and Berlin, Teubner, 1936.
[14] O. Bobrowski and M. Strom Borman, "Euler Integration of Gaussian Random Fields and Persistent Homology," `arXiv:1003.5175`, 2011.
[15] J. Bradley and C. Brislawn, "The wavelet/scalar quantization compression standard for digital fingerprint images," *ISCAS: IEEE Symp. on Circuits and Systems*, vol. 3, 1994, 205–208.
[16] G. Bredon, *Sheaf Theory*, 2nd ed., GTM 25, Springer-Verlag, 1997.
[17] R. Brinks, "On the convergence of derivatives of B-splines to derivatives of the Gaussian function," *Comp. Appl. Math.* 27:1, 2008, 79–92.
[18] L. Bröcker, "Euler integration and Euler multiplication," *Adv. Geom.* 5(1), 2005, 145–169.
[19] L. Bröcker and M. Kuppe, "Integral geometry of tame sets," *Geom. Dedicata* 82, 2000, 285–323.
[20] G. Carlsson, "Topology and Data", *Bull. Amer. Math. Soc.*, 46:2, 2009, 255–308.
[21] F. Chazal and A. Lieutier, "Weak feature size and persistent homology: computing homology of solids in $\mathbb{R}^n$ from noisy data samples," in *Proceedings of the Twenty-First Annual Symposium on Computational Geometry*, 2005, 255-262.
[22] B. Chen, "On the Euler measure of finite unions of convex sets," *Discrete and Computational Geometry* 10, 1993, 79–93.

[23] R. Cluckers and M. Edmundo, "Integration of positive constructible functions against Euler characteristic and dimension," *J. Pure Appl. Algebra*, 208, no. 2, 2006, 691–698.
[24] R. Cluckers and F. Loeser, "Constructible motivic functions and motivic integration," *Invent. Math.*, 173(1), 2008, 23–121.
[25] O. Cornea, , G. Lupton, J. Oprea, and D. Tanré, *Lusternik-Schnirelmann Category*, Amer. Math. Soc., 2003.
[26] I. Daubechies, *Ten lectures on wavelets*, SIAM, 1992.
[27] A. Dimca, *Sheaves in Topology*, Springer, 2004.
[28] K. Dohmen, *Improved Bonferroni Inequalities via Abstract Tubes*, Lecture Notes in Mathematics vol. 1826, Springer, 2003.
[29] H. Edelsbrunner and J. Harer, *Computational Topology: an Introduction*, AMS, 2010.
[30] *Eucharis: Euler Characteristic Integration Software:*
http://www.math.upenn.edu/~ghrist/eucharis/eucharis.jar
[31] M. Farber, *Invitation to Topological Robotics*, Eurpoean Mathematical Society, 2008.
[32] J. Fu, "Curvature measures of subanalytic sets," *Amer. J. Math.* 116, 1994, 819–880.
[33] S. Gal, "Euler characteristic of the configuration space of a complex," *Colloquium Mathematicum* 89:1, 2001, 61–67.
[34] J. Galambos and I. Simonelli, *Bonferroni-type Inequalities with Applications*, Springer-Verlag, 1996.
[35] S. I. Gelfand and Y. I. Manin, *Methods of Homological Algebra*, 2nd Edition, Springer, 2003.
[36] R. Ghrist, "Barcodes: The persistent topology of data," *Bull. Amer. Math. Soc.*, 45:1, 2008, 61–75.
[37] R. Ghrist and Y. Hiraoka, "Applications of sheaf cohomology and exact sequences to network coding", in *Proc. NOLTA*, 2011.
[38] R. Ghrist and M. Robinson, "Euler-Bessel and Euler-Fourier transforms," *Inverse Problems*, to appear. Preprint, 2010.
[39] R. Godement, *Topologie Algébrique et Théorie des Faisceaux*, Hermann, 1973.
[40] J. A. Goguen, "Sheaf Semantics for Concurrent Interacting Objects," *Math. Structures Comput. Sci.* 2, 1992, no. 2, 159–191.
[41] M. Goresky and R. MacPherson, *Stratified Morse Theory*, Springer-Verlag, Berlin, 1988.
[42] A. Graps, "An introduction to wavelets," *IEEE Comput. Sci. Engr*, 2, 1995, 50–61.
[43] H. Groemer, "Minkowski addition and mixed volumes," *Geom. Dedicata* 6, 1977, 141–163.
[44] A. Grothendieck, "Sur quelques points d'algèbre homologique" *Tohoku Math. J. (2)*. Part I: 9:2, 1957, 119–221; Part II: 9:3, 1957, 185–221.
[45] Guesin-Zade, "Integration with respect to the Euler characteristic and its applications," *Russ. Math. Surv.*, 65:3, 2010, 399–432.
[46] L. Guibas. "Sensing, Tracking and Reasoning with Relations," *IEEE Signal Processing Magazine*, 19(2), Mar 2002.
[47] L. Guibas, L. Ramshaw, and J. Stolfi, "A kinetic framework for computational geometry," in *Proc. IEEE Sympos. Found. Comput. Sci.*, 1983, 100–111.
[48] A. Haar, "Zur theorie der orthogonalen funktionensysteme," *Math. Ann.*, 69, 1910, 331–371.
[49] H. Hadwiger, "Integralsätze im Konvexring," *Abh. Math. Sem. Hamburg*, 20, 1956, 136–154.
[50] A. Hatcher, *Algebraic Topology*, Cambridge University Press, 2002.
[51] B. Hubbard, *The world according to wavelets: The story of a mathematical technique in the making*, AK Peters Ltd., 1998.
[52] B. Iversen, *Cohomology of Sheaves*, Springer, 1986.
[53] M. Kashiwara, "Index theorem for maximally overdetermined systems of linear differential equations," *Proc. Japan Acad. Ser. A Math. Sci.*, 49 (10), 1973, 803–804.
[54] M. Kashiwara, "Index theorem for constructible sheaves," *Astérisque* 130, 1985, 193–209.
[55] M. Kashiwara and P. Schapira, *Sheaves on Manifolds*, Springer-Verlag, 1994.
[56] V. Kiritchenko, "A Gauss-Bonnet theorem for constructible sheaves on reductive groups," *Math. Res. Lett.*, 9, 2002, 791–800.
[57] D. Klain, K. Rybnikov, K. Daniels, B. Jones, C. Neacsu, "Estimation of Euler Characteristic from Point Data," technical report, U. Mass. Lowell, 2005.
[58] D. Klain and G.-C. Rota, *Introduction to Geometric Probability*, Cambridge University Press, 1997.
[59] D. Koslov, *Combinatorial Algebraic Topology*, Springer, 2008.

[60] R. MacPherson, "Chern classes for singular algebraic varieties," *Ann. of Math.* 100, 1974, 423–432.
[61] C. McCrory and A. Parusiński, "Algebraically constructible functions: real algebra and topology," Panoramas & Synthèses 24, Soc. Math. France 2007, 69–85.
[62] J. Milnor, *Morse Theory*, Princeton University Press, 1963.
[63] R. Morelli, "A Theory of Polyhedra," *Adv. Math.* 97, 1993, 1–73.
[64] S. Oudot, L. Guibas, J. Gao, and Y. Wang, "Geodesic Delaunay triangulation in bounded planar domains," *ACM Trans. Algorithms*, 6(4): article 67, 2010.
[65] G. Ramkumar, "Tracings and Their Convolutions: Theory and Applications," Ph.D. thesis, Stanford University, 1998.
[66] M. Robinson, Inverse problems in geometric graphs using internal measurements, arXiv:1008.2933v1, August 2010.
[67] M. Robinson, Asynchronous logic circuits and sheaf obstructions, arXiv:1008.2729v1, August 2010.
[68] G.-C. Rota, "On the combinatorics of the Euler characteristic," Studies in Pure Mathematics, Academic Press, London, 1971, 221–233.
[69] J. Rotman, *An Introduction to Homological Algebra*, Springer, 2008.
[70] S. Schanuel, "Negative sets have Euler characteristic and dimension," in *Lecture Notes In Mathematics* 1488, Springer, 1991.
[71] P. Schapira, "Operations on constructible functions," *J. Pure Appl. Algebra*, 72, 1991, 83–93.
[72] P. Schapira, "Tomography of constructible functions," in *11th Intl. Symp. on Applied Algebra, Algebraic Algorithms and Error-Correcting Codes*, 1995, 427–435.
[73] R. Schneider and W. Weil, *Stochastic and Integral Geometry*, Springer, 2009.
[74] J. Schürmann, *Topology of Singular Spaces and Constructible Sheaves*, Birkhäuser, 2003.
[75] M. Shiota, *Geometry of Subanalytic and Semialgebraic Sets*, Birkhäuser, 1997.
[76] B. Usevitch,"A tutorial on modern lossy wavelet image compression: foundations of JPEG 2000," *IEEE Sig. Proc. Mag.*, 18:5, 2001, 22–35.
[77] L. Van den Dries, *Tame Topology and O-Minimal Structures*, Cambridge University Press, 1998.
[78] O. Viro, "Some integral calculus based on Euler characteristic," Lecture Notes in Math., vol. 1346, Springer-Verlag, 1988, 127–138.
[79] J. Wang and K. Huang, "Medical image compression by using three-dimensional wavelet transformation," *IEEE Transactions on Medical Imaging*, 15:4, 1996, 547–554.
[80] K. Worsley, "Local Maxima and the Expected Euler Characteristic of Excursion Sets of χ^2, F and t Fields," *Advances in Applied Probability*, 26(1), 1994, 13–42.
[81] M. Wright, "Hadwiger Integration of Definable Functions," Ph.D. thesis, Univ. Penn., May 2011.
[82] A. Zomorodian and G. Carlsson , "Computing persistent homology," *Disc. Comput. Geom.* 33:2, 2005, 247–274.

DEPARTMENT OF MATHEMATICS, UNIVERSITY OF PENNSYLVANIA
E-mail address: jucurry@math.upenn.edu

DEPARTMENTS OF MATHEMATICS AND ELECTRICAL/SYSTEMS ENGINEERING, UNIVERSITY OF PENNSYLVANIA
E-mail address: ghrist@math.upenn.edu

DEPARTMENT OF MATHEMATICS, UNIVERSITY OF PENNSYLVANIA
E-mail address: mrobin@math.upenn.edu

On the Topology of Discrete Planning with Uncertainty

Michael Erdmann

ABSTRACT. This chapter explores the topology of planning with uncertainty in discrete spaces. The chapter defines the *strategy complex* of a finite discrete graph as the collection of all plans for accomplishing all tasks specified by goal states in the graph. Transitions in the graph may be nondeterministic or stochastic. One key result is that a system can attain any state in its graph despite control uncertainty if and only if its strategy complex is homotopic to a sphere of dimension two less than the number of states in the graph.

1. Planning with Uncertainty in Robotics

The goal of Robotics is to animate the inanimate, so as to endow machines with the ability to act purposefully in the world. Roboticists, working in the subfield of planning, create software by which robots reason about future outcomes of potential actions. Using such planning software, robots combine individual actions into collections that together accomplish particular tasks in the world [29, 30].

Two fundamental and intertwined issues confound this seemingly straightforward approach. One is world complexity, the other is uncertainty.

1.1. Discrete Modeling. Modeling a seemingly continuous world with finite discrete symbols is a well-known problem at the heart of algorithmic computability [32, 49]. Practical robotics largely avoids the uncomputability questions by focusing on fixed levels of granularity deemed to be appropriate for the robot tasks at hand. The ensuing models may take the form of *a priori* shape families for describing the objects that a robot might encounter [33, 16], commensurate geometric representations of sensor data (e.g., visual, tactile, proprioceptive, auditory, laser) [45, 34, 13, 38, 5, 31], and partitions of configuration or control space into regions within which the dynamics of interaction are invariant [50, 9]. For instance, in [12] two robot palms manipulate objects based on prior geometric models of the objects and their frictional contact mechanics. A robot planner subdivides the configuration space of the palms and the object being manipulated into volumes within which the relative sliding motions at the contacts have invariant sign. These

---

2010 *Mathematics Subject Classification.* Primary 68T37, 68T40; Secondary 55U05, 55U10.
Key words and phrases. topology, graph, complex, strategy, robotics, planning, uncertainty.
This work was sponsored by DARPA under contract HR0011-07-1-0002. This work does not necessarily reflect the position or the policy of the Government. No official endorsement should be inferred.

volumes form the states in a discrete graph whose edges represent connectivity as a function of changing palm orientations. The planner constructs a strategy for reorienting the object by searching this graph for a path leading from the object's initial configuration to some desired final configuration.

1.2. Uncertainty. Uncertainty arises in modeling, control, and sensing. Models are inaccurate, control is errorful, and sensors are noisy. Despite uncertainty, roboticists seek to create planners that allow robots to operate purposefully and successfully in the world. The preimage methodology of [35] describes a general approach for planning in the presence of control and sensing uncertainty, generalized to model uncertainty in [11]. Specialized to physical systems, the planners often take the form of discrete graph searches. The states in these graphs need not simply be the configurations of the physical system. Instead, the graph states represent the information available to the robot at any given instant during plan execution [30, 6]. For instance, a common task in manufacturing systems is to reduce the entropy of small parts. These parts arrive in large numbers, jumbled together. An automatic system must orient and localize each part, so a robot can then, with little or no sensing, pick up the part and assemble it onto some product. The SONY SMART system [43] is a wonderful real world example. The robot systems described in [15, 46, 24] show how to construct planners for similar tasks, using the mechanics of the problem to reduce uncertainty. Each discrete state within the graphs for these planners is in fact a collection of underlying contact states, describing the extent to which the system has localized a part at runtime. Thus sensing uncertainty contributes to the definition of state.

As the previous discussion suggests, uncertainty and granularity are intertwined. A coarse world model produces relative certainty at the expense of expressive power. A coarse controller reduces search branching factors at the expense of local motion precision. A coarse sensor reduces hardware requirements at the expense of instantaneous localization. Tradeoffs between different levels of granularity and uncertainty are possible. Sequences of accurate motions may be traded for careful sensing in orienting parts, for instance. An open question is fine-tuning such tradeoffs so as to optimize system capabilities. It is a problem merely to describe these tradeoffs precisely.

1.3. Planners. There are two basic modes by which planners generate plans. In *backchaining*, a planner starts from a desired goal (possibly a set of configurations in some state space). The planner determines one or more actions (e.g., robot motions) that achieve the goal directly. The preconditions to those actions (e.g., the initial configurations of motions leading to the goal) combined with the original goal, then define a new subgoal. The planner repeats this process until it produces a subgoal that includes the current configuration of the system or until it determines that no such subgoal exists. The set of actions produced by this process constitutes a plan for attaining the original goal from the current configuration, and in fact, from any configuration satisfying the preconditions of any of the actions.

In *forward-chaining*, the planner starts from the current configuration. The planner determines the outcomes of all possible actions whose preconditions are currently satisfied. Then the planner repeats this process starting from each of the outcomes just determined, and so forth, until it produces a frontier of outcomes that satisfies the goal conditions or determines that the goal cannot be attained.

There exist numerous variations of these basic modes, for instance, simultaneously forward-chaining and backchaining [4, 3, 30].

1.4. Collections of Plans. Whatever the precise planning process, the result of planning is a plan for attaining some particular goal, perhaps from some particular initial configuration. Often when planning with robots, such unique plans are all one needs. In manufacturing, for instance, one may only need to know how to assemble, not disassemble. Yet, in principle, one could run the planner for all possible goals that are describable, not just those one explicitly needs. Doing so would reveal global system capabilities.

Plans fit together like jigsaw pieces. One can remove an action from a particular plan and obtain a new plan, with perhaps a slightly different goal. Sometimes one can add an action to a given plan, perhaps as a redundant backup, without changing the plan's outcomes. In other cases, adding an action might change the possible outcomes, possibly creating an infinite loop. By studying this jigsaw puzzle one gains insight into the dependence of a system's capabilities on design choices.

Moving up a level, one may thus view any one system as a point in a larger design space. If one understands the capabilities of a system and how those capabilities change when one alters the underlying system model, available actions, or sensing capabilities, one can begin to address the granularity tradeoffs mentioned earlier. For instance, in a directed graph one can readily decide whether the graph is strongly connected [1], modifying it accordingly if one desires different connectivity. No such straightforward tools exist currently for describing the capabilities of uncertain systems. This chapter proposes methods to help fill that deficit.

1.5. Topology of Plans. The aim of this chapter is to explore the topology of the collection of all plans that exist for an uncertain system and in so doing to characterize the system's capabilities. The many robotics results Ghrist found via algebraic topology [10, 22, 20, 21] motivate this exploration.

Our exploration focuses on systems that may be modeled as finite discrete spaces. As discussed, the states in such discrete spaces may represent fairly complicated system properties, so the tools presented here should have broad applicability. The core of this work has appeared previously as a robotics paper [14]. This chapter expands the topological perspective, generalizing several of the earlier results.

Beyond robotics applications, the research presented here is inspired topologically by the study of the collection of all partial orders on n items [8, 26, 27]. Two types of uncertain actions appear in this chapter, nondeterministic and stochastic. In the nondeterministic setting, as the earlier description of backchaining suggests, one may view a single plan as a particular partial order on a system's state space. When executing the plan, the system will visit states in some order consistent with this partial order. There will never be any cycling and the system will eventually wind up at the goal.

The collection of all plans is therefore related to the collection of all partial orders on the state space. A difference between our work and previous work is that the primitive motions in our collection of plans are nondeterministic actions rather than directed edges, producing slightly more complicated primitive partial orders (called *atoms* in the language of partial orders) than the single comparators $\{x > y\}$ that directed edges produce. By allowing these additional types of atoms, one can generate homotopy types of any finite simplicial complex, not just the spheres and points that are possible when atoms are single comparators [26]. From that perspective, nondeterministic planning is a natural "physical" realization of simplicial complexes, and, indeed, is forced upon us as soon as motions in a graph become uncertain or adversarial. In the stochastic setting, one cannot necessarily think of a plan as defining a partial order in the manner just described. The system may cycle between states. However, as long as cycling is transient, one may again view a plan as defining a stochastic ordering based on reachable states and convergence times.

The connection between planning and simplicial complexes holds categorically as well. The complexes defined in this chapter may be viewed, homeomorphically, as the classifying spaces of various planning categories, related to the forward-chaining and backchaining planners described earlier.

1.6. Chapter Outline. Section 2 introduces nondeterministic graphs and their strategy complexes. Section 3 introduces loopback complexes, using these to characterize goal attainability. Section 4 introduces stochastic graphs and their strategy complexes. Section 5 characterizes full controllability by the existence of a certain sphere, homotopically. Section 6 introduces source and dual complexes, indicating how these are useful for design and in assessing adversarial capabilities. Section 7 shows how a strategy complex factors into a part modeling full controllability on subgraphs and a part modeling obstructions to controllability. Section 8 examines the topology of links of actions. Section 9 develops a topological test to decide whether a set of actions is essential for accomplishing a goal. Section 10 uses decision trees to reveal further structure in loopback complexes. Section 11 examines the categorical foundations of strategy complexes and source complexes. Section 12 ends the chapter with a brief discussion.

2. Nondeterministic Graphs and Strategy Complexes

2.1. Nondeterministic Graphs. We model systems with uncertainty using discrete states and discrete actions with multiple outcomes. As suggested by the earlier robotics examples, a state may encapsulate not just the configuration of the robot but also the information known to the robot at runtime, such as that provided by sensors. Actions at a state represent the control choices available to the robot. The outcomes of an action describe the various state changes possible upon execution of that action. Following [14], we make the following definitions:

DEFINITION 2.1. A *nondeterministic graph* $G = (V, \mathfrak{A})$ is a set of *states* V and a collection of *(nondeterministic) actions* $\mathfrak{A}$. V is also known as G's *state space*. Each $A \in \mathfrak{A}$ consists of a *source* state v and a nonempty set T of *targets*, with $v \in V$ and $T \subseteq V$. We may write action A as $v \to T$. If T consists of a single state, A is also said to be *deterministic*. In that case, with $T = \{u\}$, we may write A more simply as $v \to u$. All graphs in this chapter are finite.

INTERPRETATION: Action A may be executed whenever the system is at state v. When action A is executed, the system moves from state v to one of the target states in T. If T contains multiple targets, the precise target attained is not known to the system before executing A, but is known after. Different execution instances of action A could attain different target states within T. (For instance, nature might choose a different target.) In order to model worst-case behaviors, we may imagine a potentially malevolent *adversary* who chooses the precise target attained.

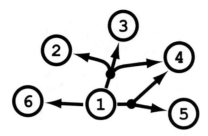

FIGURE 1. Three actions with source state 1. One action is a deterministic transition, with target state 6. The other two actions are both nondeterministic, one with targets $\{2, 3, 4\}$, the other with targets $\{4, 5\}$.

REMARKS 2.2. (1) Distinct actions may have the same source state and overlapping or identical target sets. (2) In particular, distinct actions may have the same written representation $v \to T$. (3) A nondeterministic graph in which each action is deterministic and no two actions have the same written representation is equivalent to a standard directed graph. (4) We allow the *null graph*, $(\emptyset, \emptyset)$.

EXAMPLE 2.3. In figures, we will draw an action $v \to T$ as a possibly bifurcated directed arrow from v to T. For example, Figure 1 shows three actions, each with source state 1, with different target sets. The written representations of these actions are $1 \to 6$, $1 \to \{2, 3, 4\}$, and $1 \to \{4, 5\}$.

2.2. Simplicial Complexes. Throughout this chapter we use the following definition of simplicial complex:

DEFINITION 2.4. An *(abstract) simplicial complex* Σ with underlying vertex set X is a collection of finite subsets of X, such that if σ is in Σ then so is every subset of σ (including the empty set $\emptyset$). The elements of Σ are *simplices*. We refer both to the elements of a simplex and to singleton simplices as *vertices*. Although traditionally many authors require that every element of X appear as a vertex in Σ, we do *not* impose this requirement. This broader definition is useful for modeling systems in which underlying states may or may not satisfy some monotone Boolean property. The set of vertices that actually appear in Σ is denoted by $\Sigma^{(0)}$, called the *zero-skeleton of* Σ. The *dimension* of a simplex is one less than its cardinality. All simplicial complexes and underlying vertex sets in this chapter are finite.

DEFINITION 2.5. The simplicial complex consisting solely of the empty simplex is the *empty complex*. The simplicial complex consisting of no simplices is the *void complex* [**27**].

REMARK 2.6. Every nonvoid finite abstract simplicial complex has a geometric realization in some finite-dimensional Euclidean space with relative topology the same as its weak/polytope topology [**37**]. Thus we may view any such complex as a topological space. The empty complex corresponds to the empty space. We also think of it as $\mathbb{S}^{-1}$, the sphere of dimension -1. The void complex does not seem to have such a nice topological interpretation, but is nonetheless convenient combinatorially. Viewing simplicial complexes as monotone Boolean functions which are TRUE for subsets of the underlying vertex set outside the complex, the void complex represents the constant function TRUE. The void complex is considered to be contractible. This is consistent with viewing the void complex as a collapse of any complex $\{\emptyset, \{v\}\}$ that represents a single point v.

DEFINITION 2.7. Suppose Σ and Γ are simplicial complexes with disjoint underlying vertex sets. The *simplicial join* [47] of Σ and Γ is the simplicial complex

$$\Sigma * \Gamma = \{\sigma \cup \gamma \mid \sigma \in \Sigma \text{ and } \gamma \in \Gamma\}.$$

The underlying vertex set of $\Sigma * \Gamma$ is the union of the underlying (disjoint) vertex sets of Σ and Γ.

2.3. Strategy Complexes Arising From Nondeterministic Graphs. We will define a simplicial complex for modeling the space of all plans on a nondeterministic graph. As the discussion of Section 1 suggests, we view a plan as a type of control law, that specifies what action the system should execute when it finds itself in a given state. Executing the plan should move the system from its current state to some desired state or set of states. In a nondeterministic graph, the result is a partial order on the system's state space. For this reason, we draw many of our techniques from [8, 26, 27].

We make one variation to the previous plan structure: We allow a plan to specify multiple possible actions at a given state. One may view multiple actions as additional permissible nondeterminism. At runtime, the system can execute any of the actions available at its current state or even leave the choice to an adversary. To emphasize this distinction from traditional plans and control laws, we generally speak of *strategies*. We capture the essence of a strategy via the following definitions:

DEFINITION 2.8. Suppose $G = (V, \mathfrak{A})$ is a nondeterministic graph and $\mathcal{A} \subseteq \mathfrak{A}$ is some set of actions. We say $\mathcal{A}$ *contains a circuit* if $\mathcal{A}$ contains a sequence of actions $v_1 \to T_1, \ldots, v_k \to T_k$, such that $v_{i+1} \in T_i$, for $i = 1, \ldots, k$, with $k \geq 1$ and $k+1$ meaning 1. We say $\mathcal{A}$ *converges* or *is convergent* if $\mathcal{A}$ does not contain a circuit.

If $\mathcal{A}$ contains a circuit, then an adversary could select action transitions to keep the system looping forever within the directed cycle $\{v_1, \ldots, v_k\}$, so we would not want to view $\mathcal{A}$ as a strategy. If $\mathcal{A}$ converges, then we may view $\mathcal{A}$ as a strategy.

DEFINITION 2.9. Suppose $G = (V, \mathfrak{A})$ is a nondeterministic graph, with $V \neq \emptyset$. The *strategy complex* of G, denoted Δ_G, is the simplicial complex with underlying vertex set $\mathfrak{A}$ whose simplices are all the convergent subsets $\mathcal{A}$ of $\mathfrak{A}$. Every simplex of Δ_G is a *(nondeterministic) strategy*. If $V = \emptyset$, we let Δ_G be the void complex.

REMARKS 2.10. (1) If V is nonempty, then Δ_G always contains the empty simplex. Intuitively, the empty simplex represents the strategy "DO NOT MOVE". (2) A nondeterministic action $v \to T$ with a self-loop, meaning $v \in T$, cannot appear in any strategy/simplex. (3) As outlined earlier, we view each strategy as a type of control law. In particular, to say that a system *executes strategy* σ means the following: Suppose the current state of the system is v. Strategy σ may contain zero, one, or several actions with source v. The system *stops* moving precisely when σ contains no action with source v. Otherwise, the system *must* execute some action $v \to T \in \sigma$. If there are several such actions, the strategy leaves open the method for choosing between those actions. From a worst-case perspective, an adversary may make the choice. Upon execution of action $v \to T$, the system finds itself at one of the targets $t \in T$. The process repeats, with t the system's new current state.

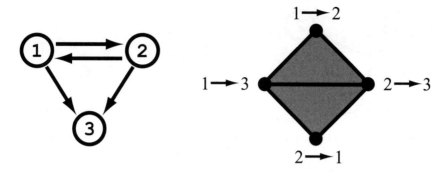

FIGURE 2. A directed graph and its associated strategy complex.

EXAMPLE 2.11. Figure 2 shows a standard directed graph along with its strategy complex. There are four directed edges in the graph, each representing a deterministic action. The strategy complex therefore could be as large as the complex generated by a tetrahedron. However, the two actions $1 \to 2$ and $2 \to 1$ together contain a circuit. Consequently, the strategy complex is actually generated by two triangles touching at an edge.

We may interpret each of the simplices in the resulting strategy complex as a strategy for accomplishing some goal, much like a traditional control law. For instance, the edge $\{1 \to 2,\ 2 \to 3\}$ represents the control law:

 – When at state 1, execute the action $1 \to 2$.
 – When at state 2, execute the action $2 \to 3$.
 – Otherwise, stop moving.

Together, the two actions $1 \to 2$ and $2 \to 3$ constitute a strategy for attaining state 3 from anywhere in the graph.

The triangle $\{1 \to 2,\ 1 \to 3,\ 2 \to 3\}$ represents another strategy for attaining state 3 from anywhere in the graph. It happens to have two actions with source 1, indicating a purposeful disinterest: whichever of these two actions executes at runtime, the system will ultimately converge to state 3.

The edge common to both triangles, $\{1 \to 3,\ 2 \to 3\}$, represents the strategy one would obtain by traditional backchaining from state 3. This strategy moves to state 3 as directly as possible.

The edge $\{1 \to 2,\ 1 \to 3\}$ represents a strategy for attaining the goal set $\{2, 3\}$. Effectively, this strategy says: "Move *away* from state 1; I do not care whereto."

EXAMPLE 2.12. In contrast, Figure 3 again shows three states, with the same *possible* transitions as in Example 2.11. However, in this example there are actually only two actions, $1 \to \{2, 3\}$ and $2 \to \{1, 3\}$, each of which is nondeterministic with two possible targets. Together, these actions contain a circuit; an adversary could force the system into an infinite loop between states 1 and 2. Consequently, the strategy complex consists of two isolated vertices, one for each action.

NOTATION 2.13. For any $m \geq -1$, $\mathbb{S}^m$ denotes the sphere of dimension m.

EXAMPLE 2.14. Figure 4 shows two strongly connected directed graphs on three states, along with their strategy complexes. The graphs are not isomorphic, but their strategy complexes are both homotopic to $\mathbb{S}^1$, the circle.

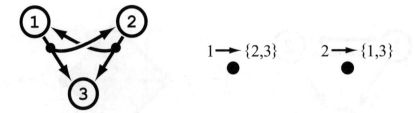

FIGURE 3. Left: A nondeterministic graph with two actions that together could produce a directed cycle. Right: The graph's strategy complex; it consists of two isolated vertices, one for each action.

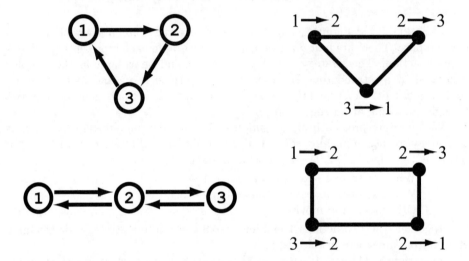

FIGURE 4. The left column shows two different strongly connected graphs on three states. The right column shows their respective strategy complexes.

Indeed, by [**26**], the strategy complex of any directed graph is homotopic either to a sphere or to a point. If the graph has n states and can be written as the disjoint union of k strongly connected subgraphs, then the strategy complex is homotopic to $\mathbb{S}^{n-k-1}$. Otherwise, the strategy complex is homotopic to a point. This result will re-appear in more general form for uncertain graphs, as Theorem 7.9.

Observe further that the graph in the lower left of Figure 4 may be viewed as the overlapping union of two strongly connected graphs on two states (the two graphs touch at state 2). Each of the subgraphs therefore has $\mathbb{S}^0$ as a strategy complex. One $\mathbb{S}^0$ is formed from the two actions $1 \to 2$ and $2 \to 1$, the other from the two actions $2 \to 3$ and $3 \to 2$. Mirroring this graph decomposition, observe that the join $\mathbb{S}^0 * \mathbb{S}^0$ is homotopic to $\mathbb{S}^1$ in general and in fact isomorphic in this case to the strategy complex of the overall graph.

Our research generalizes from directed graphs to nondeterministic (and stochastic) graphs. Doing so leads to a much larger class of strategy complexes than

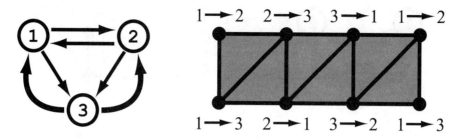

FIGURE 5. Left panel: The loopback graph $G_{\leftarrow 3}$ formed from the graph of Figure 2 by adding loopback actions at state 3 (indicated by thick arrows). Right panel: The associated loopback complex $\Delta_{G_{\leftarrow 3}}$. It is a polygonal cylinder, homotopic to $\mathbb{S}^1$.

just spheres and points, as the next theorem shows. One conclusion is that precise control is very much a special case in motion planning and that planning for uncertain systems is both topologically interesting and physically natural.

There is another observation. Directed graphs represent locally certain connectivity. Strongly connected directed graphs represent globally certain connectivity. That global property is reflected by the spherical nature of the graph's strategy complex. The connection between spheres and globally certain connectivity turns out to be significant as well for graphs in which local motions are uncertain. Much of the remainder of this chapter explores that connection.

NOTATION 2.15. Let Γ and Σ be simplicial complexes. (a) We write $\Gamma \cong \Sigma$ to mean that Γ and Σ are isomorphic, disregarding underlying vertex sets. (b) We let $\mathrm{sd}(\Sigma)$ denote the *first barycentric subdivision* of Σ. See [44, 37, 42].

THEOREM 2.16. *For any finite simplicial complex Σ, there exists a nondeterministic graph G such that $\mathrm{sd}(\Sigma) \cong \Delta_G$.*

PROOF. We give the basic construction and point to [14] for further details. Let $G = (V, \mathfrak{A})$, with V consisting of $(\mathrm{sd}(\Sigma))^{(0)}$ plus one additional state, and $\mathfrak{A}$ containing exactly one action $v \to T_v$ for each $v \in (\mathrm{sd}(\Sigma))^{(0)}$. The target set T_v consists of all states of V not adjacent to or equal to v in $\mathrm{sd}(\Sigma)$. □

3. Topological Characterization of Goal Attainability

The question of whether a nondeterministic graph contains a strategy for attaining some particular goal state may be rephrased as the problem of deciding whether the strategy complex associated with a variation of the graph is homotopic to a sphere or to a point. We first illustrate this property with two examples, then state the property as a theorem.

Consider again the graph of Figure 2. We seek to construct a topological space whose homotopy type tells us whether the graph contains a strategy for attaining state 3 from anywhere in the graph. (Of course, we know the graph contains such a strategy, by inspection, but we want a topological characterization.) Imagine that we add to the graph two deterministic actions at state 3, transitioning back to the other two states. We call these actions *loopback actions*, we call the resulting graph a *loopback graph*, and we call the associated strategy complex a *loopback complex*. See Figure 5. Observe that the loopback complex is homotopic to $\mathbb{S}^1$.

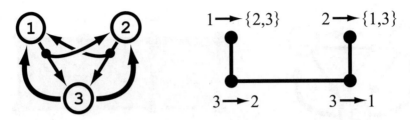

FIGURE 6. Left panel: The loopback graph $G_{\leftarrow 3}$ formed from the graph of Figure 3 by adding loopback actions at state 3. Right panel: The associated loopback complex $\Delta_{G_{\leftarrow 3}}$. It is contractible.

In contrast, suppose we ask whether the graph of Figure 3 contains a strategy for attaining state 3 from anywhere in the graph. (Again, we know it does not, by inspection, but again we seek a topological characterization.) As before, we add loopback actions at state 3 and compute the associated loopback complex, as shown in Figure 6. This time the loopback complex is homotopic to a point.

It turns out that loopback complexes are always homotopic either to a point or to $\mathbb{S}^{n-2}$, with n being the number of states in the graph. Moreover, the complex is a sphere precisely when the graph contains a strategy for moving from anywhere in the graph to the state at which we have added loopback actions. We now make this statement precise with the following definitions and theorem.

DEFINITION 3.1. Let $G = (V, \mathfrak{A})$ be a nondeterministic graph and s a desired *stop state* in V. A *complete strategy for attaining s in G* is a convergent set $\sigma \subseteq \mathfrak{A}$, such that σ contains at least one action with source v for every v in $V \setminus \{s\}$. If such a σ exists, we also say *G contains a complete strategy for attaining s*. (Later, we will use the same terminology for stochastic graphs.)

REMARK 3.2. In the previous definition, σ cannot contain any action with source s, as otherwise σ would not be convergent.

DEFINITION 3.3. With notation as above, define $G_{\leftarrow s}$ to be the nondeterministic graph constructed from G by first removing all actions of G that have source s, then adding all possible *loopback actions* at state s, that is, all deterministic actions $s \to v$, with $v \in V \setminus \{s\}$. Call $G_{\leftarrow s}$ the *loopback graph formed from G and s*. Define $\Delta_{G_{\leftarrow s}}$ to be the strategy complex associated with $G_{\leftarrow s}$ and call it the *loopback complex formed from G and s*. (Later, we will use the same terminology for stochastic graphs.)

REMARK 3.4. In the previous definition, we could simply have *added* loopback actions at s, without first *removing* any existing actions. There would be no difference in the homotopy type of the resulting complex, as Example 3.10 will show.

NOTATION 3.5. The rest of this chapter employs the following notation:
- $\mathbf{x} \in \mathbb{R}^n$ means the point $(x_1, \ldots, x_n)$ in n-dimensional Euclidean space.
- $X \simeq Y$ means that X and Y are homotopic as topological spaces (this is the same as saying that X and Y have the same homotopy type). The notation makes sense for simplicial complexes by Remark 2.6. Later, by Section 8.2, the notation will make sense for partially ordered sets.

- $\diagup_{=}^{n}$ denotes the *diagonal* in $\mathbb{R}^n$ and $\odot_{\neq}^{n}$ denotes its complement. So $\diagup_{=}^{n} = \{\mathbf{x} \in \mathbb{R}^n \mid x_1 = \cdots = x_n\}$, $\odot_{\neq}^{n} = \{\mathbf{x} \in \mathbb{R}^n \mid x_i \neq x_j, \text{ for some } i, j\}$.

THEOREM 3.6. *Let G be a nondeterministic graph with state space V, $s \in V$, and $n = |V|$. If G contains a complete strategy for attaining s, then $\Delta_{G \leftarrow s} \simeq \mathbb{S}^{n-2}$. Otherwise, $\Delta_{G \leftarrow s}$ is contractible.*

PROOF. This proof is motivated by the techniques in [**8, 26**]. The proof given here appears in similar though not quite identical form in [**14**].

If $n = 1$, then $\Delta_{G \leftarrow s}$ is necessarily the empty complex, which we view as $\mathbb{S}^{-1}$. The empty simplex is a complete strategy for attaining the only state there is in the graph. This shows the theorem holds for $n = 1$. So we may assume that $V = \{1, \ldots, n\}$, with $s = n > 1$.

I. Suppose σ is a complete strategy for attaining s in G. Let $\mathfrak{A}$ be the actions of $G_{\leftarrow s}$. For each action $A \in \mathfrak{A}$, with $A = i \rightarrow T$, define the following open polyhedral cone, which we refer to as a *nondeterministic covering set*:

$$U_A = \left\{ \mathbf{x} \in \mathbb{R}^n \mid x_i > \max_{j \in T} x_j \right\}.$$

(Looking ahead, nondeterministic covering sets constitute a special case of the covering sets to appear in Def. 5.2.)

The nerve of the cover $\{U_A\}_{A \in \mathfrak{A}}$ conveys information. In particular, a set of actions $\{A_1, \ldots, A_k\}$ is convergent if and only if $U_{A_1} \cap \cdots \cap U_{A_k}$ is not empty. When nonempty, the intersection is contractible. By the Nerve Lemma [**7, 25**], $\Delta_{G \leftarrow s}$ therefore has the homotopy type of $\bigcup_{A \in \mathfrak{A}} U_A$. We will show that $\bigcup_{A \in \mathfrak{A}} U_A = \odot_{\neq}^{n}$. Consequently, $\Delta_{G \leftarrow s}$ is homotopic to $\mathbb{S}^{n-2}$.

If $\mathbf{x}$ is a point on the diagonal $\diagup_{=}^{n}$, then $\mathbf{x}$ cannot lie in any nondeterministic covering set U_A, by construction. If $\mathbf{x} \in \mathbb{R}^n$ with $x_n > x_i$ for some index i, then $\mathbf{x}$ lies in the nondeterministic covering set associated with the loopback action $n \rightarrow i$.

Otherwise, $\mathbf{x} \in \mathbb{R}^n$ with $x_i > x_n$ for some index i. Suppose that $\mathbf{x}$ does not lie in any U_A. Since σ is a complete strategy for attaining n, σ contains an action B with source i. Write this action as $i \rightarrow T$, for some set of targets T. Action B is an element of $\mathfrak{A}$, so $\mathbf{x}$ does not lie in U_B. Consequently, for some target $j \in T$, $x_i \leq x_j$, implying that $x_j > x_n$. One may now repeat the argument with index j. Continuing in this manner, one obtains an unbounded sequence of actions $v_1 \rightarrow T_1, v_2 \rightarrow T_2, \ldots$, all in σ, such that $v_{k+1} \in T_k$ for all $k = 1, 2, \ldots$. Since G is finite, this means σ contains a circuit, establishing a contradiction.

II. Suppose G contains no complete strategy for attaining s. Let Σ be the subcomplex of $\Delta_{G \leftarrow s}$ consisting of all simplices that contain no loopback actions and let σ_0 be a maximal simplex of Σ. Suppose $\sigma_0 \neq \emptyset$ and consider the collection (not a simplicial complex) $\Gamma = \{\gamma \in \Delta_{G \leftarrow s} \mid \sigma_0 \subseteq \gamma\}$.

Since σ_0 is not a complete strategy for attaining $s = n$, there must be some state $i \neq n$ such that σ_0 contains no action with source i. Suppose γ is in Γ and does not contain the loopback action $n \rightarrow i$. By maximality of σ_0 in Σ, γ also does not contain an action with source i. Consequently, $\gamma \cup \{n \rightarrow i\}$ is convergent and thus in Γ. Suppose γ is in Γ and does already contain the loopback $n \rightarrow i$. Then removing that loopback produces as well an element of Γ. By Lemma 7.6 of [**8**], this

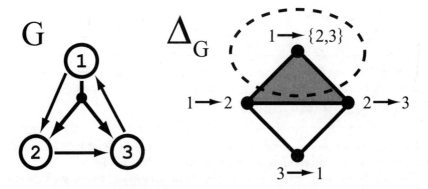

FIGURE 7. Action $1 \to 2$ is more precise than action $1 \to \{2,3\}$, in G. Consequently, all simplices containing the action $1 \to \{2,3\}$ may collapse away without changing the homotopy type of Δ_G.

means the complex $\Delta_{G_{\leftarrow s}}$ collapses to the complex $\Delta_{G_{\leftarrow s}} \setminus \Gamma$, preserving homotopy type. Repeating this process, one may collapse away all nonempty simplices of Σ along with their supersets in $\Delta_{G_{\leftarrow s}}$, leaving only the loopback actions. All the loopback actions together are convergent. So, we have shown how to collapse the complex $\Delta_{G_{\leftarrow s}}$ to a single nonempty simplex, which in turn collapses to a point. □

REMARKS 3.7. (1) The contradiction argument in part I of the proof is much like planning, now from an adversary's perspective. (2) Part II of the proof actually establishes that $\Delta_{G_{\leftarrow s}}$ is collapsible when G fails to contain a complete strategy for attaining s. Later, Corollary 10.7 will establish the yet stronger property that $\Delta_{G_{\leftarrow s}}$ is nonevasive. (3) Allowing $\sigma_0 = \emptyset$ in part II would be fine though less explicit.

COROLLARY 3.8. *With notation as above, G contains a complete strategy for attaining s if and only if $\Delta_{G_{\leftarrow s}}$ contains an odd number of simplices (counting $\emptyset$).*

PROOF. The reduced Euler characteristic is 0 for points and ± 1 for spheres. □

COROLLARY 3.9. *Let G be a nondeterministic graph with state space V. Let $s \in V$. The number of complete strategies for attaining s in G is either zero or odd.*

EXAMPLE 3.10. Understanding the homotopy types of strategy complexes in general is an open question. Globally, we have seen the significance of spheres of a certain dimension. Locally, homotopy collapse ignores imprecise actions in favor of more precise actions, as follows: Whenever a nondeterministic graph contains two actions with the same source state and comparable target sets, then all simplices containing the less precise action may collapse away. Imagine actions $v \to T$ and $v \to S$ with $S \subseteq T$. If $v \to T \in \sigma \in \Delta_G$, then $\sigma \cup \{v \to S\} \in \Delta_G$. So all simplices containing $v \to T$ may collapse away without changing homotopy type. See Figure 7.

4. Stochastic Graphs and Strategy Complexes

EXAMPLE 4.1. Consider again the graph of Figure 3. Now suppose that the transitions of each action are not so uncertain as to be nondeterministic, but instead have associated probabilities, as indicated in Figure 8. The probabilities mean that

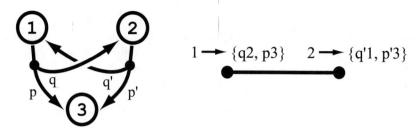

FIGURE 8. The graph on the left contains two stochastic actions. Although the actions together may cause the system to cycle between states 1 and 2, the cycling will be transient. As a result, the graph's strategy complex on the right contains not only the vertices representing the individual stochastic actions, but the edge between them. Compare with Figure 3.

during each execution of an action, a particular transition occurs with the indicated probability, independent of the past. Although such uncertainty in the actions might cause the system to cycle between states 1 and 2 for a while, the probability that the system would cycle forever is zero. With every action execution there is some minimum nonzero probability that the system will exit the cycle and move to state 3. Consequently, we should consider the set consisting of both actions to be a simplex in the graph's strategy complex, as shown in Figure 8.

In order to make this intuition precise, we need to generalize the notion of circuit given in Def. 2.8 from the nondeterministic setting to the stochastic setting. The earlier definition models an adversary who selects action transitions in such a way that the system finds itself stuck in some set of states, moving endlessly between those states. In the nondeterministic setting, the adversary can choose the transitions so as to create a cyclic path, but that is almost incidental; the key idea is that the system is stuck in a subspace. That idea generalizes readily to the stochastic setting: instead of a cyclic path, one obtains a *recurrent class* [17, 28] in a Markov chain created by the adversary. We formalize these concepts as follows:

DEFINITION 4.2. A *stochastic action* A consists of a *source* state v and a nonempty set T of *target* states, along with a strictly positive probability distribution $p : T \to (0,1]$. We may write action A as $v \to pT$. If $T = \{u_1, \ldots, u_k\}$ and $p_i = p(u_i)$, for $i = 1, \ldots, k$, then we may also write A as $v \to \{p_1 u_1, \ldots, p_k u_k\}$. In this representation, $p_i > 0$, for $i = 1, \ldots, k$, and $\sum_{i=1}^{k} p_i = 1$. If T consists of a single state, then A is deterministic. In that case, with $T = \{u\}$, we may write A in several different ways, including $v \to u$, $v \to \{1u\}$, and $v \to \{u\}$.

INTERPRETATION: As in the nondeterministic case, action A may be executed whenever the system is at state v. When action A is executed, the system moves from state v to one of the targets u in T, selected from all the targets with probability $p(u)$. This process is Markovian, that is, independent of how or when the system arrived at state v. If T contains multiple targets, the precise target attained is not known to the system before executing A, but is known after.

DEFINITION 4.3. A *stochastic graph* $G = (V, \mathfrak{A})$ is a set of *states* V and a collection of *actions* $\mathfrak{A}$, such that each action of $\mathfrak{A}$ is either nondeterministic or stochastic, with its source and targets in V. We may refer to V as G's *state space*.

(**NB:** A stochastic graph may contain *both* nondeterministic and stochastic actions. Of course, this also includes the special case of deterministic actions.)

In the remaining definitions of this section, let $G = (V, \mathfrak{A})$ be a stochastic graph.

DEFINITION 4.4. Define src : $\mathfrak{A} \to V$ so $\text{src}(A)$ is the source of A. Extend to sets of actions. If $\mathcal{A} \subseteq \mathfrak{A}$, we say $\text{src}(\mathcal{A})$ is the *start region* of $\mathcal{A}$.

DEFINITION 4.5. Let $W \subseteq V$ and $A \in \mathfrak{A}$. Action A *moves off* W if $\text{src}(A) \in W$ and one of the following is true: (i) A is stochastic with at least one of its targets in $V \setminus W$, or (ii) A is nondeterministic with all of its targets in $V \setminus W$.

DEFINITION 4.6. Let $\mathcal{A} \subseteq \mathfrak{A}$ be some set of actions in G. We say $\mathcal{A}$ *contains a stochastic circuit* if, for some nonempty subset $\mathcal{B}$ of $\mathcal{A}$, no action of $\mathcal{B}$ moves off $\text{src}(\mathcal{B})$. We say $\mathcal{A}$ *converges stochastically* or *is stochastically convergent* if $\mathcal{A}$ does not contain a stochastic circuit.

REMARK 4.7. Suppose $\mathcal{A}$ contains a stochastic circuit. Then an adversary could select some nonempty subset of actions $\mathcal{B}$ in $\mathcal{A}$ and some nonempty subset of states W in V, such that: (i) $W = \text{src}(\mathcal{B})$, (ii) $\mathcal{B}$ contains exactly one action with source w for every $w \in W$, and (iii) no B in $\mathcal{B}$ moves off W. Now consider an action B in $\mathcal{B}$. If B is stochastic, then every target of B lies in W. If B is nondeterministic, then the adversary could further select one target of B lying in W. The complete selection process just described amounts to the construction of a Markov chain on state space W. Since the chain is finite, it must contain a recurrent class [**17, 28**]. This means that there is a nonempty subset R of W such that the probability of the chain eventually moving from any given state of R to any other state of R is 1, while the probability of ever leaving R is 0. Restricting the Markov chain from W to R defines a new Markov chain. This new chain has state space R and is irreducible; it is the stochastic analogue of any irreducible directed cycle appearing via Def. 2.8 for the nondeterministic setting. (With this generalization in mind, we usually omit the explicit "stochastic" designation in the terms of Def. 4.6.)

We may now define a strategy complex much as we did earlier for nondeterministic graphs, but now allowing both stochastic and nondeterministic actions:

DEFINITION 4.8. If $V \neq \emptyset$, the *strategy complex* Δ_G of G is the simplicial complex whose underlying vertex set is $\mathfrak{A}$ and whose simplices are all the stochastically convergent subsets $\mathcal{A}$ of $\mathfrak{A}$. Every simplex of Δ_G is a *(stochastic) strategy*. If $V = \emptyset$, we let Δ_G be the void complex.

REMARKS 4.9. (1) Remarks 2.10 carry over to the stochastic setting with small changes. For instance, a stochastic action $v \to pT$ with a self-loop may appear in some simplices, so long as $|T| > 1$. (2) Theorem 3.6 holds as well in the stochastic setting. The proof carries over with small changes. See for example the covering sets of Section 5. The decision trees of Section 10 will provide yet a different type of proof.

REMARK 4.10. The Markov chain construction of Remark 4.7 suggests an alternate definition of stochastic convergence. Intuitively, no matter what Markov chain the adversary constructs, all states at which the strategy specifies actions should be transient states of the Markov chain, not recurrent states. The actions of a strategy should eventually move the system to some set of states at which the strategy specifies no actions. The next two definitions and lemma rephrase this intuition algebraically.

DEFINITION 4.11. Define the "adversity" function $\mathrm{adv} : \mathfrak{A} \times \mathbb{R}^{|V|} \to \mathbb{R}$ by

$$\mathrm{adv}(A, \{x_v\}_{v \in V}) = \begin{cases} \sum_{u \in T} p(u) x_u, & \text{if } A = w \to pT; \\ \max_{u \in T} x_u, & \text{if } A = w \to T; \end{cases} \quad \text{(for some } w\text{)}.$$

INTERPRETATION: Given a family $\{x_v\}_{v \in V}$ of real numbers indexed by V and given an action A, the expression $\mathrm{adv}(A, \{x_v\}_{v \in V})$ computes an expectation based on A's targets when A is stochastic and a maximization based on A's targets when A is nondeterministic. This definition will help us model expected outcomes arising from worst-case adversarial choices. Notation: We frequently drop the index set V to write $\mathrm{adv}(A, \{x_v\})$, or use vector notation to write $\mathrm{adv}(A, \mathbf{x})$.

DEFINITION 4.12. If $\mathcal{A}$ is a collection of actions and w a state, let $\mathcal{A}|w$ denote all actions in $\mathcal{A}$ that have source w.

LEMMA 4.13. Let $G = (V, \mathfrak{A})$ be a stochastic graph with $V \neq \emptyset$ and let $\mathcal{A} \subseteq \mathfrak{A}$. Then $\mathcal{A}$ is stochastically convergent if and only if the following system of equations in the real variables $\{x_v\}_{v \in V}$ has a unique finite solution, identically zero:

(4.1) $$x_w = \max_{A \in \mathcal{A}|w} \mathrm{adv}(A, \{x_v\}), \quad \text{for all } w \in V.$$

(We take any maximization over the empty set to be 0.)

PROOF. Follows by standard techniques for Markov chains [**17, 28**]. □

REMARK 4.14. Suppose $V \neq \emptyset$ and $\mathcal{A}$ is stochastically convergent. Then $\mathrm{src}(\mathcal{A})$ must be a proper subset of V, so System (4.1) has at least one explicit equation of the form $x_w = 0$. We may view strategy $\mathcal{A}$ as attaining the goal set consisting of all states w for which (4.1) contains the explicit equation $x_w = 0$, that is, all states in $V \setminus \mathrm{src}(\mathcal{A})$.

REMARK 4.15. Now imagine that we associate to each action A of $\mathfrak{A}$ a nonnegative *action transition time* δ_A. Let $\mathcal{A} \subseteq \mathfrak{A}$. Then the following system of equations in the real variables $\{t_v\}_{v \in V}$ again has a unique finite solution if and only if $\mathcal{A}$ is stochastically convergent, in which case each of the t_v is nonnegative:

(4.2) $$t_w = \max_{A \in \mathcal{A}|w} (\mathrm{adv}(A, \{t_v\}) + \delta_A), \quad \text{for all } w \in V.$$

We may interpret the unique finite solution $\{t_v\}_{v \in V}$, when it exists, as *worst-case expected convergence times*. Intuitively, an adversary can choose actions and nondeterministic transitions from $\mathcal{A}$ in such a way that the expected time for the system to enter the set of states $V \setminus \mathrm{src}(\mathcal{A})$, when started at state w, can be as great as t_w.

DEFINITION 4.16. Given a stochastically convergent set of actions $\mathcal{A}$, let $t_{\max}(\mathcal{A})$ be the maximum t_w obtained as a solution to System (4.2). For any nonnegative $\mathcal{T}$, let $\Delta_G^\mathcal{T}$ be the subcomplex of Δ_G consisting of all simplices σ for which $t_{\max}(\sigma) \leq \mathcal{T}$.

5. Topological Characterization of Full Controllability

DEFINITION 5.1. (a) Let $G = (V, \mathfrak{A})$ be a stochastic graph and suppose I and S are nonempty subsets of V. We say that a simplex σ of Δ_G is a *stochastic strategy for attaining S from I* if, with probability 1, the system eventually stops at *some* state of S whenever it starts at *any* state of I and executes strategy σ. In the case of singleton sets $I = \{i\}$ and $S = \{s\}$, we may simply say that σ is a *stochastic strategy for attaining s from i*.

(b) A nonempty set of states S is *certainly attainable (in G)* if there is some stochastic strategy $\sigma \in \Delta_G$ for attaining S from all of V.

(c) A graph G is *fully controllable* if, for any initial state i and any stop state s in G's state space, G contains a stochastic strategy for attaining s from i.

This section characterizes full controllability by the condition that Δ_G be homotopic to $\mathbb{S}^{n-2}$. This condition and its proof are a generalization of Theorem 3.6 and its proof. We will employ a generalization of nondeterministic covering sets and show how transitivity of actions translates to unions of open sets even in the stochastic setting. Later (see Remark 6.7) we will see the basis for a more combinatorial proof.

In order to simplify the discussion, throughout the rest of this section we assume that G is a stochastic graph with actions $\mathfrak{A}$ and state space $V = \{1, \ldots, n\}$, $n \geq 1$.

DEFINITION 5.2. Let $\{\delta_A\}_{A \in \mathfrak{A}}$ be nonnegative action transition times, associated to the actions of G. For each action A in $\mathfrak{A}$, define the *covering set* of A to be the following open subset of $\mathbb{R}^n$:

$$U_{A,\delta_A} = \{\mathbf{x} \in \mathbb{R}^n \mid x_i > \mathrm{adv}(A, \mathbf{x}) + \delta_A\}, \quad \text{with } i = \mathrm{src}(A).$$

For the special case in which $\delta_A = 0$, we may write U_A in place of U_{A,δ_A}. This notation is consistent with the notation for nondeterministic covering sets that appeared earlier in the proof of Theorem 3.6.

REMARK 5.3. Suppose we write stochastic action A as $i \to_p T$, with $\emptyset \neq T \subseteq V$, as per Def. 4.2. States are now integers. Then, letting $p_j = p(j)$ for $j \in T$,

$$U_{A,\delta_A} = \left\{\mathbf{x} \in \mathbb{R}^n \mid x_i > \sum_{j \in T} p_j x_j + \delta_A\right\}.$$

Similarly, if we write nondeterministic action A as $i \to T$, then

$$U_{A,\delta_A} = \left\{\mathbf{x} \in \mathbb{R}^n \mid x_i > \max_{j \in T} x_j + \delta_A\right\}.$$

In particular, for stochastic A with $\delta_A = 0$, the set U_A is an open homogeneous halfspace, whose defining hyperplane normal is determined by A's transition probabilities. This hyperplane includes the diagonal $\diagup^n$. For nondeterministic A with $\delta_A = 0$, the set U_A is the intersection of several such open homogeneous halfspaces, as specified by A's possible source-to-target transitions, just as in the proof of Theorem 3.6.

LEMMA 5.4. *Suppose G contains a stochastic strategy σ for attaining state k from state ℓ. Let t_ℓ be the worst-case expected convergence time starting from state ℓ, obtained by solving System (4.2) with $\mathcal{A} = \sigma$. Then*

$$\{\mathbf{x} \in \mathbb{R}^n \mid x_\ell > x_k + t_\ell\} \subseteq \bigcup_{A \in \mathfrak{A}} U_{A,\delta_A}.$$

PROOF. We can assume without loss of generality that $k = n$ and that σ is a complete strategy for attaining n. If $\ell = n$, then there is nothing to prove, since the set on the left is empty, so we may assume that $1 \leq \ell < n$.

Let $\{t_i\}_{i=1}^n$ be the solution to System (4.2) with $\mathcal{A} = \sigma$.

Suppose there is some $\mathbf{x}^* \in \mathbb{R}^n$ such that $x_\ell^* > x_n^* + t_\ell$, but $\mathbf{x}^*$ lies in no U_{A,δ_A}. Since the sets U_{A,δ_A} are invariant with respect to translation along the diagonal $\underline{\mathbb{Z}}^n$, we may assume that $x_n^* = 0$.

When System (4.2) has a unique finite solution, one can obtain that solution by iteration, that is by computing
$$t_w^{(m+1)} = \max_{A \in \mathcal{A}|w} (\mathrm{adv}(A, \{t_v^{(m)}\}) + \delta_A), \quad \text{for all } w \in V, \text{ for } m = 0, 1, 2, \ldots,$$
starting from any finite initial seed values for $\{t_v^{(0)}\}_{v \in V}$. As $m \to \infty$, the iteration will converge to the solution of (4.2).

In the present case, if we set $t_i^{(0)} = x_i^*$, then in the limit, by induction on m, we see that $t_i \geq x_i^*$, for all $1 \leq i \leq n$. Key in the induction is the observation that, for each action A of σ with source i, we have the inequality $\mathrm{adv}(A, \mathbf{x}^*) + \delta_A \geq x_i^*$, by the contrary assumption on $\mathbf{x}^*$. Thus $t_\ell \geq x_\ell^* > x_n^* + t_\ell = t_\ell$, a contradiction. $\square$

LEMMA 5.5. *A nonempty set of actions $\mathcal{A}$ is stochastically convergent if and only if*
$$\bigcap_{A \in \mathcal{A}} U_A \neq \emptyset.$$

PROOF. I. Suppose $\bigcap_{A \in \mathcal{A}} U_A \neq \emptyset$, and choose $\mathbf{x}^*$ to lie in the intersection.

For each $A \in \mathcal{A}$, define $\delta_A = x_{\mathrm{src}(A)}^* - \mathrm{adv}(A, \mathbf{x}^*)$. By definition of U_A, $\delta_A > 0$. Writing $\mathbf{t}$ for $\{t_j\}_{j=1}^n$, consider the following variant of System (4.2):

(5.1) $$t_i = \max_{A \in \mathcal{A}|i} (\mathrm{adv}(A, \mathbf{t}) + \delta_A) + b_i, \quad \text{for all } i \in V,$$

with $b_i = 0$ for all $i \in \mathrm{src}(\mathcal{A})$ and $b_i = x_i^*$ for all $i \notin \mathrm{src}(\mathcal{A})$.

By construction of the $\{\delta_A\}_{A \in \mathcal{A}}$, this system has at least one finite solution, given by $t_i = x_i^*$ for all i in V.

If $\mathcal{A}$ contains a stochastic circuit, then by Remark 4.7, we can construct from $\mathcal{A}$ an irreducible Markov chain on some nonempty subset R of V. Without loss of generality, $R = \{1, \ldots, k\}$, for some $1 \leq k \leq n$. Let (p_{ij}) denote the stochastic matrix of this Markov chain. Combining with System (5.1), we obtain
$$x_i^* \geq \sum_{j=1}^k p_{ij} x_j^* + \delta_i, \quad i = 1, \ldots, k,$$
where $\delta_i = \delta_A$, A being the particular action used to construct the transition(s) at state i as per Remark 4.7. Since (p_{ij}) is a stochastic matrix, this is only possible if $\delta_i \leq 0$ for at least one i, establishing a contradiction.

II. Suppose $\mathcal{A}$ is stochastically convergent. For each $A \in \mathcal{A}$, let $\delta_A = 1$. System (4.2) has a unique finite solution, call it $\mathbf{t}^*$. Pick some arbitrary $B \in \mathcal{A}$. Suppose B has source i. Then
$$t_i^* = \max_{A \in \mathcal{A}|i} (\mathrm{adv}(A, \mathbf{t}^*) + 1) \geq \mathrm{adv}(B, \mathbf{t}^*) + 1 > \mathrm{adv}(B, \mathbf{t}^*),$$
implying that $\mathbf{t}^* \in U_B$. So the intersection of all the U_A, with $A \in \mathcal{A}$, contains $\mathbf{t}^*$ and thus is not empty. $\square$

COROLLARY 5.6. *Let $G = (V, \mathfrak{A})$ be a stochastic graph, with $V \neq \emptyset$. Then*

$$\Delta_G \simeq \bigcup_{A \in \mathfrak{A}} U_A.$$

PROOF. If V consists of a single state, then Δ_G is the empty complex, which corresponds to the empty space. Any covering set is empty as well. So the corollary holds.

If V contains multiple states, then each covering set U_A is open and convex, so intersections of such covering sets are contractible when nonempty. Together, the Nerve Lemma and Lemma 5.5 establish the corollary. □

THEOREM 5.7. *Let $G = (V, \mathfrak{A})$ be a stochastic graph, with $V \neq \emptyset$. G is fully controllable if and only if $\Delta_G \simeq \mathbb{S}^{n-2}$, with $n = |V|$.*

PROOF. If V consists of a single state, then G is fully controllable and Δ_G is $\mathbb{S}^{-1}$, so the theorem holds. So assume that $n > 1$ in what follows.

I. Suppose G is fully controllable. By Lemma 5.4, with all $\delta_A = 0$, the union of all the covering sets U_A is $\odot_{\neq}^n$. Combining with Corollary 5.6, we see that

$$\Delta_G \simeq \bigcup_{A \in \mathfrak{A}} U_A = \odot_{\neq}^n \simeq \mathbb{S}^{n-2}.$$

II. Suppose $\Delta_G \simeq \mathbb{S}^{n-2}$. If G is not fully controllable, then there must be some state $s \in V$ such that G does not contain a complete strategy for attaining s. Let $G_{+s} = (V, \mathfrak{A}^+)$ be the graph obtained from G by *adding* all possible loopbacks at s, that is, all actions $s \to v$, with $v \in V \setminus \{s\}$. We see

$$\mathbb{S}^{n-2} \simeq \Delta_G \simeq \bigcup_{A \in \mathfrak{A}} U_A \subseteq \bigcup_{A \in \mathfrak{A}^+} U_A \simeq \Delta_{G_{+s}}.$$

Since each covering set U_A is homogeneous and invariant with respect to translation along the diagonal $\angle^n$, and since no proper subset of $\mathbb{S}^{n-2}$ is homotopic to $\mathbb{S}^{n-2}$, it must be that the covering sets U_A arising from G cover all of $\odot_{\neq}^n$. The subset relation above therefore implies that $\Delta_{G_{+s}} \simeq \mathbb{S}^{n-2}$. On the other hand, a collapsibility argument nearly identical to that appearing in the proof of Theorem 3.6 shows that $\Delta_{G_{+s}}$ is contractible, establishing a contradiction. □

REMARK 5.8. A similar result follows from Lemma 5.4 for time-bounded strategies: For any $\mathcal{T} \geq 0$, G is fully controllable using only strategies whose worst-case expected convergence times are bounded by $\mathcal{T}$ if and only if $\Delta_G^{\mathcal{T}} \simeq \mathbb{S}^{n-2}$. See [14].

EXAMPLE 5.9. Figure 9 shows a three-state graph in which every action is uncertain: two actions are stochastic, one is nondeterministic. The strategy complex is homotopic to $\mathbb{S}^1$, suggesting the graph is fully controllable. Indeed, for every state there is a strategy for moving to any other state. For instance, the actions at states 1 and 3 together will with certainty move the system to state 2. What is uncertain is the precise time this will take (one may of course compute a worst-case expected convergence time) and the precise route taken. The system may move directly to state 2 or it may cycle for a while between states 1 and 3. We see in this example how strategy complexes and the topological characterization of Theorem 5.7 have abstracted away detailed trajectory information, while preserving a description of

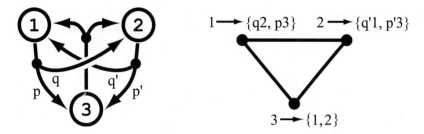

FIGURE 9. The graph on the left has a stochastic action at each of states 1 and 2 and a nondeterministic action at state 3. Its strategy complex on the right is the boundary of a triangle.

the system's overall capabilities. — We mention in passing that changing either or both of the stochastic actions to nondeterministic actions would dissipate full controllability. For instance, if all three actions were nondeterministic, then the strategy complex would consist of three isolated vertices.

6. Source and Dual Complexes

Strategy complexes model control laws for accomplishing tasks specified by goal states in nondeterministic or stochastic graphs. Each simplex (aka strategy) is a collection of actions. The source of an action describes the conditions under which the action is applicable (modeled as a state in the graph). The targets of the action describe the possible outcomes (again modeled as states in the graph). The emphasis in strategies is on actions, yet often one cares primarily about the high-level capability of accomplishing some task, that is, moving from some set of initial states to some set of final states. This section shows how to compress the strategy complex into a smaller complex, called the *source complex*, modeling the start regions of all strategies available to the system. Moreover, this compression preserves homotopy type. Construction of the source complex leads very naturally to a dual complex that models the potentially unattainable goals. These two complexes provide a basis for analyzing and designing systems with control uncertainty.

6.1. Modeling System Capabilities. Let G be a stochastic graph with state space V. We may view any $\sigma \in \Delta_G$ as a strategy for attaining the goal set $V \setminus \mathrm{src}(\sigma)$, as suggested by Remarks 2.10(3) and 4.14: If $v \in \mathrm{src}(\sigma)$, then there is at least one action in σ with source v, possibly several. The system must execute one such action when it is at state v. If $v \notin \mathrm{src}(\sigma)$, then the system does not move when it is at state v. In short, the system moves so long as its current state lies in $\mathrm{src}(\sigma)$, and stops otherwise. Since σ is convergent, with probability 1 the system will eventually find itself in $V \setminus \mathrm{src}(\sigma)$. (Of course, in some instances one may be able to make a more precise prediction as to where the system will stop, but from a global perspective, the outcome $V \setminus \mathrm{src}(\sigma)$ is a general bound.) We model this abstraction with the following definition:

DEFINITION 6.1. The *source complex* $\overline{\Delta}_G$ of a stochastic graph $G = (V, \mathfrak{A})$ is the simplicial complex whose underlying vertex set is V and whose simplices are the start regions of all strategies in G:

$$\overline{\Delta}_G = \{\mathrm{src}(\sigma) \mid \sigma \in \Delta_G\}.$$

LEMMA 6.2. *Let $G = (V, \mathfrak{A})$ be a stochastic graph and let W be a nonempty simplex of $\overline{\Delta}_G$. Then the following subcomplex of Δ_G is contractible:*

$$\Sigma_W = \{\tau \in \Delta_G \mid \mathrm{src}(\tau) \subseteq W\}.$$

PROOF. Let $\sigma \in \Delta_G$ such that $\mathrm{src}(\sigma) = W$. Since σ is convergent, there is some action $A \in \sigma$ that moves off W. Now let $\tau \in \Sigma_W$ and suppose $A \notin \tau$. If $\tau \cup \{A\}$ were to contain a circuit, then A could not move off W. So we see that $\tau \cup \{A\} \in \Sigma_W$, establishing that Σ_W is a cone with apex A. $\square$

LEMMA 6.3. *Let $G = (V, \mathfrak{A})$ be a stochastic graph and suppose W is a nonempty subset of V such that every proper subset of W is a simplex of $\overline{\Delta}_G$. Then $W \in \overline{\Delta}_G$ if and only if some action of G moves off W.*

PROOF. I. Suppose $W \in \overline{\Delta}_G$. Some action of G moves off W, by the first part of the proof of Lemma 6.2.

II. Let A be an action of G that moves off W and let $w = \mathrm{src}(A)$. By assumption, there exists $\tau \in \Delta_G$ such that $\mathrm{src}(\tau) = W \setminus \{w\}$. Arguing as in the last part of the proof of Lemma 6.2, we see that $\tau \cup \{A\} \in \Delta_G$, establishing $W \in \overline{\Delta}_G$. $\square$

REMARKS 6.4. (1) Lemma 6.3 is backchaining topologized, abstracting a planning method known as DYNAMIC PROGRAMMING [**4**]. The connection to backchaining will appear more explicitly in the proof of Lemma 8.6.

(2) One could modify the hypotheses of the lemma, explicitly requiring merely that subsets of size $|W| - 1$ lie in $\overline{\Delta}_G$, rather than all proper subsets of W. The formulation given emphasizes the intuition that a goal is certainly attainable precisely when every subspace of the goal's complement has an exit.

THEOREM 6.5. *For any stochastic graph G, $\Delta_G \simeq \overline{\Delta}_G$.*

PROOF. If G is null, both complexes are void. Otherwise, the theorem follows from Lemma 6.2 and the Quillen Fiber Lemma [**41, 7, 8, 47**]. It is also a corollary to upcoming Theorem 8.10, so we omit further details here. $\square$

COROLLARY 6.6. *A stochastic graph G with nonempty state space V is fully controllable if and only if $\overline{\Delta}_G$ is the boundary complex of the full simplex on V.*

REMARK 6.7. Here we view Corollary 6.6 as a consequence of Theorems 5.7 and 6.5. Alternatively, we could view Theorem 5.7 as a consequence of Corollary 6.6 and Theorem 6.5, thus giving us a different, combinatorial, proof of our main controllability theorem. Indeed, Corollary 6.6 is almost self-evident from the definition of source complex and Remark 6.9 below. To prove it fully, one would also need to make a backchaining argument showing how to combine individual strategies for attaining a particular state into a complete strategy for attaining that state.

DEFINITION 6.8. The *dual complex* of a stochastic graph $G = (V, \mathfrak{A})$ is the combinatorial Alexander dual of the source complex:

$$\overline{\Delta}_G^* = \{V \setminus W \mid W \subseteq V \text{ and } W \notin \overline{\Delta}_G\}.$$

(The underlying vertex set of $\overline{\Delta}_G^*$ is again V.)

Observe that $\emptyset$ is always a simplex of $\overline{\Delta}_G^*$, since V is never a simplex of $\overline{\Delta}_G$.

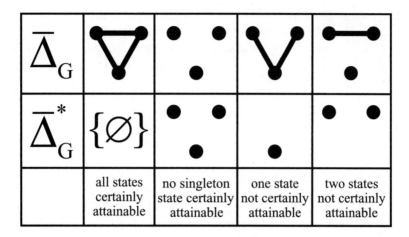

FIGURE 10. Possible source and dual complexes for a graph with three states and at least one convergent action at each state.

REMARK 6.9. The source complex $\overline{\Delta}_G$ of a graph G is the collection of all start regions of convergent sets of actions of G. The complements (relative to V) of these start regions are all the certainly attainable goals (see again Def. 5.1). The simplices of $\overline{\Delta}_G^*$ describe all *potentially unattainable* goals, that is, all sets of states that are not certainly attainable (from everywhere in the graph).

EXAMPLE 6.10. Figure 10 shows the relationship between $\overline{\Delta}_G$ and $\overline{\Delta}_G^*$, along with their meanings, for graphs on three states, assuming that, for every state, some action moves off that state. Given such actions, there are exactly four source complexes possible, ignoring state permutations, since there are three possible edges that may or may not be present in the source complex. Observe that there could be many different graphs that give rise to these complexes, but the details of these graphs are irrelevant at the level of understanding global capabilities. That observation is one interpretation of Theorem 6.5.

EXAMPLE 6.11. The source and dual complexes for the loopback graph of Figure 5 appear in the first column of complexes in Figure 10. The graph is fully controllable. The example of Figure 9 similarly maps to this same column in Figure 10. In contrast, the loopback graph of Figure 6 maps to the third column of complexes in Figure 10. Its source complex is in fact 1——3——2, with dual complex given by the singleton vertex representing state 3. This means that all goals are certainly attainable (in the loopback graph) except for state 3 alone, as we have seen in a variety of ways elsewhere.

The next theorem establishes arbitrary finite complexes as source complexes.

THEOREM 6.12. *For any finite simplicial complex Σ, there exists a nondeterministic graph G such that $\Sigma = \overline{\Delta}_G$ (disregarding underlying vertex sets).*

PROOF. We give the basic construction and point to [**14**] for further details. Let $G = (V, \mathfrak{A})$, with V consisting of $\Sigma^{(0)}$ plus one additional state, and let $\mathfrak{A}$ consist of all actions $x \to V \setminus X$, with $x \in X$ and X a maximal simplex of Σ. □

6.2. System Design.
The minimal nonfaces of $\overline{\Delta}_G$ are useful indicators of how a system loses full controllability, as Lemma 6.3 suggests. By a *minimal nonface* of $\overline{\Delta}_G$ we mean a set of states that is not a simplex of $\overline{\Delta}_G$ but all of whose proper subsets are simplices of $\overline{\Delta}_G$. An adversary can prevent the system from leaving any minimal nonface W of $\overline{\Delta}_G$, since no action moves off W. In particular, any stochastic action with source in W has all its targets in W, while any nondeterministic action with source in W has at least one target in W. For such actions, an adversary can, in effect, delete all the targets outside W (if even there are any). Doing so would produce a new graph with state space W. That graph would be fully controllable. To see this, observe that all proper subsets of W lie in $\overline{\Delta}_G$ since W is a minimal nonface. So the original system can certainly attain any goal of the form $\{w\} \cup (V \backslash W)$, with $w \in W$. Relative to the adversary preventing exit from W, this says the system has full controllability within W. This *relative controllability* is consistent with the simplices of $\overline{\Delta}_G$ inside W forming a sphere of dimension $|W| - 2$.

From a design perspective, one can treat the minimal nonfaces of an existing system as hints for improving system capabilities. The key is to fill in nonfaces by adding or modifying actions. For instance, in the example of Figure 6, since $\{1, 2\}$ is the only minimal nonface of $\overline{\Delta}_{G_{\leftarrow 3}}$, adding *any* action with source 1 or 2 that moves off $\{1, 2\}$, will establish full controllability. There are many different possibilities, including deterministic action $1 \to 3$ and stochastic action $1 \to \{p_1 1, p_2 2, p_3 3\}$.

See [**14**] for further discussion of design.

These ideas extend to improving the performance of a system by considering the complexes $\Delta_G^{\mathcal{T}}$, along with their source and dual variants, for various times $\mathcal{T}$.

6.3. Inferring Adversarial Capabilities.
The source complex also allows one to infer fairly high-level adversarial capabilities, again by looking for missing simplices. A simple observation is that

$$\overline{\Delta}_G = \Gamma * \Sigma_1 * \cdots * \Sigma_k,$$

where Γ is generated by a full simplex consisting of all states in the graph G that do not lie in any minimal nonface of $\overline{\Delta}_G$, while the $\{\Sigma_i\}$ are defined as follows: Define an equivalence relation on the states outside $\Gamma^{(0)}$ as the transitive closure of a simple relation in which two states are related whenever they lie in a common minimal nonface of $\overline{\Delta}_G$. Then Σ_i consists of all simplices of $\overline{\Delta}_G$ that lie within the ith equivalence class of this equivalence relation.

This decomposition of $\overline{\Delta}_G$ reveals some time-varying adversarial capabilities. At any state that lies within multiple minimal nonfaces of $\overline{\Delta}_G$, an adversary may select within which of these minimal nonfaces to keep the system. Consequently, an adversary has some control over an *impatient* system, meaning a system that keeps trying to escape a minimal nonface, for instance by moving to every state of the minimal nonface and by eventually executing all actions available to it. If the system starts within a particular $\Sigma_i^{(0)}$ and is impatient, then the adversary can eventually force the system to reach any particular state within $\Sigma_i^{(0)}$. (Adversarially chosen transitions between different equivalence classes may also be possible, but that information is not directly knowable from the source complex.)

Finally, one can infer some adversarial capabilities more abstractly from homology and cohomology representatives, again by finding minimal nonfaces. Here are some sample results (see [**37, 25**] for background and notation):

LEMMA 6.13. *Suppose* $0 \neq [\alpha] \in \tilde{H}_p(\overline{\Delta}_G; \mathbb{Z})$, *with* $G = (V, \mathfrak{A})$ *a stochastic graph*, $p \geq 0$, *and* α *a simplicial cycle. Write* $\alpha = \sum_i n_i \sigma_i$, *with* $\{\sigma_i\}$ *a basis of elementary p-chains for* $\overline{\Delta}_G$ *and each* n_i *an integer. Then:*

(a) *For every* v *in* V, *there is some* σ_i *with* $n_i \neq 0$ *such that* $\sigma_i \cup \{v\} \notin \overline{\Delta}_G$.
(b) *No action moves off the set of states* $\bigcup_{n_i \neq 0} \sigma_i$ *(called the* support *of* α*).*

PROOF. (a) follows from the definition of nontrivial reduced homology: if no σ_i with the claimed property existed, then α would be homologous to zero.
(b) follows from (a) and Lemma 6.3. □

COROLLARY 6.14. *With notation as above, given any initial state of the system, either an adversary can force the system into the support of* α *and keep it there or the system has no action for moving off its initial state.*

LEMMA 6.15. *Suppose* $[\alpha^*] \in \tilde{H}^p(\overline{\Delta}_G; \mathbb{Z})$, *with* $G = (V, \mathfrak{A})$ *a stochastic graph, $p \geq 0$, and* α^* *a simplicial cocycle (possibly cobounding). Write* $\alpha^* = \sum_i n_i \sigma_i^*$, *with* $\{\sigma_i\}$ *a basis of elementary p-chains for* $\overline{\Delta}_G$ *and each* n_i *an integer.*
Let $v \in V \setminus \bigcup_{n_i \neq 0} \sigma_i$. *Then, for every* σ_i *with* $n_i \neq 0$, $\sigma_i \cup \{v\} \notin \overline{\Delta}_G$.

PROOF. Write $\tau_i = \sigma_i \cup \{v\}$ and suppose $\tau_i \in \overline{\Delta}_G$ for some i such that $n_i \neq 0$. With δ as coboundary operator and ∂ as boundary operator, bearing in mind that v lies outside the support of α^*, we calculate: $<\delta\alpha^*, \tau_i> = <\alpha^*, \partial\tau_i> = \pm n_i \neq 0$. That is a contradiction, since α^* is a cocycle. □

COROLLARY 6.16. *With notation as above, given any initial state of the system outside the support of* α^*, *either the system has no action for moving off its initial state or an adversary can force the system into some subspace of* any *of the defining simplices of* α^*. *Thereafter, the adversary can hold the system to the union of that subspace and the system's initial state.*

REMARKS 6.17. Lemmas 6.13 and 6.15 infer missing simplices of $\overline{\Delta}_G$ from simplicial cycles and cocycles. The two lemmas differ primarily in quantification. Lemma 6.13 and its corollary assert that an adversary can force a system to *some* defining simplex of a nonbounding cycle, whereas Lemma 6.15 and its corollary make a similar assertion for *any* defining simplex of a (possibly cobounding) cocycle. Lemma 6.15 is a generalization of the observation that the coboundary operator implicitly reveals missing simplices. (The two lemmas and their corollaries also impose slightly different conditions on the initial state of the system and describe different subspaces within which the adversary can hold the system.)

EXAMPLE 6.18. Suppose $\overline{\Delta}_G$ is the triangulation of the torus shown in Figure 11. Each minimal nonface of $\overline{\Delta}_G$ is a triangle and a generator of some one-dimensional homology. An adversary can hold the system to any such triangle once the system is at a state defining the triangle. For instance, the oriented 1-cycle $\alpha = [a, b] + [b, c] + [c, a]$ is a generator of some homology and constitutes a minimal nonface of $\overline{\Delta}_G$, so certainly the adversary can hold the system to the support of α. Corollary 6.14 makes the stronger point that the adversary can force the system to this 1-cycle *from any* state, should the system move from that state. Figure 11 also depicts the 1-cocycle α^*, whose support consists of all states except state c. Lemma 6.15 asserts that state c does not form a triangle with any of the defining edges of α^*. Consequently, the adversary can force the system from state c *to any* of α^*'s defining edges, should the system move from state c.

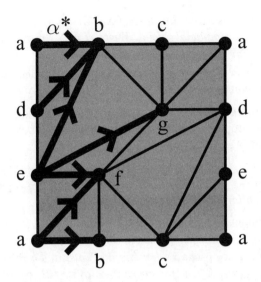

FIGURE 11. A particular triangulation of the torus. The thickened oriented edges depict a 1-cocycle α^* dual to the oriented 1-cycle $\alpha = [a,b] + [b,c] + [c,a]$.

The triangulation of Figure 11 is in fact very bad for the system. All states are equivalent by the equivalence relation defined at the beginning of this subsection, meaning an adversary can eventually force an impatient system anywhere.

7. Unions and Quotients

In this section we explore how the relationship between subgraphs and their encompassing graphs carries over to strategy complexes.

DEFINITION 7.1. A *stochastic subgraph* $H = (W, \mathfrak{B})$ of a stochastic graph $G = (V, \mathfrak{A})$ is a stochastic graph in its own right such that $W \subseteq V$ and $\mathfrak{B} \subseteq \mathfrak{A}$.

7.1. Graph Unions.

DEFINITION 7.2. If $G_1 = (V_1, \mathfrak{A}_1)$ and $G_2 = (V_2, \mathfrak{A}_2)$ are stochastic graphs with nonempty (possibly overlapping) state spaces, define their union $G_1 \cup G_2$ to be the stochastic graph $(V_1 \cup V_2, \mathfrak{A}_1 \sqcup \mathfrak{A}_2)$. Here "$\sqcup$" means "disjoint union"; we treat actions in $\mathfrak{A}_1$ and $\mathfrak{A}_2$ as distinct even if their written representations are identical.

LEMMA 7.3. *With notation and hypotheses as above, write $G = G_1 \cup G_2$.*
  (a) *Always, $\Delta_G \subseteq \Delta_{G_1} * \Delta_{G_2}$.*
  (b) *If either of the following conditions is satisfied, then $\Delta_G = \Delta_{G_1} * \Delta_{G_2}$:*
    (i) *At least one of G_1 and G_2 has no actions with sources in $V_1 \cap V_2$;*
    (ii) *$|V_1 \cap V_2| \leq 1$.*

PROOF. (a) The simplicial join $\Delta_{G_1} * \Delta_{G_2}$ is sensible since G combines actions of G_1 and G_2 with a disjoint union. The join is nonvoid since neither graph is null.

Finally, any convergent subset σ of $\mathfrak{A}_1 \sqcup \mathfrak{A}_2$ can be written as $(\sigma \cap \mathfrak{A}_1) \sqcup (\sigma \cap \mathfrak{A}_2)$, each term of which is a convergent set of actions in one of the original graphs.

(b)(i). Suppose $\operatorname{src}(\mathfrak{A}_1) \subseteq V_1 \backslash V_2$. Choose $\sigma_i \in \Delta_{G_i}$, $i = 1, 2$, and let $\sigma = \sigma_1 \sqcup \sigma_2$. If σ contains a circuit, then there is some nonempty set of actions $\tau \subseteq \sigma$ such that no action of τ moves off $\operatorname{src}(\tau)$. Since actions of G_2 have no targets in $V_1 \setminus V_2$ and since $\operatorname{src}(\tau \cap \mathfrak{A}_1)$ is disjoint from V_2, either $\tau \cap \mathfrak{A}_2$ is empty or it too defines a circuit. Since $\tau \cap \mathfrak{A}_2 \subseteq \sigma_2$, the second possibility cannot occur. The first possibility implies that $\tau \subseteq \sigma_1$, which means τ could not have defined a circuit.

(ii). If $V_1 \cap V_2 = \emptyset$, then $\Delta_G = \Delta_{G_1} * \Delta_{G_2}$ by (i). Otherwise, suppose τ is some nonempty subset of $\sigma_1 \sqcup \sigma_2$, with $\sigma_i \in \Delta_{G_i}$, $i = 1, 2$, such that no action of τ moves off $\operatorname{src}(\tau)$. One may choose τ so no two actions have the same source. Specializing (i) to a subgraph of G_1 and a subgraph of G_2, whose collections of actions are $\tau \cap \mathfrak{A}_1$ and $\tau \cap \mathfrak{A}_2$, respectively, produces a contradiction. □

7.2. Collapsing Fully Controllable Subgraphs.

DEFINITION 7.4. Suppose $G = (V, \mathfrak{A})$ is a stochastic graph with $V \neq \emptyset$. Let $\sim$ be an equivalence relation on V. Define the quotient graph $G/\sim = (V/\sim, \mathfrak{A}/\sim)$ as follows: The state space $V/\sim$ consists of one representative state for every equivalence class of $\sim$. The set of actions $\mathfrak{A}/\sim$ is nearly identical to the set $\mathfrak{A}$; the difference is we relabel source and target states by their representatives in $V/\sim$.

Three comments: (a) State relabeling may identify targets. In the case of stochastic actions, we sum the corresponding transition probabilities when such identifications occur. (b) Distinct actions of G may appear identical in $G/\sim$. We treat them as distinct. (c) Convergent actions of $\mathfrak{A}$ may become self-loops in $\mathfrak{A}/\sim$.

DEFINITION 7.5. Suppose $G = (V, \mathfrak{A})$ is a stochastic graph and W is a nonempty subset of V. Let $\sim$ be the relation in which two states are equivalent if they are identical or if they both lie in W. Write G/W for the quotient graph $G/\sim$.

LEMMA 7.6. *Suppose $H = (W, \mathfrak{B})$ is a fully controllable stochastic subgraph of stochastic graph $G = (V, \mathfrak{A})$, with $W \neq \emptyset$. Then*

$$\Delta_G \simeq \Delta_H * \Delta_{G/W}.$$

PROOF. This lemma is a special case of upcoming Lemma 8.12. □

DEFINITION 7.7. Suppose $G = (V, \mathfrak{A})$ is a stochastic graph. Let $\leftrightarrow$ be the equivalence relation in which two states u and v are equivalent if there exists some fully controllable stochastic subgraph $H = (W, \mathfrak{B})$ of G such that u and v both lie in W. (H may depend on u and v.)

REMARK 7.8. Relation $\leftrightarrow$ is not the same as attainability, by Figure 12.

THEOREM 7.9. *Let G be a non-null stochastic graph. Then*

$$\Delta_G \simeq \mathbb{S}^{n-k-1} * \Delta_{G/\leftrightarrow},$$

with n the size of G's state space and k the number of $\leftrightarrow$ equivalence classes.

PROOF. Apply Lemma 7.6 repeatedly, once for each $\leftrightarrow$ equivalence class (the order does not matter), using the fact that $\mathbb{S}^i * \mathbb{S}^j \simeq \mathbb{S}^{i+j+1}$. □

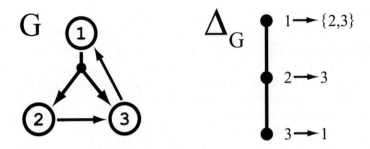

FIGURE 12. States 1 and 3 are certainly attainable from each other, but they do not together lie in a fully controllable subgraph. The nondeterminism of the action at state 1 may or may not force the system to pass through state 2 on its way to state 3; the system can neither definitely avoid state 2 nor be certain of attaining it.

REMARK 7.10. For deterministic directed graphs, $\leftrightarrow$ is the same as strong connectivity. In that case, $\Delta_{G/\leftrightarrow}$ is generated by a full simplex consisting of all directed edges between the strong components of G. If there are no such edges, then $\Delta_{G/\leftrightarrow}$ is the empty complex and $\Delta_G \simeq \mathbb{S}^{n-k-1}$. Otherwise, Δ_G is contractible. This is Hultman's result [26], mentioned earlier in Example 2.14. For nondeterministic graphs, $\Delta_{G/\leftrightarrow}$ can be much more general, as we have seen. Via the methods of Section 6, the complex $\Delta_{G/\leftrightarrow}$ reveals system limitations and adversarial power.

8. Topology of Prescribed Motions

In some situations, when creating a plan to accomplish a task, some actions may be prescribed, perhaps by earlier choices or by external constraints. For example, construction on a highway may force a local detour, or a broken finger may require a robot to perform an assembly from some specific angle. The link of the prescribed actions in the original strategy complex describes all the strategies consistent with the prescription. This section explores links in strategy complexes. We will see that many of our earlier ideas generalize. Indeed, we deferred proofs for some of the earlier results since they really are corollaries to this section. The next section will further use these results to characterize essential actions topologically.

8.1. Links and Sources. Following [27], we define deletion and link more generally than is customary, allowing deletions and links with respect to sets of elements that might not even be in the complex's underlying vertex set:

DEFINITION 8.1. Given a simplicial complex Σ and some set $\mathcal{E}$, we define the *deletion* $\mathrm{dl}(\Sigma, \mathcal{E})$ and the *link* $\mathrm{lk}(\Sigma, \mathcal{E})$ to be the following subcomplexes of Σ:

$$\mathrm{dl}(\Sigma, \mathcal{E}) = \{\tau \in \Sigma \mid \tau \cap \mathcal{E} = \emptyset\},$$
$$\mathrm{lk}(\Sigma, \mathcal{E}) = \{\tau \in \Sigma \mid \tau \cap \mathcal{E} = \emptyset \text{ and } \tau \cup \mathcal{E} \in \Sigma\}.$$

REMARKS 8.2. (1) If the underlying vertex set of Σ is X, then the underlying vertex set of both $\mathrm{dl}(\Sigma, \mathcal{E})$ and $\mathrm{lk}(\Sigma, \mathcal{E})$ is generally understood to be $X \setminus \mathcal{E}$.

(2) If $\mathcal{E}$ is not a simplex of Σ, then $\mathrm{lk}(\Sigma, \mathcal{E})$ is the void complex.

First, we generalize the definition of "moves off". To set the stage, suppose $G = (V, \mathfrak{A})$ is a stochastic graph, $\sigma \in \Delta_G$, and $W \subseteq V$. Suppose A is an action of

G whose source lies in W. We are interested in the behavior of the system starting from src(A), given that the system first executes action A and then executes strategy σ, with the provision that execution of σ stops if ever the system re-enters W. Since $\sigma \cup \{A\}$ need not lie in Δ_G, we refer to this behavior by saying that the system *moves according to* $[A; \sigma \dashv W]$.

DEFINITION 8.3. Let $G = (V, \mathfrak{A})$ be a stochastic graph, $\sigma \in \Delta_G$, $W \subseteq V$, and $A \in \mathfrak{A}$. Action A *moves off* W *subject to* σ if $A \notin \sigma$, src(A) $\in W$, and the worst-case probability of returning to W from src(A) is strictly less than 1 whenever the system moves according to $[A; \sigma \dashv W]$.

REMARKS 8.4. (1) If $\sigma = \emptyset$, Def. 8.3 is equivalent to Def. 4.5.

(2) Algebraically, we may express the probability condition of Def. 8.3 by requiring that adv($A, \{q_v\}$) < 1, where $\{q_v\}_{v \in V}$ is the solution to the following system of equations (the solution exists and is unique since σ is convergent):

(8.1) $$q_w = \begin{cases} \max_{B \in \sigma | w} \mathrm{adv}(B, \{q_v\}), & \text{if } w \in \mathrm{src}(\sigma) \setminus W; \\ 1, & \text{if } w \in W; \\ 0, & \text{otherwise.} \end{cases}$$

(The maximization at w is taken over $\sigma|w$, that is, all actions of σ with source w.)

Next, we generalize the definition of source complex:

DEFINITION 8.5. Suppose $G = (V, \mathfrak{A})$ is a stochastic graph and $\sigma \in \Delta_G$. The *source link of* σ *in* Δ_G is the subcomplex of $\overline{\Delta}_G$ given by:

$$\overline{\mathrm{Lk}}(\Delta_G, \sigma) = \{\mathrm{src}(\tau) \mid \tau \in \mathrm{lk}(\Delta_G, \sigma)\}.$$

Lemma 6.3 generalizes as follows:

LEMMA 8.6. *Let* $G = (V, \mathfrak{A})$ *be a stochastic graph and* $\sigma \in \Delta_G$. *Suppose* W *is a nonempty subset of* V *such that every proper subset of* W *is a simplex of* $\overline{\mathrm{Lk}}(\Delta_G, \sigma)$. *Then* $W \in \overline{\mathrm{Lk}}(\Delta_G, \sigma)$ *if and only if some action of* G *moves off* W *subject to* σ.

PROOF. I. Suppose $W \in \overline{\mathrm{Lk}}(\Delta_G, \sigma)$. Then there exists $\tau \in \Delta_G$ such that $\tau \cap \sigma = \emptyset$, $\tau \cup \sigma \in \Delta_G$, and src($\tau$) $= W$. We will now backchain using actions of τ and σ to define a sequence of triples $(W_1, A_1, v_1), \ldots, (W_k, A_k, v_k)$ such that $W \cup \mathrm{src}(\sigma) = \{v_1, \ldots, v_k\}$ and action A_j has source v_j and moves off W_j (in the standard sense of Def. 4.5):

- Let $W_1 = W \cup \mathrm{src}(\sigma)$ and $k = |W_1|$. (So $W_1 = \mathrm{src}(\tau) \cup \mathrm{src}(\sigma)$.)
- For $j = 1, \ldots, k$: Choose $v_j \in W_j$ so that all actions of $\tau \cup \sigma$ with source v_j move off W_j. Such a v_j exists since $\tau \cup \sigma$ is convergent and since $W_j \subseteq \mathrm{src}(\tau) \cup \mathrm{src}(\sigma)$. If $v_j \in W$, let A_j be an action of τ with source v_j. Otherwise, let A_j be an action of σ with source v_j. Finally, define $W_{j+1} = W_j \setminus \{v_j\}$. (Observe: $W_{k+1} = \emptyset$.)

Choose i to be the smallest index in $\{1, \ldots, k\}$ such that $A_i \in \tau$ (well-defined since $W \neq \emptyset$). This is the action A we seek. To verify: Action A is in τ, so not in σ. Action A moves off W in the standard sense, by minimality of i. We need to show that adv($A, \{q_v\}$) < 1, with $\{q_v\}_{v \in V}$ being the solution to System (8.1).

Suppose $i = 1$. Then A moves off $W \cup \mathrm{src}(\sigma)$. If A is nondeterministic, then all of A's targets lie outside $W \cup \mathrm{src}(\sigma)$, so adv($A, \{q_v\}$) $= 0 < 1$. If A is stochastic, then at least one target u of A lies outside $W \cup \mathrm{src}(\sigma)$, meaning $q_u = 0$, so $\mathrm{adv}(A, \{q_v\}) < 1$.

Suppose $i > 1$. By Def. 4.11, the previous conclusion holds more generally whenever $0 \leq q_u < 1$ for all relevant targets u (meaning all targets when A is nondeterministic; at least one target when A is stochastic). Recall that *all* actions of σ with source v_j move off W_j, for $j = 1, \ldots, i-1$. Inductively we therefore see that $q_{v_j} < 1$ for all $j = 1, \ldots, i-1$, and so $\mathrm{adv}(A, \{q_v\}) < 1$.

II. Let A be an action of G that moves off W subject to σ and let $w = \mathrm{src}(A)$. By assumption, there exists $\tau \in \Delta_G$ such that $\tau \cap \sigma = \emptyset$, $\tau \cup \sigma \in \Delta_G$, and $\mathrm{src}(\tau) = W \setminus \{w\}$. Write $\tau' = \tau \cup \{A\}$. Then $\mathrm{src}(\tau') = W$. Since $A \notin \sigma$, $\tau' \cap \sigma = \emptyset$. To establish that $W \in \overline{\mathrm{Lk}}(\Delta_G, \sigma)$, we therefore should show that $\tau' \cup \sigma \in \Delta_G$.

Suppose otherwise. Then, for some $\gamma \subseteq \tau' \cup \sigma$, with $A \in \gamma$, no action of γ moves off $\mathrm{src}(\gamma)$, in the standard sense of Def. 4.5. Let $\gamma' = \{B \in \gamma \mid \mathrm{src}(B) \notin W\} \subseteq \sigma$. Either γ' is empty or an adversary could select nondeterministic transitions so that, with probability 1, execution of γ' eventually moves the system into W whenever it starts in $\mathrm{src}(\gamma')$. That means $q_v = 1$ in (8.1) for every $v \in \mathrm{src}(\gamma)$, implying $\mathrm{adv}(A, \{q_v\}) = 1$, a contradiction. $\square$

Finally, we generalize Lemma 6.2:

LEMMA 8.7. *Let $G = (V, \mathfrak{A})$ be a stochastic graph, $\sigma \in \Delta_G$, and $W \in \overline{\mathrm{Lk}}(\Delta_G, \sigma)$, with $W \neq \emptyset$. Then the following subcomplex of $\mathrm{lk}(\Delta_G, \sigma)$ is contractible:*

$$\Sigma_{W,\sigma} = \{\tau \in \mathrm{lk}(\Delta_G, \sigma) \mid \mathrm{src}(\tau) \subseteq W\}.$$

PROOF. Let A move off W subject to σ, as by Lemma 8.6. We claim that $\Sigma_{W,\sigma}$ is a cone with apex A. To see this, suppose $\tau \in \Sigma_{W,\sigma}$ and $A \notin \tau$. Write $\tau' = \tau \cup \{A\}$. We know that $\tau \cap \sigma = \emptyset$, $\tau \cup \sigma \in \Delta_G$, and $\mathrm{src}(\tau) \subseteq W$. We also know that $A \notin \sigma$ and $\mathrm{src}(A) \in W$. So we should show that $\tau' \cup \sigma \in \Delta_G$. The argument is identical to that given in the last paragraph of the proof of Lemma 8.6. $\square$

8.2. Quillen Fiber Lemma.

Before generalizing the key results of Sections 6 and 7, we review the Quillen Fiber Lemma [40, 41] in a form for our proofs. See [7, 8, 47] for further details regarding partially ordered sets (posets), order complexes, and the Quillen Fiber Lemma.

Every simplicial complex Σ defines a poset $\mathfrak{F}(\Sigma)$, called the *face poset* of Σ. The elements of $\mathfrak{F}(\Sigma)$ are the nonempty simplices of Σ, partially ordered by set inclusion. Conversely, any poset P defines a simplicial complex $\Sigma(P)$, called the *order complex* of P. The simplices of $\Sigma(P)$ are given by the finite chains $p_1 < \cdots < p_k$ in P. The order complex of a face poset, $\Sigma(\mathfrak{F}(\Sigma))$, is isomorphic to $\mathrm{sd}(\Sigma)$, showing that Σ and $\Sigma(\mathfrak{F}(\Sigma))$ are homeomorphic. One may therefore speak of the topology of a poset, implicitly meaning the topology of its order complex.

DEFINITION 8.8. If Q is a poset, let $Q_{\leq q}$ denote the set $\{q' \in Q \mid q' \leq_Q q\}$, with $\leq_Q$ being the partial order on Q (set inclusion in the case of face posets derived from simplicial complexes).

THEOREM 8.9 (Quillen Fiber Lemma). *Suppose $f : P \to Q$ is an order-preserving map between two posets. If $f^{-1}(Q_{\leq q})$ is contractible for all $q \in Q$, then P and Q are homotopy equivalent.*

Quillen's original version (as Theorem A in [40]) makes a stronger category-theoretic assertion than the combinatorial version we have reproduced here. Quillen's version for posets [41] is also stronger, asserting that f actually induces a homotopy equivalence. In this chapter, we only require the version stated above.

8.3. Two Link-Based Homotopy Equivalences. We generalize Theorem 6.5 and Lemma 7.6 to links.

THEOREM 8.10. *For any stochastic graph G and any $\sigma \in \Delta_G$,*
$$\text{lk}(\Delta_G, \sigma) \simeq \overline{\text{Lk}}(\Delta_G, \sigma).$$

PROOF. Let $P = \mathfrak{F}(\text{lk}(\Delta_G, \sigma))$ and $Q = \mathfrak{F}(\overline{\text{Lk}}(\Delta_G, \sigma))$ be the face posets of the two complexes. Define $f : P \to Q$ by $f(\tau) = \text{src}(\tau)$. For any $W \in Q$, $f^{-1}(Q_{\leq W})$ is the face poset of $\Sigma_{W,\sigma}$, which is contractible by Lemma 8.7. The Quillen Fiber Lemma completes the argument. □

REMARK 8.11. For non-null G, Theorem 6.5 follows as a corollary, with $\sigma = \emptyset$.

In what follows, let tilde accents (as in $\tilde{\sigma}$) designate the image of G's actions in G/W as per Defs. 7.4 and 7.5.

LEMMA 8.12. *Suppose $H = (W, \mathfrak{B})$ is a fully controllable stochastic subgraph of stochastic graph $G = (V, \mathfrak{A})$, with $W \neq \emptyset$. Let $\sigma \in \Delta_G$ with $W \cap \text{src}(\sigma) = \emptyset$. Then*
$$\text{lk}(\Delta_G, \sigma) \simeq \Delta_H * \text{lk}(\Delta_{G/W}, \tilde{\sigma}).$$

REMARK 8.13. Lemma 7.6 follows as a corollary, with $\sigma = \emptyset$.

PROOF OF LEMMA 8.12. Observe that $\tilde{\sigma} \in \Delta_{G/W}$, since $\sigma \in \Delta_G$ and σ contains no actions at states in W, so the lemma's statement makes sense. Moreover, by Theorems 6.5 and 8.10, we only need to prove that
$$\overline{\text{Lk}}(\Delta_G, \sigma) \simeq \overline{\Delta}_H * \overline{\text{Lk}}(\Delta_{G/W}, \tilde{\sigma}).$$

Let $P = \mathfrak{F}(\overline{\Delta}_H * \overline{\text{Lk}}(\Delta_{G/W}, \tilde{\sigma}))$ and $Q = \mathfrak{F}(\overline{\text{Lk}}(\Delta_G, \sigma))$ be the associated face posets. The elements of P are the nonempty simplices of $\overline{\Delta}_H * \overline{\text{Lk}}(\Delta_{G/W}, \tilde{\sigma})$. We may write any such simplex uniquely as $X \cup Y$, with $X \in \overline{\Delta}_H$ and $Y \in \overline{\text{Lk}}(\Delta_{G/W}, \tilde{\sigma})$, not both X and Y empty.

Let $\diamond$ be the state in G/W representing W identified to a point and define $f : P \to Q$ by
$$f(X \cup Y) = \begin{cases} X \cup Y, & \text{if } \diamond \notin Y; \\ W \cup Y \setminus \{\diamond\}, & \text{if } \diamond \in Y; \end{cases}$$

with $X \in \overline{\Delta}_H$ and $Y \in \overline{\text{Lk}}(\Delta_{G/W}, \tilde{\sigma})$. Observe that $X \subseteq W$ and $Y \subseteq (V \setminus W) \cup \{\diamond\}$.

Establishing the following conditions will complete the proof, by the Quillen Fiber Lemma:

(i) f is well-defined, meaning that $f(X \cup Y)$ really is a simplex of $\overline{\text{Lk}}(\Delta_G, \sigma)$;
(ii) f is order-preserving;
(iii) the fibers $f^{-1}(Q_{\leq q})$ are contractible.

There are several cases to verify. We will prove the most interesting cases, leaving the remainder for the reader.

(i). We assume here that $\diamond \in Y$, leaving to the reader the case for which $\diamond \notin Y$.

We know there exists a set of actions $\tau \subseteq \mathfrak{A}$ such that $\tilde{\tau} \cap \tilde{\sigma} = \emptyset$, $\tilde{\tau} \cup \tilde{\sigma} \in \Delta_{G/W}$, and $\text{src}(\tilde{\tau}) = Y$. (Recall again that, for example, $\tilde{\tau}$ is the image of G's actions τ in the quotient graph G/W.) We can assume that $\tilde{\tau}$ contains exactly one action $\tilde{A}$

with source $\diamond$. Let $w \in W$ be the source of the corresponding action A of τ. Since H is fully controllable, there exists $\gamma \in \Delta_H$ such that $\text{src}(\gamma) = W \setminus \{w\}$. Since $W \cap \text{src}(\sigma) = \emptyset$, $(\gamma \cup \tau) \cap \sigma = \emptyset$. We know $\text{src}(\gamma \cup \tau) = (W \setminus \{w\}) \cup (Y \setminus \{\diamond\}) \cup \{w\} = W \cup Y \setminus \{\diamond\}$. So it remains to establish that $\gamma \cup \tau \cup \sigma \in \Delta_G$.

The set of actions $(\tau \setminus \{A\}) \cup \sigma$ is convergent since the set $(\tilde{\tau} \setminus \{\tilde{A}\}) \cup \tilde{\sigma}$ is convergent and since the only difference, if any, between these two sets of actions is in the labeling of targets at which neither set has sources. Since γ has no sources or targets outside W and $(\tau \setminus \{A\}) \cup \sigma$ has no sources in W, Lemma 7.3(b)(i) tells us $\gamma \cup (\tau \setminus \{A\}) \cup \sigma$ is convergent. Any circuit contained in $\gamma \cup \tau \cup \sigma$ would therefore necessarily involve the action A. Mapping that circuit to the quotient graph G/W then tells us $\tilde{\tau} \cup \tilde{\sigma}$ must contain a circuit, which is impossible.

(ii). Easy.

(iii). Every $q \in Q$ is a nonempty simplex of $\overline{\text{Lk}}(\Delta_G, \sigma)$, so we will write U in place of q, with $U \subset V$. Let such a nonempty U be given. We need to show that $f^{-1}(Q_{\leq U})$ is contractible. We assume here that $U \cap W = W$, leaving to the reader the case in which $U \cap W$ is a proper subset of W.

To establish contractibility, we will show that $f^{-1}(Q_{\leq U})$ is the face poset of a cone with apex $\diamond$. Observe that $f^{-1}(Q_{\leq U})$ is indeed the face poset of some simplicial complex, since f is order-preserving.

Let $X \cup Y \in f^{-1}(Q_{\leq U})$, with $X \in \overline{\Delta}_H$ and $Y \in \overline{\text{Lk}}(\Delta_{G/W}, \tilde{\sigma})$. Suppose $\diamond \notin Y$. We need to show that $X \cup Y \cup \{\diamond\}$ is an element of $f^{-1}(Q_{\leq U})$. Observe:

$$f(X \cup Y) = X \cup Y \subseteq U, \quad \text{since } X \cup Y \in f^{-1}(Q_{\leq U});$$
$$f(X \cup Y \cup \{\diamond\}) = W \cup Y \subseteq U, \quad \text{since } W \subseteq U.$$

To complete the proof, we will show that $Y \cup \{\diamond\} \in \overline{\text{Lk}}(\Delta_{G/W}, \tilde{\sigma})$.

Suppose otherwise. Then there must be some set of states $Y' \subseteq Y$, such that every proper subset of $Y' \cup \{\diamond\}$ is a simplex of $\overline{\text{Lk}}(\Delta_{G/W}, \tilde{\sigma})$ but $Y' \cup \{\diamond\}$ is not. By Lemma 8.6, no action of G/W moves off $Y' \cup \{\diamond\}$ subject to $\tilde{\sigma}$. Consequently, no action of G moves off $Y' \cup W$ subject to σ. Lemma 8.6 therefore implies that $Y' \cup W \notin \overline{\text{Lk}}(\Delta_G, \sigma)$. That establishes a contradiction, since $Y' \cup W \subseteq Y \cup W \subseteq U \in \overline{\text{Lk}}(\Delta_G, \sigma)$.

The same argument shows that $\{\diamond\}$ is itself in $f^{-1}(Q_{\leq U})$. □

9. Essential Actions

A graph may contain redundant actions. For instance, in the graph of Figure 7, the action $1 \to \{2, 3\}$ is completely unessential. We can remove the action without changing the overall capabilities of the system: with or without the action, the system can move from any state to any other state. The link of this action in the graph's strategy complex is the edge appearing in the right panel of Figure 13. It is contractible. That observation is generally true, as the next lemma shows.

DEFINITION 9.1. Given stochastic graph $G = (V, \mathfrak{A})$ and actions $\mathcal{A} \subseteq \mathfrak{A}$, let $G \setminus \mathcal{A}$ denote the graph $(V, \mathfrak{A} \setminus \mathcal{A})$ formed from G by removing the actions $\mathcal{A}$.

REMARK 9.2. Viewing $G \setminus \mathcal{A}$ as a subgraph of G, we may think of $\Delta_{G \setminus \mathcal{A}}$ as a subcomplex of Δ_G. In fact, $\Delta_{G \setminus \mathcal{A}} = \text{dl}(\Delta_G, \mathcal{A})$.

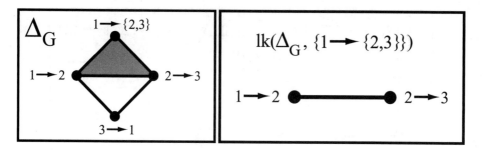

FIGURE 13. Left panel: The strategy complex Δ_G of Figure 7. Right panel: The link of the action $1 \to \{2,3\}$ in Δ_G. The link is contractible since $1 \to \{2,3\}$ is not essential for full controllability.

LEMMA 9.3. *Let $G = (V, \mathfrak{A})$ be a fully controllable stochastic graph. Suppose $\mathcal{A}$ is a nonempty set of actions in $\mathfrak{A}$ such that $G \backslash \mathcal{A}$ is also fully controllable. Then $\mathrm{lk}(\Delta_G, \mathcal{A})$ is contractible.*

PROOF. If $\mathcal{A}$ is not a simplex in Δ_G, then $\mathrm{lk}(\Delta_G, \mathcal{A})$ is the void complex, which is considered to be contractible. So we may assume that $\mathcal{A} \in \Delta_G$.

The covering set homotopy equivalence of Corollary 5.6 carries over to links of strategies as follows:

$$\mathrm{lk}(\Delta_G, \mathcal{A}) \simeq \left(\bigcup_{A \in \mathfrak{A} \backslash \mathcal{A}} U_A \right) \cap \left(\bigcap_{A \in \mathcal{A}} U_A \right).$$

Since $G \backslash \mathcal{A}$ is fully controllable, $\Delta_{G \backslash \mathcal{A}} \simeq \mathbb{S}^{n-2}$, with $n = |V|$. As a result, the union of all the covering sets U_A, with $A \in \mathfrak{A} \backslash \mathcal{A}$, is all of $\bigodot_{\neq}^n$. So

$$\mathrm{lk}(\Delta_G, \mathcal{A}) \simeq \bigcap_{A \in \mathcal{A}} U_A.$$

That last intersection is nonempty and convex, hence contractible. □

DEFINITION 9.4. Let $G = (V, \mathfrak{A})$ be a stochastic graph and let $s \in V$. Suppose G contains a complete strategy for attaining s. Let $\mathcal{A}$ be a set of actions in $\mathfrak{A}$. We say *$\mathcal{A}$ is essential for attaining s in G* if $G \backslash \mathcal{A}$ does not contain a complete strategy for attaining s.

THEOREM 9.5. *Let $G = (V, \mathfrak{A})$ be a stochastic graph and let $s \in V$. Suppose G contains a complete strategy for attaining s. Let $\mathcal{A}$ be a nonempty subset of $\mathfrak{A}$.*

(a) *If $\mathcal{A}$ is not essential for attaining s in G, then $\mathrm{lk}(\Delta_{G_{\leftarrow s}}, \mathcal{A})$ is contractible.*
(b) *If $\mathcal{A}$ is essential, but no proper subset of $\mathcal{A}$ is essential, for attaining s in G, then $\mathrm{lk}(\Delta_{G_{\leftarrow s}}, \mathcal{A}) \simeq \mathbb{S}^{n-3}$, with $n = |V|$.*

PROOF SKETCH. Part (a) follows from Lemma 9.3. For part (b), we first observe that $\mathbb{S}^{n-3}$ is sensible, since $\mathcal{A}$'s existence means n is at least 2. We next outline the key steps of a proof, leaving the details to the reader:

(1) One may replace any stochastic action A of $\mathfrak{A} \setminus \mathcal{A}$ with a collection of deterministic actions, one for each stochastic transition of A, without changing the homotopy types of these complexes: Δ_G, $\Delta_{G_{\leftarrow s}}$, $\mathrm{lk}(\Delta_G, \mathcal{A})$, $\mathrm{lk}(\Delta_{G_{\leftarrow s}}, \mathcal{A})$. See Remark 9.7 below for more detail.

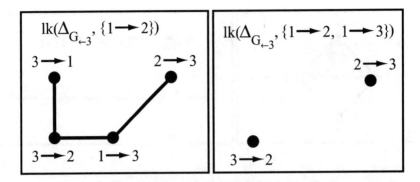

FIGURE 14. The link shown in the left panel is contractible since action $1 \to 2$ is not essential for attaining state 3 in the graph of Figure 2. The link shown in the right panel is homotopic to $\mathbb{S}^0$ since together the actions $1 \to 2$ and $1 \to 3$ are essential for attaining state 3 while neither is individually. See also Figure 5.

(2) Let G_{ndet} be G with all stochastic actions of $\mathfrak{A} \setminus \mathcal{A}$ replaced as in Step (1). Define W to be all states w in V for which G_{ndet} contains a strategy that attains s from w without requiring any of the actions $\mathcal{A}$. Observe $s \in W$.

(3) The loopback graph formed from G_{ndet} and s contains a fully controllable subgraph with state space W, disjoint from $\mathrm{src}(\mathcal{A})$. Prereasoning some of Step (4) shows that $\mathcal{A}$ is convergent. Lemma 8.12 then factors $\mathbb{S}^{|W|-2}$ out of the link of $\mathcal{A}$ in the loopback complex formed from G_{ndet} and s.

(4) Consider G again. Steps (1)–(3), along with the hypotheses regarding $\mathcal{A}$, allow us to assume without loss of generality that every action in $\mathcal{A}$ moves off $V \setminus \{s\}$ and that no action in $\mathfrak{A} \setminus \mathcal{A}$ moves off $V \setminus \{s\}$.

(5) Let $H = (V', \mathfrak{A}')$, with $V' = V \setminus \{s\}$ and $\mathfrak{A}' = \{A \in \mathfrak{A} \setminus \mathcal{A} \mid \mathrm{src}(A) \in V'\}$. This construction makes sense for all stochastic actions of $\mathfrak{A} \setminus \mathcal{A}$ by Step (4). There could, however, be nondeterministic actions in $\mathfrak{A} \setminus \mathcal{A}$ that have transitions both to s and to one or more states in $V \setminus \{s\}$. In constructing $\mathfrak{A}'$, remove from any such action the transition to s.

(6) Observe that for any $v \in \mathrm{src}(\mathcal{A})$, H contains a complete strategy for attaining state v, by the hypotheses regarding $\mathcal{A}$.

(7) Let H_+ designate H with all possible loopbacks added at every state of $\mathrm{src}(\mathcal{A})$. Using Theorems 6.5 and 8.10, along with the Quillen Fiber Lemma, one sees that $\Delta_{H_+} \simeq \mathrm{lk}(\Delta_{G_{\leftarrow s}}, \mathcal{A})$.

(8) Since H_+ is fully controllable and contains $n - 1$ states, Theorem 5.7 establishes that $\mathrm{lk}(\Delta_{G_{\leftarrow s}}, \mathcal{A}) \simeq \mathbb{S}^{n-3}$.

□

EXAMPLE 9.6. The three-state graph of Figure 2 contains a complete strategy for attaining state 3. The associated loopback graph and complex appear in Figure 5. The action $1 \to 2$ is not essential for attaining state 3. Topologically, the link $\mathrm{lk}(\Delta_{G_{\leftarrow 3}}, \{1 \to 2\})$ is contractible, as shown in the left panel of Figure 14. Action $1 \to 3$ is not essential by itself either. However, the two actions $1 \to 2$ and $1 \to 3$ together clearly are essential for attaining state 3. Topologically, the right panel of Figure 14 shows that $\mathrm{lk}(\Delta_{G_{\leftarrow 3}}, \{1 \to 2, 1 \to 3\})$ is indeed homotopic to $\mathbb{S}^0$.

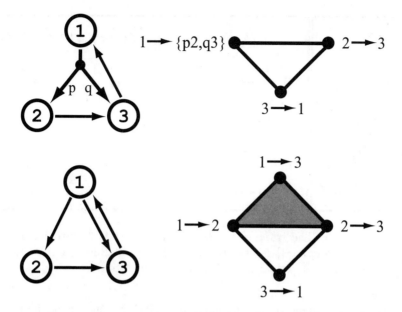

FIGURE 15. The graph in the top left panel contains an action with stochastic transitions. These transitions appear as separate deterministic actions in the graph shown in the bottom left panel. The strategy complexes for the two graphs, shown in the right two panels, have the same homotopy type.

REMARK 9.7. In Step (1) of the proof of Lemma 9.5, we observed that one may replace any stochastic action in a graph G by a collection of deterministic actions, one such action for each of the original stochastic transitions, without changing the homotopy type of Δ_G. In fact, the corresponding source complexes are identical, as a straightforward application of Lemma 6.3 shows. See Figure 15 for an example. The argument generalizes with the aid of Lemma 8.6, showing that replacement of stochastic transitions by deterministic actions does not change the homotopy type of $\mathrm{lk}(\Delta_G, \mathcal{A})$, so long as one does not make such replacements for any actions of $\mathcal{A}$. The same reasoning applies to $\Delta_{G \leftarrow s}$ and $\mathrm{lk}(\Delta_{G \leftarrow s}, \mathcal{A})$.

EXAMPLE 9.8. Lemma 9.3 is phrased in terms of a fully controllable graph, whereas Lemma 9.5 is specialized to loopback graphs. One wonders whether part (b) of Lemma 9.5 holds as well for fully controllable graphs. In fact, it need not hold when stochastic actions appear in the essential set $\mathcal{A}$, as Figure 16 shows.

10. Decision Trees

This section re-examines the structure of loopback complexes. We have seen that strategy complexes and source complexes can have the topology of any finite simplicial complex (Theorems 2.16 and 6.12), but loopback complexes are either spheres or points, homotopically (Theorem 3.6). Decision trees allow us to be more specific. We are motivated in this exploration by the extensive results Jonsson obtained for directed graph complexes via decision trees [27] as well as the connections Forman established between decision trees and discrete Morse theory [18, 19].

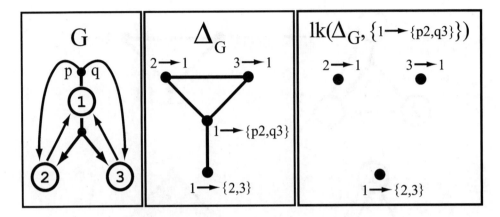

FIGURE 16. The graph on the left is fully controllable. Stochastic action $1 \to \{p2, q3\}$ is essential for full controllability. However, the link of that action in Δ_G is not homotopic to $\mathbb{S}^0$.

Intuitively, a decision tree is a variant of the "20 Questions" game, for determining whether an unknown set lies in some known collection of sets. Let Σ (perhaps a simplicial complex) be such a collection, drawn from an underlying vertex set X. We can view any subset σ of X as a bit vector over X. Suppose we know Σ exactly and someone has a secret σ that may or may not lie in Σ. We may ask whether individual bits in σ's bit vector are on or off. Any question we ask may depend on the answers to earlier questions. Our goal is to ask as few questions as possible in order to decide whether $\sigma \in \Sigma$. For example, suppose $x_0 \in X$ is some specific point, and suppose Σ consists of all subsets of X that do not contain x_0. Then the answer to one question, "Is x_0 in σ?", is sufficient to establish whether $\sigma \in \Sigma$. (We do not need to figure out what σ is exactly, merely whether it lies in Σ.)

In the worst case, one may need to ask $|X|$-many questions. Such sets σ are called *evasive* (relative to Σ and the questions being asked). Simplicial complexes for which one can structure the questions in such a way that no simplex is evasive are called *nonevasive*. This is a strong property. For finite simplicial complexes it is well-known [7] that the following proper inclusions hold:

$$\text{Cones} \subset \begin{array}{c}\text{Nonevasive}\\ \text{Complexes}\end{array} \subset \begin{array}{c}\text{Collapsible}\\ \text{Complexes}\end{array} \subset \begin{array}{c}\text{Contractible}\\ \text{Complexes}\end{array}.$$

This section shows that a contractible loopback complex is in fact nonevasive. Similarly, for any loopback complex $\Delta_{G \leftarrow s}$ homotopic to a sphere, one can pose the membership questions in an order such that exactly one simplex is evasive, corresponding to a complete strategy for attaining goal state s.

We now define decision trees recursively, much as one would in a functional programming language such as SML [39]. A decision tree is an object containing some data, with the object assuming one of two forms. One form of object, which we designate NODE below, contains a simplicial complex, a vertex, and two subtrees. By containing two subtrees, a NODE spawns two structural recursions. The other form of object, which we designate LEAF below, stops such recursions. A LEAF contains only a simplicial complex. There are further restrictions on the data in each object, made explicit in the next definition.

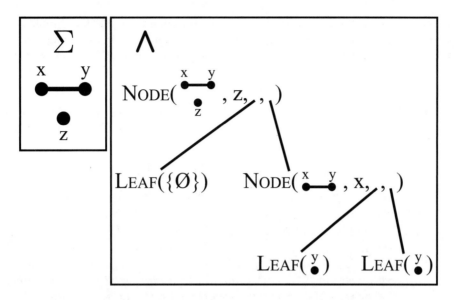

FIGURE 17. Left panel: A simplicial complex Σ generated by an edge $\{x, y\}$ and a point z. Right panel: One possible decision tree $\wedge$ for Σ. (Notation: Formally, a NODE contains its subtrees. Pictorially, it is convenient to draw edges to the subtrees. For ease of viewing, the figure depicts nonempty complexes geometrically rather than algebraically.)

DEFINITION 10.1. Suppose Σ is a finite simplicial complex with finite underlying vertex set X. A *decision tree for Σ (with underlying vertex set X)* is defined recursively as follows:
- Suppose Σ is one of the following complexes: $\emptyset$, $\{\emptyset\}$, $\{\emptyset, \{x\}\}$ (void complex, empty complex, single point x, respectively). Then LEAF(Σ) is a decision tree for Σ.
- Suppose $x \in X$. Define the underlying vertex set for both $\mathrm{lk}(\Sigma, \{x\})$ and $\mathrm{dl}(\Sigma, \{x\})$ to be $X \setminus \{x\}$. Suppose $\wedge_{\mathrm{lk}}$ is a decision tree for $\mathrm{lk}(\Sigma, \{x\})$ and $\wedge_{\mathrm{dl}}$ is a decision tree for $\mathrm{dl}(\Sigma, \{x\})$. Then NODE($\Sigma, x, \wedge_{\mathrm{lk}}, \wedge_{\mathrm{dl}}$) is a decision tree for Σ.
- Nothing else is a decision tree for Σ.

REMARKS 10.2. (1) One may think of NODE and LEAF as functions that take data and produce a decision tree, or, equivalently, as the actual objects containing that data. (2) Figure 17 shows one possible decision tree for a simplicial complex generated by an edge and a point. (3) Decision trees need not be unique. For instance, the complex consisting of a single point x with underlying vertex set $\{x\}$ gives rise to these two possible decision trees:

$$\text{LEAF}(\{\emptyset, \{x\}\}) \qquad \text{NODE}(\{\emptyset, \{x\}\}, x, \text{LEAF}(\{\emptyset\}), \text{LEAF}(\{\emptyset\})).$$

DEFINITION 10.3. A decision tree $\wedge$ for a finite simplicial complex Σ and a simplex σ of Σ together define a *descent path* through the tree, as follows: The path starts at $\wedge$. Suppose $\wedge = \text{NODE}(\Sigma, x, \wedge_{\mathrm{lk}}, \wedge_{\mathrm{dl}})$. If $x \in \sigma$, then the path continues via tree $\wedge_{\mathrm{lk}}$ and simplex $\sigma \setminus \{x\}$; otherwise, the path continues via tree

$\wedge_{\mathrm{dl}}$ and simplex σ. The path stops upon encountering a LEAF. The simplex σ is *evasive with respect to* $\wedge$ if the LEAF attained contains the empty complex, and *nonevasive with respect to* $\wedge$ otherwise. A simplicial complex Σ is *nonevasive* if there is some decision tree $\wedge$ for Σ such that every simplex of Σ is nonevasive with respect $\wedge$. See [18, 27] for further details.

REMARK 10.4. The void complex is nonevasive, as is the complex representing any full nonempty simplex. The empty simplex is evasive with respect to any decision tree for the empty complex.

The following lemma is immediate from the definitions:

LEMMA 10.5. *Suppose* $\wedge = \mathrm{NODE}(\Sigma, x, \wedge_{\mathrm{lk}}, \wedge_{\mathrm{dl}})$ *is a decision tree for finite simplicial complex* Σ.

  (a) *If* $\wedge_{\mathrm{lk}}$ *and* $\wedge_{\mathrm{dl}}$ *establish that* $\mathrm{lk}(\Sigma, \{x\})$ *and* $\mathrm{dl}(\Sigma, \{x\})$, *respectively, are nonevasive, then* $\wedge$ *establishes that* Σ *is nonevasive.*
  (b) *If exactly one simplex* τ *of* $\mathrm{lk}(\Sigma, \{x\})$ *is evasive with respect to* $\wedge_{\mathrm{lk}}$ *and every simplex of* $\mathrm{dl}(\Sigma, \{x\})$ *is nonevasive with respect to* $\wedge_{\mathrm{dl}}$, *then exactly one simplex* σ *of* Σ *is evasive with respect to* $\wedge$, *given by* $\sigma = \tau \cup \{x\}$.
  (c) *If every simplex of* $\mathrm{lk}(\Sigma, \{x\})$ *is nonevasive with respect to* $\wedge_{\mathrm{lk}}$ *and exactly one simplex* τ *of* $\mathrm{dl}(\Sigma, \{x\})$ *is evasive with respect to* $\wedge_{\mathrm{dl}}$, *then exactly one simplex* σ *of* Σ *is evasive with respect to* $\wedge$, *given by* $\sigma = \tau$.

The following technical lemma regarding strategy complexes is also immediate:

LEMMA 10.6. *Let* $G = (V, \mathfrak{A})$ *be a stochastic graph,* $v \in V$, *and* $\tau \in \Delta_G$. *Suppose* $\mathcal{B}$ *is some collection of actions, all with source* v, *such that* $\tau \cup \{B\} \in \Delta_G$ *for each single action* $B \in \mathcal{B}$. *Then* $\tau \cup \mathcal{B} \in \Delta_G$.

We now strengthen the contractibility claim of Theorem 3.6:

COROLLARY 10.7. *Let* $G = (V, \mathfrak{A})$ *be a stochastic graph and* $s \in V$. *Suppose* G *does not contain a complete strategy for attaining* s. *Then* $\Delta_{G \leftarrow s}$ *is nonevasive.*

PROOF. Let $\mathcal{A}_{\not{s}}$ denote the set of all actions of $G_{\leftarrow s}$ except for those with source s. Define a partial decision tree for $\Delta_{G_{\leftarrow s}}$ by arranging the actions $\mathcal{A}_{\not{s}}$ in any order and recursively constructing $\mathrm{NODE}(\Sigma, A, \wedge_{\mathrm{lk}}, \wedge_{\mathrm{dl}})$, with A ranging over $\mathcal{A}_{\not{s}}$, starting from $\Sigma = \Delta_{G_{\leftarrow s}}$, until all actions of $\mathcal{A}_{\not{s}}$ have been used along every possible descent path. Consider the (yet to be instantiated) subtrees $\{\wedge_i\}$ at the frontier of this partial decision tree. The "lk" branches on the descent path from the overall decision tree for $\Delta_{G_{\leftarrow s}}$ to subtree $\wedge_i$ define a subset $\mathcal{A}_i$ of $\mathcal{A}_{\not{s}}$. Subtree $\wedge_i$ is a decision tree for a subcomplex Σ_i of $\Delta_{G_{\leftarrow s}}$, consisting of all simplices σ such that $\sigma \cap \mathcal{A}_{\not{s}} = \emptyset$ and $\sigma \cup \mathcal{A}_i \in \Delta_{G_{\leftarrow s}}$. Subcomplex Σ_i could be void and thus nonevasive. Otherwise, its zero-skeleton is some subset of the loopback actions. That subset is nonempty. (To see this, observe that $\mathcal{A}_i$ must be a simplex of $\Delta_{G_{\leftarrow s}}$ when Σ_i is nonvoid, but it cannot be a complete strategy for attaining s. As in the proof Theorem 3.6, $\mathcal{A}_i$ can therefore join with at least one loopback action.) By Lemma 10.6, Σ_i must in fact represent a full nonempty simplex, so is nonevasive. Lemma 10.5(a) then implies that $\Delta_{G_{\leftarrow s}}$ is nonevasive. □

The next lemma will be a useful stepping stone to two results. By a *minimal strategy* σ we mean a strategy with exactly one action at every state in $\mathrm{src}(\sigma)$.

LEMMA 10.8. *Let $G = (V, \mathfrak{A})$ be a stochastic graph and let $s \in V$. Suppose σ_0 is a minimal complete strategy for attaining s in G. Let $\mathcal{A}$ be some nonempty subset of $\mathfrak{A} \setminus \sigma_0$. Then $\mathrm{lk}(\Delta_{G_{\leftarrow s}}, \mathcal{A})$ is nonevasive.*

PROOF. By Lemma 10.5(a), along with Remarks 8.2(2), 9.2, and 10.4, the proof reduces to the case in which $\mathcal{A}$ is convergent and consists of *all* actions in $\mathfrak{A} \setminus \sigma_0$ that do not have source s.

Let $\tau_0 = \{B \in \sigma_0 \mid \mathrm{src}(B) \notin \mathrm{src}(\mathcal{A})\}$ and write $\tau_0 = \{B_1, \ldots, B_k\}$, for some $k \geq 0$ ($k = 0$ means $\tau_0 = \emptyset$). Again by Lemma 10.5(a), the proof further reduces to showing that the complexes $\mathrm{dl}(\mathrm{lk}(\Delta_{G_{\leftarrow s}}, \mathcal{A} \cup \{B_1, \ldots, B_{j-1}\}), \{B_j\})$, for $j = 1, \ldots, k$, and the complex $\mathrm{lk}(\Delta_{G_{\leftarrow s}}, \mathcal{A} \cup \tau_0)$ are all nonevasive. We will show that the last of these complexes, call it Σ, is nonevasive, leaving the first k complexes to the reader.

For every $v \in V \setminus \{s\}$, now let B_v designate the action of σ_0 with source v. By convergence of σ_0, the following system of equations has a unique finite solution:

$$(10.1) \qquad \begin{aligned} t_v &= \mathrm{adv}(B_v, \{t_u\}) + 1, \quad \text{for } v \in V \setminus \{s\}; \\ t_s &= 0. \end{aligned}$$

Observe that if B_v is nondeterministic, then $t_v > t_u$ for all targets u of B_v. If B_v is stochastic, then $t_v > t_u$ for at least one target u.

Let $t_0 = \min\{t_v \mid v \in \mathrm{src}(\mathcal{A})\}$, let v_0 be a state in $\mathrm{src}(\mathcal{A})$ at which the minimum t_0 is attained, and let C_0 be the action of σ_0 with source v_0. (At least one v_0 exists since $\mathcal{A} \neq \emptyset$. If more than one v_0 is possible, pick any one.)

We will show that Σ, if not void, is a cone with apex C_0. Thus Σ is nonevasive. Let $\sigma \in \Sigma$ with $C_0 \notin \sigma$. Define $\sigma' = \sigma \cup \{C_0\}$. By construction, $\sigma \cap (\mathcal{A} \cup \tau_0) = \emptyset$ and $\sigma \cup \mathcal{A} \cup \tau_0 \in \Delta_{G_{\leftarrow s}}$. By definition, $C_0 \notin \mathcal{A}$. Since $\mathrm{src}(C_0) \in \mathrm{src}(\mathcal{A})$, $C_0 \notin \tau_0$. So $\sigma' \cap (\mathcal{A} \cup \tau_0) = \emptyset$. We need to establish that $\sigma' \cup \mathcal{A} \cup \tau_0 \in \Delta_{G_{\leftarrow s}}$, for then $\sigma' \in \Sigma$.

Suppose $\sigma' \cup \mathcal{A} \cup \tau_0$ is not convergent. Then there is some nonempty set of actions $\gamma \subseteq \sigma' \cup \mathcal{A} \cup \tau_0$ such that no action of γ moves off $\mathrm{src}(\gamma)$. Necessarily, $C_0 \in \gamma$. Some target u of C_0 with $t_0 > t_u$ lies in $\mathrm{src}(\gamma)$. Let $v_1 = u$, $t_1 = t_u$, and let C_1 be an action of γ with source u. Suppose $u \neq s$. By definition of t_0, this means C_1 must be an action of σ_0. Now repeat the construction. We obtain a sequence of distinct states $v_0, v_1, \ldots, v_i$, all in $\mathrm{src}(\gamma)$. By finiteness, we must eventually find that $v_i = s$. So $s \in \mathrm{src}(\gamma)$. On the other hand, observe that $\mathrm{src}(\mathcal{A} \cup \tau_0) = V \setminus \{s\}$, implying $\mathrm{src}(\sigma \cup \mathcal{A} \cup \tau_0) = V$. That is a contradiction, since $\sigma \cup \mathcal{A} \cup \tau_0$ is convergent. □

REMARK 10.9. The method of decision trees is a combinatorial approach for inferring topology, seemingly different from the covering set approach we used earlier. A connection to covering sets appears via System (10.1).

We can now strengthen as well Lemma 9.5(a):

COROLLARY 10.10. *Let $G = (V, \mathfrak{A})$ be a stochastic graph and let $s \in V$. Suppose G contains a complete strategy for attaining s. Let $\mathcal{A}$ be a nonempty subset of $\mathfrak{A}$ that is not essential for attaining s in G. Then $\mathrm{lk}(\Delta_{G_{\leftarrow s}}, \mathcal{A})$ is nonevasive.*

The following result is useful in a decision-tree-theoretic proof that loopback complexes are spheres when complete strategies exist:

COROLLARY 10.11. *Let $G = (V, \mathfrak{A})$ be a stochastic graph and let $s \in V$. Suppose σ_0 is a minimal complete strategy for attaining s in G. There exists a decision tree for $\Delta_{G_{\leftarrow s}}$ with exactly one evasive simplex, given by σ_0.*

PROOF. By using Lemma 10.5(c) and specializing Lemma 10.8 to cases in which $\mathcal{A}$ consists of a single action, we reduce to the case in which $\mathfrak{A} = \sigma_0$. Writing $\sigma_0 = \{A_1, \ldots, A_{n-1}\}$, with $n = |V|$, we may construct a partial decision tree for $\Delta_{G \leftarrow s}$ whose frontier consists of decision trees for the following n complexes:

$$\Sigma_i = \mathrm{dl}(\mathrm{lk}(\Delta_{G \leftarrow s}, \{A_1, \ldots, A_{i-1}\}), A_i), \quad \text{for } i = 1, \ldots, n-1;$$
$$\Sigma_n = \mathrm{lk}(\Delta_{G \leftarrow s}, \{A_1, \ldots, A_{n-1}\}).$$

For $i = 1, \ldots, n-1$, complex Σ_i is a cone with apex given by the loopback action $s \to \mathrm{src}(A_i)$. Complex Σ_n is the empty complex. Repeated application of Lemma 10.5(b) finishes the proof. □

We may now obtain the sphere result of Theorem 3.6 using decision trees:

COROLLARY 10.12. *Let $G = (V, \mathfrak{A})$ be a stochastic graph, $s \in V$, and $n = |V|$. If G contains a complete strategy for attaining s, then $\Delta_{G \leftarrow s} \simeq \mathbb{S}^{n-2}$.*

PROOF. We may assume that $n > 1$, as otherwise the claim is trivially true. Now recall the following general result from discrete Morse theory [18]:

Suppose $\wedge$ is a decision tree for a simplicial complex Σ, such that the empty simplex is nonevasive with respect to $\wedge$. Then Σ is homotopy equivalent to a CW complex consisting of exactly one p-cell for every p-dimensional simplex of Σ that is evasive with respect to $\wedge$, along with one additional 0-cell. In particular, if Σ is nonempty and admits a decision tree with exactly one evasive simplex, then the associated CW complex consists of a 0-cell and a k-cell, with k the dimension of the evasive simplex. The complex Σ is therefore homotopic to a sphere of dimension k.

In our case, Corollary 10.11 produces exactly one evasive simplex. That simplex has dimension $n - 2$. □

11. Category Connections

This section explores connections between strategy complexes and category theory. For an introduction to category theory, see [2]. For more advanced treatments, see [36, 23, 48]. Then recall that the *nerve* of a small category is the simplicial set whose simplices are the diagrams of composable morphisms and that the *classifying space* of a small category is the geometric realization of its nerve. A strategy complex looks almost like the nerve of some category, particularly since actions look like morphisms. Moreover, any finite simplicial complex is a small category, via its face poset. In that setting, categorical nerve amounts to barycentric subdivision.

These observations suggest viewing strategy and source complexes as homeomorphic to the classifying spaces of planning processes, each described by a poset. Informally, "Plans are the nerve of planning." This section explores the foundations for that statement. We focus on nondeterministic graphs and give a sampling of key results. The constructions extend to stochastic graphs, with some technical modifications to account for probabilities and quantification differences.

11.1. Each Graph as a Category. We may view a given nondeterministic graph $G = (V, \mathfrak{A})$ as a category in several distinct but related ways:

(1) We *view G directly as a category* as follows: The objects are the individual states v of V plus every subset T of V that is the target set of some action in $\mathfrak{A}$. The morphisms are the actions $\mathfrak{A}$ plus all required identities. There are no compositions

except between an action $v \to T$ of $\mathfrak{A}$ and identities at v and T. In particular, here we view a source state v and a singleton target set $\{v\}$ as different objects.

(2) We *view G as a category with subsets* as follows: Much as in (1), but with additional objects and morphisms. The objects now are the individual states v plus *all* nonempty subsets S of target sets T. The morphisms now are *all* labeled arrows $v \xrightarrow{A} S$, with $A \in \mathfrak{A}$, $A = v \to T$, and $\emptyset \neq S \subseteq T$, plus all required identities.

(3) We *view G as a category with supersets* as follows: Much as in (2), except that we include supersets S of T instead of subsets.

Intuitively, one might interpret the original G in category (2) as an upper bound on an adversary's actual choices. One might interpret the original G in category (3) as a benign instantiation of more evil adversaries.

11.2. All Graphs as a Category. We may view all nondeterministic graphs as a category, again in several distinct but related ways. The objects in all cases are the graphs themselves. We include the null graph; it is an initial object in each category. A morphism $f : G \to H$ between two nondeterministic graphs is simultaneously a function on states and on actions that induces a functor from G to H, viewed as categories via Section 11.1. Composition of morphisms is composition of the underlying functions. To be explicit, we write out the analogue for case (2):

(2) In the *category of nondeterministic graphs with cycle-preserving morphisms*, a morphism $f : G \to H$, with $G = (V, \mathfrak{A})$ and $H = (W, \mathfrak{B})$, is simultaneously a function $f : V \to W$ and a function $f : \mathfrak{A} \to \mathfrak{B}$ such that: if $v \to S \in \mathfrak{A}$ and $f(v \to S) = w \to T \in \mathfrak{B}$, then $f(v) = w$ and $f(S) \subseteq T$. One can verify that identities and composition are well-defined in this category. The name for this category comes from Lemma 11.1 below.

(3) The *category of nondeterministic graphs with goal-preserving morphisms* is much like (2), except that we replace subset with superset. The name for this category comes from Lemma 11.2 below.

LEMMA 11.1. *If $f : G \to H$ is a cycle-preserving morphism, then $f(\mathcal{A})$ contains a circuit in H whenever $\mathcal{A}$ contains a circuit in G.*

PROOF. Suppose $\mathcal{A}$ contains a sequence of actions $v_1 \to S_1, \ldots, v_k \to S_k$, such that $v_{i+1} \in S_i$, for $i = 1, \ldots, k$, with $k \geq 1$ and $k+1$ meaning 1. For each i, $f(v_i \to S_i) = w_i \to T_i$ (some action of H), with $f(v_i) = w_i$ and $f(S_i) \subseteq T_i$. So $w_{i+1} \in T_i$, telling us $f(\mathcal{A})$ contains a circuit in H. $\square$

LEMMA 11.2. *If $f : G \to H$ is a goal-preserving morphism, with $G = (V, \mathfrak{A})$ and $H = (W, \mathfrak{B})$, and if Z is certainly attainable in G, then $f(Z) \cup (W \setminus f(V))$ is certainly attainable in H.*

PROOF. We know $V \setminus Z \in \overline{\Delta}_G$. We need to show that $f(V) \setminus f(Z) \in \overline{\Delta}_H$. Using Lemma 6.3, it is enough to show that for every nonempty subset Y of $f(V) \setminus f(Z)$, some action of H moves off Y. Pick any such Y and let $X = V \setminus f^{-1}(f(V) \setminus Y)$. Observe that $\emptyset \neq X \subseteq V \setminus Z$, so $X \in \overline{\Delta}_G$. By Lemma 6.3, some action $v \to S$ in $\mathfrak{A}$ moves off X. In particular, $v \in X$ and $S \subseteq V \setminus X$. Since f is a goal-preserving morphism, $f(v \to S)$ is an action $w \to T$ in $\mathfrak{B}$, with $w = f(v) \in f(X) \subseteq Y$ and $f(S) \supseteq T$. Thus $T \subseteq f(V \setminus X) \subseteq f(V) \setminus Y$. So $f(v \to S)$ moves off Y. $\square$

11.3. Functoriality of Strategy Complexes. Let JOIN$^+$ be the following category: The objects are finite join semi-lattices with two restrictions: (1) Each semi-lattice must contain a distinguished top element $\hat{1}$. (2) A semi-lattice with more than one element must contain a distinguished bottom element $\hat{0}$. Morphisms are functions between semi-lattices viewed as sets, that further respect the join structure, send $\hat{1}$ to $\hat{1}$, and send $\hat{0}$ to $\hat{0} \neq \hat{1}$ whenever $\hat{0} \neq \hat{1}$ exists in the domain.

COMMENT: Let J, K be objects of JOIN$^+$, with $|J| = 1$ and $|K| > 1$. Then JOIN$^+$ contains identity morphisms at J and K, a unique morphism $J \to K$, but *no* morphism $K \to J$.

DEFINITION 11.3. Given a nondeterministic graph G, we may view its strategy complex Δ_G as an object $L(G)$ in JOIN$^+$ by adjoining a top element $\hat{1}$. Intuitively, $\hat{1}$ represents circuits. The join operation is defined by saying that $\tau \vee \sigma = \tau \cup \sigma$ whenever τ, σ, and $\tau \cup \sigma$ are all simplices of Δ_G, and is $\hat{1}$ otherwise.

COMMENT: If G is null, then $L(G)$ consists just of $\hat{1}$. Otherwise, $L(G)$ also contains $\hat{0} = \emptyset \in \Delta_G$, distinct from $\hat{1}$.

LEMMA 11.4. *The assignment operator L from Def. 11.3 is a functor from the category of nondeterministic graphs with cycle-preserving morphisms to* JOIN$^+$.

PROOF. To define the functor, let $L(G)$ be as in Def. 11.3, then extend to morphisms as follows: If $f : G \to H$ is a cycle-preserving morphism, define the morphism $L(f) : L(G) \to L(H)$ in JOIN$^+$ as follows:

$$L(f)(\hat{1}) = \hat{1};$$
$$L(f)(\tau) = f(\tau), \quad \text{if } \tau \in \Delta_G \text{ and } f(\tau) \in \Delta_H;$$
$$L(f)(\tau) = \hat{1}, \quad \text{otherwise.}$$

We now assume that all graphs are non-null; the null cases are straightforward.

To verify that $L(f)$ is a morphism, first observe that $L(f)$ sends $\hat{1}$ to $\hat{1}$ and $\hat{0}$ to $\hat{0}$ (since $\hat{0} = \emptyset$). Moreover, if x and y are elements in $L(G)$, with at least one of them being $\hat{1}$, then $L(f)(x \vee y) = \hat{1} = L(f)(x) \vee L(f)(y)$. So suppose $\tau, \sigma \in \Delta_G$. If $\tau \cup \sigma \notin \Delta_G$, then $f(\tau \cup \sigma) \notin \Delta_H$, by Lemma 11.1. So $L(f)(\tau \vee \sigma) = L(f)(\hat{1}) = \hat{1} = L(f)(\tau) \vee L(f)(\sigma)$. The last equality holds either because one of the terms is $\hat{1}$ already or because $f(\tau) \cup f(\sigma) = f(\tau \cup \sigma) \notin \Delta_H$. If $\tau \cup \sigma \in \Delta_G$ but $f(\tau \cup \sigma) \notin \Delta_H$, then $L(f)(\tau \vee \sigma) = L(f)(\tau \cup \sigma) = \hat{1} = L(f)(\tau) \vee L(f)(\sigma)$, with the last equality holding for the same reasons as before. Finally, if $\tau \cup \sigma \in \Delta_G$ and $f(\tau \cup \sigma) \in \Delta_H$, then $L(f)(\tau \vee \sigma) = f(\tau \cup \sigma) = f(\tau) \cup f(\sigma) = L(f)(\tau) \vee L(f)(\sigma)$.

To verify that L is a functor:

Identities. If $G \xrightarrow{i} G$ is an identity morphism, then $L(G) \xrightarrow{L(i)} L(G)$ is as well.

Composition. Suppose $G \xrightarrow{g} H \xrightarrow{h} K$. Consider $L(G) \xrightarrow{L(g)} L(H) \xrightarrow{L(h)} L(K)$. $L(h \circ g)$ and $L(h) \circ L(g)$ both send $\hat{1}$ to $\hat{1}$ and $\hat{0}$ to $\hat{0}$. Let $\tau \in \Delta_G$. If $g(\tau) \notin \Delta_H$, then $h(g(\tau)) \notin \Delta_K$ by Lemma 11.1. Thus $L(h \circ g)(\tau) = \hat{1}$ and $(L(h) \circ L(g))(\tau) = L(h)(L(g)(\tau)) = L(h)(\hat{1}) = \hat{1}$. If $g(\tau) \in \Delta_H$ but $h(g(\tau)) \notin \Delta_K$, then $L(h \circ g)(\tau) = \hat{1}$ and $(L(h) \circ L(g))(\tau) = L(h)(g(\tau)) = \hat{1}$. Finally, if $g(\tau) \in \Delta_H$ and $h(g(\tau)) \in \Delta_K$, then $L(h \circ g)(\tau) = h(g(\tau)) = (L(h) \circ L(g))(\tau)$, by definition of $\circ$. □

REMARK 11.5. The join structure shows how Δ_G is homeomorphic to a planner's classifying space. Imagine a forward-chaining planner that takes unions of convergent sets of actions, retaining only those unions that remain convergent. This planner defines a category whose objects are the nonempty convergent sets of actions. The morphisms are induced by other sets of convergent actions, which may be adjoined without creating a circuit. The categorical nerve of this category is $\text{sd}(\Delta_G)$.

11.4. Functoriality of Source Complexes. It is harder to see source complexes as functors. The reason is that source complexes discard information about precise paths and targets attained, retaining only the start regions of strategies. For functoriality, some contextual information appears to be necessary. This subsection explores one possible approach.

DEFINITION 11.6. Let $G = (V, \mathfrak{A})$ be a nondeterministic graph. If $\mathcal{A}$ is any set of actions in $\mathfrak{A}$, let $\text{trg}(\mathcal{A})$ denote all the targets of those actions. Formally, $\text{trg}(\mathcal{A}) = \bigcup_{v \to T \in \mathcal{A}} T$.

DEFINITION 11.7. Let $G = (V, \mathfrak{A})$ be a nondeterministic graph. Define the *poset of local strategies of* G, denoted $P(G)$, to consist of all pairs (W, X) such that: (i) $X \subseteq W \subseteq V$, and (ii) there exists a $\sigma \in \Delta_G$ with $\text{src}(\sigma) = W \setminus X$ and $\text{trg}(\sigma) \subseteq W$. Define a partial order on $P(G)$ by $(U, Y) \leq (W, X)$ if and only if $X \subseteq Y \subseteq U \subseteq W$.

REMARKS 11.8. (1) If X is a nonempty proper subset of W, then (W, X) is in $P(G)$ precisely when G contains a nonempty strategy for attaining goal set X from start region $W \setminus X$, moving wholly within *ambient subspace* W. (2) An element of $P(G)$ "less" than another has a reduced ambient subspace and/or a looser goal set. (3) If G is the null graph, then $P(G)$ is the empty poset. Otherwise, $P(G)$ includes, for each subset W of G's state space, the trivial element (W, W).

LEMMA 11.9. *The assignment operator P from Def. 11.7 is a functor from the category of nondeterministic graphs with goal-preserving morphisms to the category of finite posets.*

PROOF. To define the functor, let $P(G)$ be as in Def. 11.7, then extend to morphisms as follows: If $f : G \to H$ is a goal-preserving morphism, define the poset morphism $P(f) : P(G) \to P(H)$ by $P(f)(W, X) = (f(W), f(X))$. One should verify that: (a) $P(f)(W, X) \in P(H)$ whenever $(W, X) \in P(G)$, (b) $P(f)$ is a poset morphism, and (c) P preserves identities and composition.

We will prove (a) and leave verification of (b) and (c) to the reader.

Let $(W, X) \in P(G)$. So there exists $\sigma \in \Delta_G$ such that $\text{src}(\sigma) = W \setminus X$ and $\text{trg}(\sigma) \subseteq W$. Consider any $v \to S \in \sigma$. Let $u \to T = f(v \to S)$. Since f is goal-preserving, $T \subseteq f(S) \subseteq f(W)$. So, we see that $\text{trg}(f(\sigma)) \subseteq f(W)$.

Now let U be any nonempty subset of $f(W) \setminus f(X)$. We will show that some action of $f(\sigma)$ moves off this set. As a result, by applying Lemma 6.3 repeatedly, we see that there is some $\tau \in \Delta_H$ for which $\text{src}(\tau) = f(W) \setminus f(X)$ and $\text{trg}(\tau) \subseteq f(W)$.

Let $Y = W \setminus f^{-1}(f(W) \setminus U)$. Observe $\emptyset \neq Y \subseteq W \setminus X$. So σ contains an action $v \to S$ that moves off Y. Let $u \to T = f(v \to S)$. Then $u = f(v) \in f(Y) \subseteq U$ and $T \subseteq f(S) \subseteq f(W \setminus Y) \subseteq f(W) \setminus U$. This says the action $u \to T$ moves off U. □

DEFINITION 11.10. Given a nondeterministic graph $G = (V, \mathfrak{A})$ and a nonempty subset W of V, let $G|W = (W, \mathfrak{A}_W)$ with $\mathfrak{A}_W = \{v \to T \in \mathfrak{A} \mid v \in W \text{ and } T \subseteq W\}$. So $G|W$ consists of state space W and all actions of G whose motions lie within W. Also let $P|W = \{(W, X) \in P(G)\}$ (W fixed, X varying), inheriting $P(G)$'s partial order. So $P|W$ models the certainly attainable goals of $G|W$, suggesting the next lemma.

LEMMA 11.11. *Let $G = (V, \mathfrak{A})$ be a nondeterministic graph. Suppose $\emptyset \neq W \subseteq V$. Then $P|W$ and $\overline{\Delta}_{G|W}$ are isomorphic when viewed as partially ordered sets.*

PROOF. Define $P|W \xrightarrow{f} \overline{\Delta}_{G|W}$ and $\overline{\Delta}_{G|W} \xrightarrow{g} P|W$ by $f(W, X) = W \setminus X$ and $g(Y) = (W, W \setminus Y)$. The reader may verify that f and g are well-defined and that they are poset maps. They are inverses. □

11.5. A Planner's Classifying Space. Lemma 11.11 suggests how one may view $\overline{\Delta}_{G|W}$ as homeomorphic to a planner's classifying space. In particular, $P|W$ defines a poset category which one may interpret as a planner's search space. The category's objects are the elements of $P|W$. The category's morphisms are arrows of the form $(W, Y) \to (W, X)$, exactly one such arrow for each comparison $(W, Y) \leq (W, X)$ in $P|W$. This subsection elaborates the interpretation of $P|W$ as a planner.

LEMMA 11.12. *Let $G = (V, \mathfrak{A})$ be a nondeterministic graph. Suppose $\emptyset \neq X \subseteq W \subseteq V$. Then $(W, X) \in P(G)$ if and only if there exists a sequence of triples $(X_0, A_0, v_0), \ldots, (X_{k-1}, A_{k-1}, v_{k-1})$, with $k \geq 0$, such that:*

(i) $W = X_0 \supset X_1 \supset \cdots \supset X_{k-1} \supset X_k \stackrel{\text{def}}{=} X$;
(ii) $X_j \setminus X_{j+1} = \{v_j\}$, *for* $j = 0, \ldots, k - 1$;
(iii) A_j *is an action of $G|W$ with source v_j that moves off $W \setminus X_{j+1}$, for $j = 0, \ldots, k - 1$.*

PROOF. The lemma holds when $W = X$, by letting $k = 0$, so we may assume that X is a (nonempty) proper subset of W.

I. Suppose $(W, X) \in P(G)$. Let $\sigma \in \Delta_G$ be a minimal strategy satisfying $\text{src}(\sigma) = W \setminus X$ and $\text{trg}(\sigma) \subseteq W$. In particular, all actions of σ are actions of $G|W$. Now backchain from X, starting by letting $k = |W \setminus X|$ and $X_k = X$. Inductively, suppose $X_{j+1} \supset \cdots \supset X_k$ (and the corresponding triples) have been defined, for some $j \in \{0, \ldots, k - 1\}$. Since σ is convergent, some action A_j of σ must move off $W \setminus X_{j+1}$. Let $v_j = \text{src}(A_j)$ and $X_j = X_{j+1} \cup \{v_j\}$. Decrement j by one. Repeat this process until reaching (X_0, A_0, v_0), at which point $X_0 = W$.

II. Suppose the specified sequence of triples exists. Then the set of actions $\sigma = \{A_0, \ldots, A_{k-1}\}$ is convergent. Moreover, $\text{src}(\sigma) = W \setminus X$ and $\text{trg}(\sigma) \subseteq W$. So $(W, X) \in P(G)$. □

REMARK 11.13. Viewing $P|W$ as a poset category, Lemma 11.12 says that for every object (W, X) in this category, there is a diagram of composable morphisms

(11.1) $\quad (W, W) = (W, X_0) \to (W, X_1) \to \cdots \to (W, X_k) = (W, X),$

with the $\{X_i\}$ satisfying the conditions stated in the lemma. This type of diagram defines a forward-chaining planner, as described next.

The planner starts its search at (W, W) and thereafter regards (W, W) as *visited*. The planner iterates as follows:

(a) Suppose the planner has visited $(W, Z) \in P|W$.
(b) Suppose there is some state $z \in Z$ and some action A of $G|W$ such that A has source z and A moves off $W \setminus Z'$, with $Z' = Z \setminus \{z\}$.
(c) This means (W, Z') is an object in $P|W$ and thus $(W, Z) \to (W, Z')$ is a morphism in $P|W$. If (W, Z') has not yet been visited, the planner *traverses the arrow* and visits (W, Z').

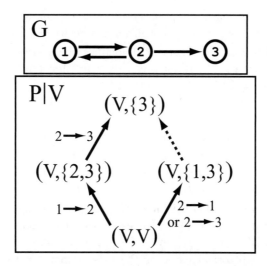

FIGURE 18. Top panel: A graph with state space $V = \{1, 2, 3\}$. Bottom panel: The category $P|V$, drawn with its four generating morphisms. (Not shown are four identity morphisms and the morphism $(V, V) \to (V, \{3\})$ arising from composition.) Three of the four morphisms (solid arrows) are labeled by the actions that the forward-chaining planner of Remark 11.13 might produce. The fourth morphism (dashed arrow) amounts to a strategy switch; the planner would not traverse this arrow merely by forward-chaining, since no action of G with source 1 moves off $\{1, 2\}$.

The planner loops over all possible choices of z and A at all visited objects, visiting more objects in the process, until it has visited every possible object in $P|W$. Some further comments:

(1) The conclusion that $(W, Z') \in P(G)$ in step (c) is independent of whatever strategy establishes that $(W, Z) \in P(G)$. This observation means that the planner needs to keep track merely of the objects (W, X) it encounters not of the actions leading to those objects.

(2) If desired, one may also view the planner as an output device. As such, the planner reports each arrow $(W, Z) \to (W, Z')$ that it traverses, labeled with the action A found in step (b). Implicitly, this linear output defines a strategy for attaining X from W for each object (W, X) visited by the planner. For any given (W, X), one may recover that strategy by scanning the output in reverse. (Of course, backchaining directly in G would be more efficient.)

(3) The category $P|W$ contains non-identity morphisms beyond the arrows traversed by the basic forward-chaining search just described, either because an arrow points to a previously visited object or because no action can label the arrow. See Figure 18. Viewed as planning operations, these additional morphisms constitute *strategy switches*: If the planner discovers both (W, X) and (W, Y) via forward-chaining, with $X \subset Y$, then the planner may, at least implicitly, traverse the arrow $(W, Y) \to (W, X)$,

thereby instantiating a preference for strategies with tighter goals over those with looser goals. (Morphism composition is a particular instance.)

SUMMARY: The poset category $P|W$ defines the search space of a forward-chaining planner augmented with strategy switches that improve goal attainment. The objects (W, X) of $P|W$ constitute the planner's search states; the morphisms $(W, Y) \to (W, X)$ constitute the planner's possible state transitions.

COROLLARY 11.14. *Let $G = (V, \mathfrak{A})$ be a nondeterministic graph. Suppose $\emptyset \neq W \subseteq V$. Let $P^\circ|W$ be the poset subcategory of $P|W$ formed by removing the object (W, W). Then the categorical nerve of $P^\circ|W$ is isomorphic to $\mathrm{sd}(\overline{\Delta}_{G|W})$. Consequently, $\overline{\Delta}_G$ is homeomorphic to the classifying space of a category which one may interpret as a planner for finding all nontrivial certainly attainable goals in G.*

PROOF. By Lemma 11.11, $P|W$ and $\overline{\Delta}_{G|W}$ are isomorphic posets. The bottom elements, (W, W) and $\emptyset$, respectively, correspond via this isomorphism. As a result, $P^\circ|W$ (which is $P|W \setminus \{(W, W)\}$) and $\overline{\Delta}_{G|W} \setminus \{\emptyset\}$ are isomorphic poset categories, implying that the categorical nerve of $P^\circ|W$ is isomorphic to $\mathrm{sd}(\overline{\Delta}_{G|W})$. Using $W = V$ in Remark 11.13, then dropping the planner's trivial initial state (V, V), establishes the lemma's planning assertion. □

REMARK 11.15. The diagram of morphisms (11.1) gives rise to a $(k-1)$-simplex in $P^\circ|W$'s categorical nerve, of the form $(W, X_1) \to \cdots \to (W, X_k)$, with $k \geq 1$.

The following lemma provides an alternative perspective:

LEMMA 11.16. *Let $G = (V, \mathfrak{A})$ be a nondeterministic graph. Suppose $\emptyset \neq W \subseteq V$. Then $(W, X_1) < (W, X_2) < \cdots < (W, X_k)$ is a chain in $P(G)$ if and only if there exist minimal strategies $\sigma_1 \subset \sigma_2 \subset \cdots \subset \sigma_k$ in $\Delta_{G|W}$, with all inclusions proper, such that $\mathrm{src}(\sigma_j) = W \setminus X_j$, with $X_j \subseteq W$, for $j = 1, \ldots, k$.*

PROOF. I. Suppose $(W, X_1) < (W, X_2) < \cdots < (W, X_k)$ is a chain in $P(G)$. Let $\sigma_k \in \Delta_G$ be a minimal strategy satisfying $\mathrm{src}(\sigma_k) = W \setminus X_k$ and $\mathrm{trg}(\sigma_k) \subseteq W$. So $\sigma_k \in \Delta_{G|W}$. For $j = 1, \ldots, k-1$, let $\sigma_j = \{A \in \sigma_k \mid \mathrm{src}(A) \in W \setminus X_j\}$. Observe that $\mathrm{src}(\sigma_j) = W \setminus X_j$ since $X_k \subset X_j$. Moreover, σ_j is minimal since σ_k is minimal. By definition of $P(G)$, $X_j \subseteq W$. Finally, $\sigma_1 \subset \sigma_2 \subset \cdots \subset \sigma_k$ since $X_1 \supset X_2 \supset \cdots \supset X_k$, with all inclusions proper.

II. Suppose the specified strategies exist. Then $(W, X_j) \in P(G)$, for $j = 1, \ldots, k$. Since the strategies are minimal, proper inclusion of strategies implies proper inclusion of start regions. Consequently, $(W, X_1) < \cdots < (W, X_k)$. □

REMARK 11.17. Consider a planner that constructs all possible strategies of $\Delta_{G|W}$ in the manner described by Remark 11.5. Suppose further that the planner reports the status of its search not by outputting full strategies but simply their start regions, along with corresponding arrows. Passing to complements in W, Lemma 11.16 tells us that this planner will traverse exactly all the arrows of the category $P|W$ (some arrows perhaps more than once). Omitting again the trivial object of $P|W$, one may therefore view $\overline{\Delta}_{G|W}$ as homeomorphic to the classifying space of a category derived from a planner. The difference between this planner and that described in Remark 11.13 lies in each planner's search space. The planner of Remark 11.13 never needs to remember actions; its search space truly is $P|W$. In contrast, the planner suggested by Lemma 11.16 operates in the join semi-lattice $L(G|W)$, then produces $P|W$ as a trace.

12. Discussion

12.1. Uncertainty, Geometry, and Topology. This research shows how the following are equivalent topologically: (a) the convergent sets of motions in finite graphs with control uncertainty, (b) finite simplicial complexes, (c) certain families of polyhedral cones in $\mathbb{R}^n$. Similar results were known for braid arrangements in $\mathbb{R}^n$ via earlier work on directed graph complexes and partially ordered sets [7, 8, 26, 47, 27]. Those ideas extend readily from directed graphs to graphs with nondeterministic transitions. The ideas extend as well to graphs with stochastic transitions, by allowing the hyperplanes comprising the $\mathbb{R}^n$ arrangements to rotate more freely about the line $\{x_1 = \cdots = x_n\}$. The usefulness of these hyperplanes is their ability to cast geometrically the expected convergence time equations of adversarially chosen Markov chains. That geometry provides a stepping stone to topology via the Nerve Lemma.

The significance of the step to topology is in showing how to infer global system capabilities from local graph connectivity. Traditionally, in order for a system with uncertainty to know that it can attain a goal, it tries to exhibit a strategy for doing so. Exhibiting a strategy entails combining uncertain actions in a manner that converges to the desired goal. Our theorems provide an alternative: instead of creating a specific strategy, one merely needs to show that the available actions cover a sphere. Exhibiting a specific strategy is one way to cover a sphere, but not the only way. For instance, imagine a collection of strategies $\{\sigma_v\}$, such that σ_v converges to goal s in time $\mathcal{T}$ when started from state v. The strategies need not be consistent with each other; their union may contain a circuit or simply take too long. Nonetheless, we know there is some strategy $\sigma \subseteq \bigcup_v \sigma_v$ that will converge to goal s in time $\mathcal{T}$ from all relevant v, simply because the union of the original strategies along with loopbacks from s must cover a sphere of the correct dimension. Section 6.2 suggests related applications in system design.

12.2. Higher-Order Interactions. The compression of strategy complexes to source complexes discards detailed action information while preserving knowledge of the system's global capabilities. This permits higher-level reasoning about interactions with an adversary. In effect, the start regions of strategies now become much like actions, facilitating reasoning about time-varying goals and tactics. This perspective holds as well when there are prescribed motions, via the source complexes of links. An interesting direction for future exploration is to vary these links and see how the complexes vary. From a practical perspective, this may be useful in disaster preparation.

12.3. Computational Complexity. Our recent robotics paper [14] provides explicit algorithms for computing many of the structures presented in this chapter, such as the source complex and the fully controllable subgraphs of any stochastic graph. That paper also discusses several approaches for computing the strategy complex of a stochastic graph. The subroutine used within many of these algorithms is a form of backchaining, much as it appeared in several proofs throughout this chapter. The backchaining algorithm starts with a desired goal set S. It searches for some action A that moves off the complement $V \setminus S$, then enlarges the goal to a new subgoal S' by adding the source of A to S. The algorithm repeats this process, now with S' in place of S, enlarging to a new S', and so forth. If and when an enlarged subgoal S' engulfs all of V, the algorithm terminates successfully; the actions it found constitute a strategy for attaining the original goal S. Otherwise, for some S', the algorithm is unable to find an action that moves off $V \setminus S'$, meaning that no strategy exists for attaining the original goal S from all of V. The worst-case runtime complexity of this backchaining algorithm is $O(|V|^2|\mathfrak{A}|)$, with $G = (V, \mathfrak{A})$ being the underlying stochastic graph. Depending on the particular graph G and the particular goal S, faster versions may be possible.

While backchaining has low polynomial time complexity, computing the source or strategy complex of a graph may require exponential time. At first glance, one suspects a representational defect, since the number of simplices in a simplicial complex can be exponential in the size of the complex's zero-skeleton. However, the underlying reasons appear to be intrinsic to the questions we are investigating rather than merely an artifact of the methods. Observe that the maximal simplices of a graph's source complex correspond to the minimal certainly attainable goals in the graph (as complements with respect to the state space V). It turns out that the problem of finding the size of the smallest certainly attainable goal in a nondeterministic graph is *NP*-complete [**14**]. Consequently, it is unlikely that faster than exponential time algorithms exist for determining the global capabilities of an uncertain system (as via a source complex), in the worst case.

From a robotics perspective, discovering that a problem is *NP*-complete is good news. The problem is nontrivial enough to be interesting yet probably not so complicated as to be intractable in all settings. Indeed, in applications one may be fortunate to have small upper bounds on the sizes of the minimal certainly attainable goals. Backchaining then allows one to construct the maximal simplices of the source complex reasonably quickly. The maximal simplices fully determine the complex. Other practical efficiency improvements are possible. For instance, in some cases the dual complex has a compact representation, as for a fully controllable graph. More generally, one may find the fully controllable subgraphs of a stochastic graph in fairly low polynomial time [**14**]. The results of Section 7 then simplify the source and strategy complexes by collapsing each such subgraph to a single state.

12.4. Complex Structure and Future Work. Loopback complexes are highly specialized structures, as analysis via decision trees shows, yet source and strategy complexes can be fairly arbitrary. Full controllability appears as homotopy equivalence to a sphere of a particular dimension. Dually, subcomplexes arising from certain subgraphs in a fully controllable graph must be cones, as we saw in several proofs. Yet, we do not understand in and of itself what contractibility of a strategy complex implies. This gap, between specific structure on the one hand and almost arbitrary structure on the other, suggests a spectrum of potential results. Future research should further classify the homotopy types of uncertain systems in order to facilitate effective design. A robot encountering a novel scenario should be able to abstract from it a topological hash, index into a table of applicable strategies, then select one such strategy optimized with respect to attendant objectives, much like a robot hand grasping an object today will select from a collection of forces in the grasp's null space. The research discussed in this chapter provides one step in that direction.

Acknowledgments

The author is grateful to Ben Mann, Rob Ghrist, and the entire SToMP group for making this work possible, and to Afra Zomorodian for the opportunity to present the results at the Computational Topology Short Course during JMM2011.

References

1. A. V. Aho, J. E. Hopcroft, and J. D. Ullman, *The Design and Analysis of Computer Algorithms*, Addison-Wesley, Reading, MA, 1974.
2. S. Awodey, *Category Theory*, second ed., Oxford University Press, Oxford, 2010.
3. A. Barr, P. Cohen, and E. Feigenbaum (eds.), *The Handbook of Artificial Intelligence*, William Kaufmann, Los Altos, California, 1981–1989.
4. R. Bellman, *Dynamic Programming*, Princeton University Press, Princeton, 1957.
5. T. Bergquist, C. Schenck, U. Ohiri, J. Sinapov, S. Griffith, and A. Stoytchev, *Interactive object recognition using proprioceptive feedback*, Proc. IEEE IROS Workshop: Semantic Perception for Mobile Manipulation, 2009.
6. D. P. Bertsekas, *Dynamic Programming: Deterministic and Stochastic Models*, Prentice-Hall, Englewood Cliffs, N.J., 1987.
7. A. Björner, *Topological methods*, Handbook of Combinatorics (R. L. Graham, M. Grötschel, and L. Lovász, eds.), vol. II, Elsevier, Amsterdam, 1995, pp. 1819–1872.
8. A. Björner and V. Welker, *Complexes of directed graphs*, SIAM J. Discrete Math **12(4)** (1999), 413–424.
9. M. Brady, J. M. Hollerbach, T. L. Johnson, T. Lozano-Pérez, and M. T. Mason, *Robot Motion: Planning and Control*, MIT Press, Cambridge, MA, 1982.
10. V. de Silva and R. Ghrist, *Coordinate-free coverage in sensor networks with controlled boundaries via homology*, Intl. J. Robotics Research **25(12)** (2006), 1205–1222.
11. B. R. Donald, *Error Detection and Recovery in Robotics*, Lecture Notes in Computer Science, No. 336. Springer Verlag, Berlin, 1989.
12. M. A. Erdmann, *An exploration of nonprehensile two-palm manipulation*, Intl. J. Robotics Research **17(5)** (1998), 485–503.
13. _____, *Shape recovery from passive locally dense tactile data*, Robotics: The Algorithmic Perspective (The Third Workshop on the Algorithmic Foundations of Robotics) (P. K. Agarwal, L. E. Kavraki, and M. T. Mason, eds.), A K Peters, Natick, MA, 1998, pp. 119–132.
14. _____, *On the topology of discrete strategies*, Intl. J. Robotics Research **29(7)** (2010), 855–896.
15. M. A. Erdmann and M. T. Mason, *An exploration of sensorless manipulation*, IEEE J. Robotics and Automation **4(4)** (1988), 369–379.
16. R. S. Fearing, *Tactile sensing for shape interpretation*, Dexterous Robot Hands (S. T. Venkataraman and T. Iberall, eds.), Springer Verlag, New York, 1990, pp. 209–238.
17. W. Feller, *An Introduction to Probability Theory and Its Applications*, (revised printing) third ed., vol. 1, John Wiley & Sons, New York, 1968.
18. R. Forman, *Morse theory and evasiveness*, Combinatorica **20(4)** (2000), 489–504.
19. _____, *A user's guide to discrete Morse theory*, Séminaire Lotharingien de Combinatoire **48** (2002), B48c.
20. R. Ghrist and S. LaValle, *Nonpositive curvature and Pareto-optimal coordination of robots*, SIAM J. Control & Optimization **45(5)** (2006), 1697–1713.
21. R. Ghrist, J. O'Kane, and S. LaValle, *Computing Pareto optimal coordinations on roadmaps*, Intl. J. Robotics Research **24(11)** (2005), 997–1010.
22. R. Ghrist and V. Peterson, *The geometry and topology of reconfiguration*, Advances in Applied Mathematics **38(3)** (2007), 302–323.
23. P. G. Goerss and J. F. Jardine, *Simplicial Homotopy Theory*, Birkhäuser Verlag, Basel, 1999.
24. K. Y. Goldberg, *Orienting polygonal parts without sensors*, Algorithmica **10(2–4)** (1993), 201–225.
25. A. Hatcher, *Algebraic Topology*, Cambridge University Press, Cambridge, 2002.
26. A. Hultman, *Directed subgraph complexes*, Elec. J. Combinatorics **11(1)** (2004), R75.
27. J. Jonsson, *Simplicial complexes of graphs*, Ph.D. thesis, Department of Mathematics, KTH, Stockholm, Sweden, 2005.

28. S. Karlin and H. M. Taylor, *A Second Course in Stochastic Processes*, Academic Press, New York, 1981.
29. J.-C. Latombe, *Robot Motion Planning*, Kluwer Academic Publishers, Boston, 1991.
30. S. M. LaValle, *Planning Algorithms*, Cambridge University Press, New York, 2006.
31. M. Levoy, *The digital Michelangelo project*, Proc. Second Intl. Conf. on 3-D Digital Imaging and Modeling, 1999, pp. 2–11.
32. H. R. Lewis and C. H. Papadimitriou, *Elements of the Theory of Computation*, Prentice-Hall, Englewood Cliffs, New Jersey, 1981.
33. T. Lozano-Pérez, *The design of a mechanical assembly system*, Tech. Report AI-TR-397, S.M. thesis, MIT, Cambridge, MA, 1976.
34. T. Lozano-Pérez, J. L. Jones, E. Mazer, and P. A. O'Donnell, *HANDEY: A Robot Task Planner*, MIT Press, Cambridge, MA, 1992.
35. T. Lozano-Pérez, M. T. Mason, and R. H. Taylor, *Automatic synthesis of fine-motion strategies for robots*, Intl. J. Robotics Research **3(1)** (1984), 3–24.
36. J. P. May, *Simplicial Objects in Algebraic Topology*, University of Chicago Press, Chicago, 1967.
37. J. R. Munkres, *Elements of Algebraic Topology*, Addison-Wesley, Menlo Park, CA, 1984.
38. D. K. Pai, *Multisensory interaction: Real and virtual*, Robotics Research: The Eleventh International Symposium (P. Dario and R. Chatila, eds.), Springer Verlag, Berlin, 2005, pp. 489–498.
39. L. C. Paulson, *ML for the Working Programmer*, second ed., Cambridge University Press, Cambridge, 1996.
40. D. Quillen, *Higher algebraic K-theory: I*, Lecture Notes in Mathematics, vol. 341, Springer Verlag, Berlin, 1973, pp. 85–147.
41. _____, *Homotopy properties of the poset of nontrivial p-subgroups of a group*, Advances in Math. **28(2)** (1978), 101–128.
42. J. J. Rotman, *An Introduction to Algebraic Topology*, Springer Verlag, Berlin, 1988.
43. M. Shirai and A. Saito, *Parts supply in SONY's general-purpose assembly system SMART*, Japanese Journal of Advanced Automation Technology **1** (1989), 108–111.
44. E. H. Spanier, *Algebraic Topology*, McGraw-Hill, San Francisco, 1966.
45. S. A. Stansfield, *Robotic grasping of unknown objects: A knowledge-based approach*, Intl. J. Robotics Research **10(4)** (1991), 314–326.
46. R. H. Taylor, M. T. Mason, and K. Y. Goldberg, *Sensor-based manipulation planning as a game with nature*, Robotics Research: The Fourth International Symposium (R. Bolles and B. Roth, eds.), MIT Press, Cambridge, MA, 1988, pp. 421–429.
47. M. L. Wachs, *Poset Topology: Tools and Applications*, IAS/Park City Mathematics Institute, Summer 2004.
48. G. Warner, *Topics in Topology and Homotopy Theory*, University of Washington, Seattle, https://digital.lib.washington.edu/researchworks/handle/1773/2641, 1999.
49. S. Weinberger, *Computers, Rigidity, and Moduli: The Large-Scale Fractal Geometry of Riemannian Moduli Space*, Princeton University Press, Princeton, 2005.
50. D. E. Whitney, *Quasi-static assembly of compliantly supported rigid parts*, Journal of Dynamic Systems, Measurement, and Control **104(1)** (1982), 65–77.

SCHOOL OF COMPUTER SCIENCE, CARNEGIE MELLON UNIVERSITY, PITTSBURGH, PA 15213
E-mail address: me@cs.cmu.edu

Combinatorial Optimization of Cycles and Bases

Jeff Erickson

ABSTRACT. We survey algorithms and hardness results for two important classes of topology optimization problems: computing minimum-weight cycles in a given homotopy or homology class, and computing minimum-weight cycle bases for the fundamental group or various homology groups.

1. Introduction

Identification of topological features is an important subproblem in several geometric applications. In many of these applications, it is important for these features to be represented as compactly as possible.

For example, a common method for building geometric models is to reconstruct a surface from a set of sample points, obtained from range finders, laser scanners, or some other physical device. Gaps and measurement errors in the point cloud data induce errors in the reconstructed surface, which often take the form of spurious handles or tunnels. For example, the original model of *David*'s head constructed by Stanford's Digital Michelangelo Project [130] has 340 small tunnels, none of which are present in the original marble sculpture [101]. Because most surface simplification and parametrization methods deliberately preserve the topology of the input surface, topological noise must be identified, localized, and removed before these other algorithms can be applied [74, 101, 191, 193].

Another example arises in VLSI routing [46, 86, 129], map simplification [54, 81, 53], and graph drawing [70, 75]. Given a rough sketch of one or more paths in a planar environment with fixed obstacles—possibly representing roads or rivers near cities or other geographic features, or wires between components on a chip—we want to produce a topologically equivalent set of paths that are as short or as simple as possible, perhaps subject to some tolerance constraints.

Similar optimization problems also arise in higher-dimensional simplicial complexes. For example, several researchers model higher-order connectivity properties

---

2000 *Mathematics Subject Classification.* Primary 68W05, 68Q25; Secondary 57M05, 55U05.

Key words and phrases. computational topology, topological graph theory, algorithms.

Portions of this work were done while the author was visiting IST Austria. This work was partially supported by NSF grant CCF 09-15519.

of sensor networks using simplicial complexes. In one formulation, complete coverage of an area by a sensor network is indicated by the presence of certain nontrivial second homology classes in the associated simplicial complex [57, 58, 89, 178]; smaller generators indicate that fewer sensors are required for coverage. Moreover, if coverage is incomplete, holes in the network are indicated by nontrivial one-dimensional homology classes; localizing those holes makes repairing or routing around them easier.

Similar topological optimization problems arise in shape analysis [63, 64], low-distortion surface parametrization [99, 174, 184], probabilistic embedding of high-genus graphs into planar graphs [20, 128, 171], computing crossing numbers of graphs [117], algorithms for graph isomorphism [116], approximation of optimal traveling salesman tours [60, 88] and Steiner trees [16, 17, 18], data visualization and analysis [31, 32, 50], shape modeling [33], localization of invariant sets of differential equations [113], minimal surface computation [176, 71], image and volume segmentation [24, 92, 94, 119], and sub-sampling point cloud data for topological inference [154].

This survey gives an overview of algorithms and hardness results for four classes of topological optimization problems:

Optimal Homotopy Basis: Given a space Σ and a basepoint x, find an optimal set of loops whose homotopy classes generate the fundamental group $\pi_1(\Sigma, x)$.

Homotopy Localization: Given a cycle γ in a space Σ, find a shortest cycle in Σ that is homotopic to γ.

Optimal Homology Basis: Given a space Σ and an integer p, find an optimal set of p-cycles whose homology classes generate the homology group $H_p(\Sigma)$.

Homology Localization: Given a p-cycle c in a space Σ, find an optimal p-cycle in Σ that is homologous to c.

In the interest of finiteness, the survey is limited to *exact* and *efficient* algorithms for these problems; I do not even attempt to cover the vast literature on approximation algorithms, numerical methods, and practical heuristics for topological optimization. For similar reasons, the survey attempts only to give a high-level overview of the most important results; many crucial technical details are mentioned only in passing or ignored altogether.

I assume the reader is familiar with basic results in algebraic topology (cell complexes, surface classification, homotopy, covering spaces, homology, relative homology, Poincaré-Lefschetz duality) [103, 150, 175], graph algorithms (graph data structures, depth-first search, shortest paths, minimum spanning trees, NP-completeness) [52, 122, 179], and combinatorial optimization (flows, cuts, circulations, linear programming, LP duality) [4, 122, 166].

2. Input Assumptions

2.1. Combinatorial spaces. Our focus on efficient, exact algorithms necessarily limits the scope of the problems we can consider. With one exception, we consider only *combinatorial* topological spaces as input: finite cell complexes whose cells do not have geometry per se, but where each cell c has an associated non-negative *weight* $w(c)$. For simplicity of exposition, we restrict our attention to simplicial complexes; however, most of the algorithms we discuss can be applied with little or no modification to finite regular CW-complexes. For questions about homotopy, we consider only paths, loops, and cycles in the 1-skeleton; the weight (or 'length') of a cycle is the sum of the weights of its edges. For questions about homology, the output is either a subcomplex of the p-skeleton, whose weight is the sum of the weights of its cells, or a real or integer p-chain, whose cost is the weighted sum of its coefficients.

Unless noted otherwise, the time bounds reported here assume that the input complex is (effectively) represented by a sequence of boundary matrices, each stored explicitly in a standard two-dimensional array. (For sparse complexes, representing each boundary matrix using a sparse-matrix data structure leads to faster algorithms, at least in practice, especially in combination with simplification heuristics [10, 30, 56, 84, 113, 147, 194, 195].) For each index i, we let n_i denote the number of i-dimensional cells in the input complex, and we let n denote the total number of cells of all dimensions.

2.2. Combinatorial surfaces. For problems related to homotopy of loops and cycles, we must further restrict the class of input spaces to combinatorial 2-*manifolds*. Classical results of Markov and others [136, 137, 138] imply that most computational questions about homotopy are undecidable for general 2-complexes or manifolds of dimension 4 and higher. Thurston and Perelman's geometrization theorem [123, 145, 182] implies that the homotopy problems we consider are decidable in 3-manifolds [23, 139, 159]; however, the few explicit algorithms that are known are extremely complex, and no complexity bounds are known. The homology problems we consider are all decidable for arbitrary finite complexes, but (not surprisingly) they can be solved more efficiently in combinatorial 2-manifolds than in general spaces.

To simplify exposition, this survey explicitly considers only *orientable* surfaces; however, most of the results we describe apply to non-orientable surfaces with little or no modification.

The surface algorithms we consider are most easily described in the language of topological graph theory [90, 125, 144]. An *embedding* of an undirected graph G on an abstract 2-manifold Σ maps vertices of G to distinct points in Σ and edges of G to interior-disjoint curves in Σ. The *faces* of an embedding are maximal connected subsets of Σ that are disjoint from the image of the graph. An embedding is *cellular* (or *2-cell* [144]) if each of its faces is homeomorphic to an open disk. A combinatorial surface is simply a cellular embedding of a graph on an abstract 2-manifold.

Any cellular embedding of a graph on an orientable surface can be represented combinatorially by a *rotation system*, which encodes the cyclic order of edges around each vertex. To define rotation systems more formally, we represent each edge uv in G by a pair of directed edges or *darts* $u{\to}v$ and $v{\to}u$; we say that the dart $u{\to}v$ *leaves* u and *enters* v. A rotation system is a pair (σ, ρ) of permutations of the darts, such that $\sigma(u{\to}v)$ is the next dart leaving u after $u{\to}v$ in counterclockwise order around u (with respect to some fixed orientation of the surface), and ρ is the reversal permutation $\rho(u{\to}v) = v{\to}u$. The permutation $\sigma \circ \rho$ encodes the clockwise order of darts around each 2-cell.

Any cellularly embedded graph with n_0 vertices, n_1 edges, and n_2 faces lies on a surface Σ with Euler characteristic $\chi = n_0 - n_1 + n_2 = 2 - 2g$. Assuming without loss of generality that no vertex has degree less than 3, it follows that $n_1 = O(n_0 + g)$

and $n_2 = O(n_0 + g)$. Let $n = n_0 + n_1 + n_2$ denote the total complexity of the input surface.

Any graph G embedded on any surface Σ has an associated *dual graph* G^*, defined intuitively by giving each edge of G a clockwise quarter-turn around its midpoint. More formally, the dual graph G^* has a vertex f^* for each face f of G, and two vertices in G^* are joined by an edge e^* if and only if the corresponding faces of G are separated by an edge e. The dual graph G^* has a natural cellular embedding in the same surface Σ, determined by the rotation system $(\sigma \circ \rho, \rho)$. Each face v^* of this embedding corresponds to a vertex v of the primal graph G. Duality can be extended to the darts of G by defining $(u \to v)^* = (\ell^*) \to (r^*)$, where ℓ and r are respectively the faces on the left and right sides of $u \to v$. See Figure 1.

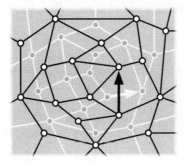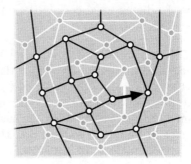

FIGURE 1. A portion of an embedded graph G and its dual G^*, with one dart and its dual emphasized.

For any subgraph H of G, we abuse notation by letting H^* denote the corresponding subgraph of the dual graph G^*, and letting $G \setminus H$ denote the subgraph of G obtained by deleting the *edges* of H. In particular, we have $(G \setminus H)^* = G^* \setminus H^*$.

For algorithms that act on combinatorial surfaces, we assume that the input is a data structure of size $O(n)$ that efficiently supports standard graph operations, such as enumerating neighbors of vertices in constant time each, both in the primal graph and in the dual graph. Several such data structures are known [**12, 100, 131, 148, 189**]. It is straightforward to determine the Euler characteristic, and therefore the genus, of a given surface in $O(n)$ time by counting vertices, edges, and faces.

To simplify our exposition, we implicitly assume that shortest vertex-to-vertex paths are unique in every weighted graph we consider. This assumption can be enforced automatically using standard perturbation techniques [**149**], but in fact, none of the algorithms we describe actually require this assumption.

2.3. Why not solve the real problem? The problems we consider are obviously well-defined for continuous geometric spaces, such as piecewise-linear or Riemannian surfaces (with some appropriate discrete or implicit representation). Unfortunately, at least with the current state of the art, these spaces permit only inefficient or approximate solutions.

Almost all the algorithms we describe here rely heavily on the ability to compute exact shortest paths. Shortest paths in combinatorial spaces can be computed in $O(n \log n)$ time using Dijkstra's algorithm in the 1-skeleton. In general Riemannian surfaces, shortest paths have no analytic representation, and therefore cannot be computed exactly even in principle. For piecewise-linear surfaces, existing shortest-path algorithms, both exact [**42, 133, 143, 158, 164, 165, 177**] and approximate [**1, 5**], are efficient only under the assumption that any shortest path crosses each edge of the input complex at most a constant number of times. (Some common numerical algorithms [**118, 170**] require even stronger geometric assumptions.) The crossing assumption is reasonable in practice—for example, it holds if the input complex is PL-embedded in some ambient Euclidean space (in which case any shortest path crosses any edge at most *once*), or if all face angles are larger than some fixed constant—but it does not hold in general.

As an elementary bad example, consider the piecewise-linear annulus defined by identifying the non-horizontal edges of the Euclidean trapezoid with vertices $(0,0)$, $(1,0)$, $(x,1)$, $(x+1,1)$, for some arbitrarily large integer x, as shown in Figure 2. (Essentially the same example appears as Figure 1 in Alexandrov's seminal paper on convex polyhedral metrics [**7**].) The shortest path in this annulus between its two vertices is a vertical segment that crosses the oblique edge $x - 1$ times. All existing shortest-path algorithm require at least constant time for each crossing; thus, their total running time is *unbounded* as a function of the combinatorial complexity of the input.

FIGURE 2. A shortest path in a piecewise-linear annulus

Even for piecewise-linear surfaces embedded in $\mathbb{R}^3$, shortest-path algorithms can only be efficient in a model of computation that supports exact constant-time real arithmetic, including square roots. Even for polyhedra with integer vertices, most vertex-to-vertex geodesics have irrational lengths, representable analytically only with deeply nested radicals. Admittedly, this does not appear to be a significant problem in practice; it is relatively easy to implement existing shortest-path algorithms to compute paths that are optimal up to floating-point precision [**177**].

Subject to those caveats, a few of the algorithms we describe for combinatorial surfaces can be extended to embedded piecewise-linear surfaces, with some loss of efficiency, by replacing Dijkstra's algorithm with a piecewise-linear shortest-path algorithm. For example, Erickson and Whittlesey's algorithm [**80**] (described in Section 3.2) can be modified to compute an optimal homotopy basis with a given basepoint, in an embedded piecewise-linear surface, in $O(n^2)$ real arithmetic operations. However, for most of the problems we consider, no efficient algorithms are known for piecewise-linear surfaces.

—— PART I. HOMOTOPY ——

3. Homotopy Bases

Suppose we are given an undirected graph G with non-negatively weighted edges, a cellular embedding of G on an orientable surface Σ of genus g, and a vertex x of G. A *homotopy basis* is a set of $2g$ loops based at x whose homotopy classes generate the fundamental group $\pi_1(\Sigma, x)$. Our goal in this section is to compute a homotopy basis of minimum total length.

The standard structure for a homotopy basis is a *system of loops*, which is a set $\{\ell_1, \ell_2, \ldots, \ell_{2g}\}$ of $2g$ loops in Σ, each with basepoint x, such that the subsurface $\Sigma \setminus (\ell_1 \cup \ell_2 \cup \cdots \cup \ell_{2g})$ is an open topological disk; see Figure 3. The resulting disk is called a *(reduced) polygonal schema*.

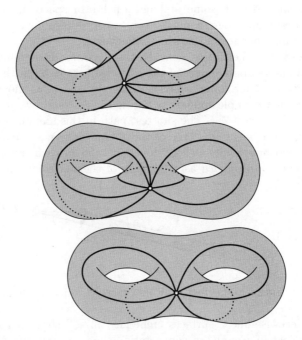

FIGURE 3. A homotopy basis that is not a system of loops, a non-canonical system of loops, and a canonical system of loops for a surface of genus 2.

Suppose we assign each loop ℓ_i an arbitrary orientation. Each loop ℓ_i appears as two paths on the boundary of the polygonal schema, once in each orientation. The cyclic sequence of these directed boundary paths is the *gluing pattern* of the polygonal schema. We obtain a one-relator presentation of $\pi_1(\Sigma, x)$, in which the loops ℓ_i are generators and the gluing pattern is the relator.

Many statements and proofs of the surface-classification theorem [6, 22, 156] rely on the existence of a *canonical* system of loops with the gluing pattern

$$\ell_1 \ell_2 \bar{\ell}_1 \bar{\ell}_2 \ell_3 \ell_4 \bar{\ell}_3 \bar{\ell}_4 \cdots \ell_{2g-1} \ell_{2g} \bar{\ell}_{2g-1} \bar{\ell}_{2g};$$

see the bottom of Figure 3. We emphasize that most of the algorithms we describe do *not* compute canonical systems of loops; indeed, it is open whether the shortest canonical system of loops can be computed in polynomial time. Fortunately, most applications of systems of loops do not require this canonical structure.

3.1. Without Optimization. If we don't care about optimization, we can compute a system of loops for any combinatorial surface in $O(n)$ time, using a straightforward extension of the textbook algorithm [168, 169] to construct a (non-minimal) presentation of the fundamental group of an arbitrary cell complex. The algorithm was first described explicitly by Eppstein [**76**], but is implicit in earlier work of several authors [**65, 98, 161, 172, 180**].

A *spanning tree* of G is a connected acyclic subgraph of G that includes every vertex of G; a *spanning cotree* of G is a subgraph C such that the corresponding dual subgraph C^* is a spanning tree of G^*.

A *tree-cotree decomposition* [**13**] is a partition of G into three edge-disjoint subgraphs: a spanning tree T, a spanning cotree C, and the leftover edges $L = G\backslash(T\cup C)$. In any tree-cotree decomposition (T, L, C) of any graph embedded on an orientable surface of genus g, the set L contains exactly $n_1 - (n_0 - 1) - (n_2 - 1) = 2g$ edges. (In particular, if $g = 0$, then $L = \varnothing$, and we recover the classical result that the complement of any spanning tree of a planar graph is a spanning cotree [**187**].) For each edge e in G, let $\ell_x(T, e)$ denote the loop obtained by concatenating the unique path in T from x to one endpoint of e, the edge e itself, and the unique path in T back to x. The set of $2g$ loops $\mathcal{L} = \{\ell_x(T, e) \mid e \in L\}$ is a system of loops.

We can construct a tree-cotree decomposition in $O(n)$ time by computing an arbitrary spanning tree T of G, for example by breadth- or depth-first search, and then computing an arbitrary spanning tree C^* of the dual subgraph $(G \setminus T)^*$. (Alternatively, we can compute a spanning tree C^* of G^* first, and then compute a spanning tree T of $G\backslash C$.) The sequence of edges traversed by any loop $\ell_x(T, e)$ can then be extracted from T and e in $O(1)$ time per edge. Thus, we can construct a system of loops in $O(n+k)$ time, where k denotes the total complexity of the output (the sum over all loops of the number of edges in each loop). Each loop $\ell_x(T, e)$ traverses each edge of the combinatorial surface at most twice, so $k = O(gn)$, and this bound is tight in the worst case.

THEOREM 3.1. *Given a combinatorial surface Σ with complexity n and genus g, we can construct a homotopy basis for Σ in $\Theta(n + k) = O(gn)$ time.*

3.2. Optimization. To construct the *minimum-length* homotopy basis with a given basepoint x, Erickson and Whittlesey [**80**] modify the previous algorithm by choosing a particular *greedy* tree-cotree decomposition (T, L, C). In this greedy decomposition,

- T is the *shortest-path tree* rooted at the basepoint x, and
- C^* is a *maximum-weight* spanning tree of the dual subgraph $(G \setminus T)^*$, where the weight of each dual edge e^* is the length of the corresponding primal loop $\ell_x(T, e)$.

We call the system of loops defined by this tree-cotree decomposition the *greedy system of loops*. Erickson and Whittlesey [**80**] proved that the greedy system of loops is the shortest system of loops with basepoint x using a complex exchange argument. Here we describe a simpler proof due to Colin de Verdière [**47**].

A *pointed homology basis* is a set of $2g$ loops with a common basepoint x whose homology classes generate the first homology group $H_1(\Sigma; \mathbb{Z}_2) \cong (\mathbb{Z}_2)^{2g}$. The Hurewicz theorem implies that any homotopy basis (and therefore any system of loops) is also a pointed homology basis. Thus, it suffices to prove that the greedy system of loops is the pointed homology basis of minimum total length.

The core of Colin de Verdière's proof is the following exchange argument, which extends a similar characterization of shortest non-contractible and non-separating cycles by Thomassen [**181**]. For any loop ℓ, let $[\ell]$ denote the *homology class* of ℓ, which we identify as a vector in $(\mathbb{Z}_2)^{2g}$. A set of $2g$ loops is a pointed homology basis if and only if the corresponding set of $2g$ homology classes is linearly independent.

LEMMA 3.2. *Every loop in a minimum-length pointed homology basis has the form $\ell_x(T, e)$ for some edge e, where T is the shortest-path tree rooted at x.*

PROOF. We regard the graph G as a continuous metric space, in which any edge of length w is isometric to the real interval $[0, w]$. Let $\overline{\alpha}$ denote the reversal of any directed path α. Let $\alpha \cdot \beta$ denote the concatenation of two directed paths α and β with matching endpoints.

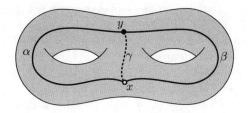

FIGURE 4. Three paths from x to y.

Fix a loop ℓ with basepoint x and a pointed homology basis $\mathcal{L}$ that contains ℓ. Let y be the midpoint of ℓ, so that ℓ can be decomposed two paths from x to y of equal length. Call these paths α and β, so that $\ell = \alpha \cdot \overline{\beta}$; see Figure 4. If we assume shortest vertex-to-vertex paths are unique, the point y lies in the interior of an edge of G.

Suppose there is a third path γ from x to y that is shorter than both α and β, as illustrated in Figure 4. Then the loops $\ell^{\flat} = \alpha \cdot \overline{\gamma}$ and $\ell^{\sharp} = \gamma \cdot \overline{\beta}$ are both shorter than $\ell = \alpha \cdot \overline{\beta}$. We immediately have $[\ell] = [\ell^{\flat}] + [\ell^{\sharp}]$, which implies that either $\mathcal{L} \cup \{\ell^{\sharp}\} \setminus \{\ell\}$ or $\mathcal{L} \cup \{\ell^{\flat}\} \setminus \{\ell\}$ is a pointed homology basis of smaller total length than $\mathcal{L}$. Thus, $\mathcal{L}$ is not a minimum-length pointed homology basis.

We conclude that if ℓ is a member of any minimum-length pointed homology basis, then α and β must be *shortest* paths from x to y. It follows that $\ell = \ell_x(T, e)$, where T is the shortest-path tree rooted at x and e is the edge of G that contains y. □

We are now faced with the following problem: From the set of $O(n)$ loops $\ell_x(T, e)$, extract a subset of $2g$ loops of minimum total length whose homology classes are linearly independent. Erickson and Whittlesey [80] observe that the loops $\ell_x(T, e_1), \ell_x(T, e_2), \ldots, \ell_x(T, e_k)$ have linearly independent homology classes if and only if deleting those loops from the surface Σ leaves a connected subsurface, or equivalently, if the dual subgraph $(G \setminus T)^* \setminus \{e_1^*, e_2^*, \ldots, e_k^*\}$ is connected. Thus, we seek the minimum-weight set of $2g$ edges in the dual subgraph $(G \setminus T)^*$ whose deletion leaves the graph connected. These are precisely the edges that are *not* in the maximum-weight spanning tree of $(G \setminus T)^*$.

The greedy tree-cotree decomposition can be constructed in $O(n \log n)$ time using textbook algorithms for shortest-path trees and minimum spanning trees. If $g = O(n^{1-\varepsilon})$ for some constant $\varepsilon > 0$, the time bound can be reduced to $O(n)$, using a more recent shortest-path algorithm of Henzinger *et al.* based on graph separators [105] and a careful implementation of Borůvka's minimum-spanning tree algorithm [21, 135]. The greedy system of loops can be extracted from this decomposition in $O(1)$ time per output edge.

THEOREM 3.3. *Given a combinatorial surface Σ with complexity n and genus g, and a vertex x of Σ, we can construct a system of loops with basepoint x, of minimum total length, in $O(n \log n + k) = O(n \log n + gn)$ time, or in $O(n + k) = O(gn)$ time if $g = O(n^{1-\varepsilon})$ for some constant $\varepsilon > 0$.*

If no basepoint is specified in advance, we can compute the globally shortest system of loops in $O(n^2 \log n + k)$ time by running the previous algorithm at every vertex. No faster algorithm is known.

3.3. Related Results. The tree-cotree algorithm can also be used to construct a system of loops in the dual graph G^*. Using this dual system of loops, one can label each edge of G with a string of length $O(g)$, in $O(gn)$ time, so that the concatenation of labels along any path encodes the homotopy type of that path. A version of this encoding can be used to quickly determine whether two paths or cycles of length k are homotopic in $O(n + k)$ time, via Dehn's algorithm [59, 61].

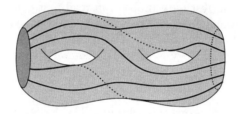

FIGURE 5. A system of five arcs for a surface with genus 2 and two boundary components

A slightly different substructure is needed to cut a surface *with boundary* into a disk. An *arc* in a surface Σ with boundary is a path whose endpoints lie on the boundary of Σ; a set of arcs $\{\alpha_1, \alpha_2, \ldots, \alpha_\beta\}$ such that $\Sigma \setminus (\alpha_1 \cup \alpha_2 \cup \cdots \cup \alpha_\beta)$ is a topological disk is called a *system of arcs*. See Figure 5. Euler's formula implies that if Σ has genus g and b boundary components, then any system of arcs for Σ has exactly $\beta = 2g + b - 1$ elements; the number β is the first *Betti number* of Σ. The minimum-length system of arcs can be constructed in $O(n \log n + k) = O(n \log n + (g + b)n)$ time using a variant of the greedy tree-cotree construction [34, 48, 79]. In fact, this greedy system of arcs is the minimum-length basis for the relative homology group $H_1(\Sigma, \partial \Sigma; \mathbb{Z}_2)$.

A *cut graph* is a subgraph X of the graph G such that $\Sigma \setminus X$ is homeomorphic to an open disk; for example, the union of all loops in a system of loops is a cut graph. Colin de Verdière [47] also described a similar greedy algorithm to compute the minimum-length cut graph with a prescribed set of vertices of degree greater than two. Computing the minimum-length cut graph with no prescribed vertices is NP-hard [78].

The algorithms described so far in this section do not necessarily construct *canonical* systems of loops. Brahana's proof of the surface classification theorem [22] describes an algorithm to transform any system of loops into a canonical system by cutting and pasting the polygonal schema; a more efficient transformation algorithm was later described by Vegter and Yap [185]. Lazarus et al. [126]

described the first direct algorithm to construct a canonical system of loops in $O(gn)$ time. The minimum-length system of loops homotopic to a given system can be computed in polynomial time by an algorithm of Colin de Verdière and Lazarus [**48, 49**]. However, it is open whether the globally shortest canonical homotopy basis can be computed in polynomial time.

4. Shortest Homotopic Paths and Cycles

The *shortest homotopic path* problem asks, given a path π in some combinatorial surface Σ, to compute the shortest path in Σ that is homotopic to π. Similarly, the *shortest homotopic cycle* problem asks, given a cycle γ in some combinatorial surface Σ, to compute the shortest cycle in Σ that is homotopic to γ. The input and output curves need not be paths and cycles in the graph-theoretic sense; they may visit vertices and edges multiple times.

The definition of the universal cover $\tilde{\Sigma}$ of Σ implies that the shortest path homotopic to any path π is the projection of the shortest path in $\tilde{\Sigma}$ between the endpoints of some lift $\tilde{\pi}$ of π. This characterization does not directly yield an algorithm, however, because the universal cover is usually infinite. Algorithms to solve the homotopic shortest path problem instead first construct a finite, simply-connected, and ideally small *relevant region* of $\tilde{\Sigma}$, and then compute a shortest path between the endpoints of $\tilde{\pi}$ within this relevant region.

A slightly different strategy is needed to compute shortest homotopic cycles, because cycles lift to *infinite* paths in the universal cover; we discuss the necessary modifications in Section 4.4.

4.1. Warm up: Polygons with Holes.
The earliest algorithms for computing shortest homotopic paths and cycles considered the Euclidean setting, where the the input path π is a chain of k line segments, and the environment is a polygon P with holes of total complexity n. Hershberger and Snoeyink [**106**] described an efficient algorithm for this special case, simplifying an earlier algorithm of Leiserson and Maley [**129**]. Hershberger and Snoeyink's algorithm proceeds in five stages, illustrated in Figures 6 and 7 on the next page.

First, in a preprocessing phase, *triangulate* the polygon P and assign each diagonal in the triangulation a unique label. The triangulation can be computed in $O(n \log n)$ time using textbook computational geometry algorithms [**55**]; faster algorithms are known when the number of holes is small [**11, 39, 167**].

Second, compute the *crossing sequence* of the input path π: the sequence of labels of diagonals crossed by the path, in order along the path. This stage is straightforward to implement in $O(k + x)$ time, where $x = O(kn)$ is the length of the crossing sequence.

Third, *reduce* the crossing sequence of π by repeatedly removing adjacent pairs of the same diagonal label; the reduced crossing sequence can be computed in $O(x)$ time. The reduction is justified by the observation that the shortest path in any homotopy class cannot cross the same diagonal e twice consecutively; otherwise, replacing the subpath between the two crossings with a sub-segment of e would yield a shorter homotopic path. The reduction of the crossing sequence mirrors a homotopy from π to the (unknown) shortest homotopic path π'. See Figure 6.

Fourth, construct the *sleeve* of triangles defined by the reduced crossing sequence of π. The sleeve is a topological disk (but not necessarily a simple Euclidean polygon) constructed by gluing together copies of the triangles in the triangulation

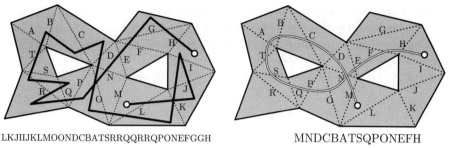

LKJIIJKLMOONDCBATSRRQQRRQPONEFGGH   MNDCBATSQPONEFH

FIGURE 6. Reducing a crossing sequence

of P. Specifically, we start with a copy the triangle containing the starting point of π, and then for each label in the reduced crossing sequence, we attach a new copy of the triangle just beyond the corresponding diagonal. See Figure 7. The sleeve can be constructed in $O(1)$ time per triangle, or $O(x)$ time altogether.

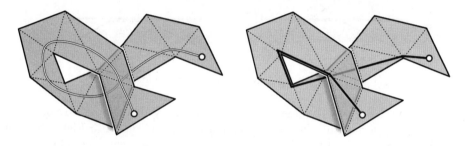

FIGURE 7. The sleeve defined by the reduced crossing sequence in Figure 6, and the shortest path within this sleeve

Finally, compute the shortest path in the sleeve between the endpoints of π; this is the shortest path in P homotopic to π. This shortest path can be computed in $O(1)$ time per sleeve triangle using the *funnel* algorithm independently proposed by Tompa [**183**], Chazelle [**38**], Lee and Preparata [**127**], and Leiserson and Maley [**129**]. Although the funnel algorithm was designed to compute shortest paths in simple Euclidean polygons, it works without modification in any simply-connected domain obtained by gluing Euclidean triangles along common edges, provided all triangle vertices lie on the boundary.

THEOREM 4.1 (Hershberger and Snoeyink [**106**]). *Let P be any polygon with holes with total complexity n, and let π be a polygonal chain with k edges. The shortest path in P homotopic to π can be computed in $O(n \log n + kn)$ time.*

4.2. Surfaces with Boundary. Colin de Verdière and Erickson [**48**] describe algorithms to compute shortest homotopic paths and cycles in arbitrary combinatorial surfaces. For surfaces with boundary, their algorithm follows almost exactly the same outline as Hershberger and Snoeyink's.

Here we sketch a variant of their algorithm for surfaces with boundary, essentially due to Colin de Verdière and Lazarus [**49**]. Suppose we are given a combinatorial surface Σ with complexity n, genus g, and b boundary components. In the preprocessing phase, we first compute a *greedy system of arcs* $\{\alpha_1, \alpha_2, \ldots, \alpha_\beta\}$ for Σ in $O(n \log n + (g+b)n)$ time, as described in Section 3.3.

The greedy system of arcs has the following important property, motivated by results of Hass and Scott [102]. We say that two paths σ and τ form a *bigon* if some subpath of σ and some subpath of τ are homotopic.

LEMMA 4.2. *For any path π, there is a shortest path homotopic to π that does not define a bigon with any arc in the greedy system of arcs.*

PROOF. Recall that the greedy system of arcs is the minimum-length collection of arcs that generates the relative homology group $H_1(\Sigma, \partial\Sigma; \mathbb{Z}_2)$. It follows that each arc α_i in the greedy system of arcs is a shortest path in its relative homology class, and therefore in its homotopy class.

Define the *crossing number* of a path π to be the total number of times π crosses the arcs α_i in the greedy system.

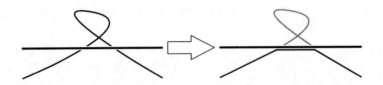

FIGURE 8. Let bigons by bygones.

If π forms a bigon with some arc α_i in the greedy system of arcs, we can replace some subpath of π with a homotopic subpath of α_i, as shown in Figure 8. This new path is homotopic to π, is no longer than π, and has smaller crossing number than π. It follows that among all shortest paths homotopic to π, the path with smallest crossing number does not define a bigon with any arc α_i. □

Now suppose we are given a path π with complexity k. We compute the *signed crossing sequence* of the directed input path π with respect to these arcs; each time π crosses an arc α_i from left to right (respectively, from right to left), the signed crossing sequence contains the symbol i^+ (respectively, i^-). The signed crossing sequence is then reduced by repeatedly removing pairs of the form $i^+ i^-$ or $i^- i^+$. As in the Euclidean setting, the reduction mirrors a homotopy of π that removes bigons one at a time; Lemma 4.2 implies that the reduced crossing sequence of π is the crossing sequence of some shortest path homotopic to π.

To construct the relevant region R of the universal cover, we glue together a sequence of $x+1$ copies of the polygonal schema $D = \Sigma \setminus (\alpha_1 \cup \alpha_2 \cup \cdots \cup \alpha_\beta)$, where $x = O((g+b)k)$ is the length of the reduced crossing sequence. The relevant region is a combinatorial *disk* with complexity $O(nx) = O((g+b)nk)$. Finally, we compute the shortest path in R between the first vertex of π in the first copy of D to the last vertex of π in the last copy of D, using a linear-time shortest-path algorithm for planar graphs [105].

THEOREM 4.3 (Colin de Verdière and Erickson [48]). *Let Σ be a combinatorial surface with genus g and $b \geq 1$ boundary components, and let π be a path of k edges in Σ. The shortest path in Σ homotopic to π can be computed in $O(n \log n + (g+b)nk)$ time.*

4.3. Surfaces without Boundary.

For orientable surfaces with no boundary, Colin de Verdière and Erickson require a different decomposition of the surface into disks. For surfaces with genus at least 2, their preprocessing algorithm computes a *tight octagonal decomposition* of the given surface Σ in $O(gn \log n)$ time. This is a set of $O(g)$ cycles $\{\gamma_1, \gamma_2, \ldots\}$, each as short as possible in its homotopy class, that decompose the surface into octagons meeting four at a vertex; see Figure 9. The algorithm to construct this decomposition is fairly technical and relies on several other results [29, 49, 78, 85].

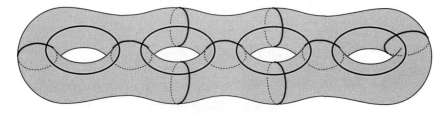

FIGURE 9. A tight octagonal decomposition.

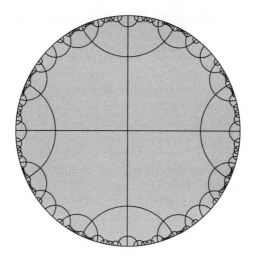

FIGURE 10. The universal cover of a tight octagonal decomposition.

The universal cover of a tight octagonal decomposition is combinatorially isomorphic to a regular tiling of the hyperbolic plane by right-angled octagons. In particular, each cycle in the decomposition lifts to a family of infinite geodesics in the universal cover $\tilde{\Sigma}$, each of which crosses any other shortest path in $\tilde{\Sigma}$ at most once. Evoking the regular hyperbolic structure, Colin de Verdière and Erickson call these infinite geodesics *lines*. These lines are drawn as circular arcs in Figure 10, following the Poincaré disk model of the hyperbolic plane.

Now let π be a given path in Σ, and let $\tilde{\pi}$ be an arbitrary lift of π to $\tilde{\Sigma}$. Let X denote the set of lines (lifts of cycles γ_i) that $\tilde{\pi}$ crosses, and let X' be the set of lines that $\tilde{\pi}$ crosses an odd number of times. Any path in $\tilde{\Sigma}$ between the endpoints in $\tilde{\pi}$ must cross every line in X' an odd number of times. Moreover, an easy variant of Lemma 4.2 implies that the shortest such path crosses each line in X' *exactly* once and does not cross any line not in X'. For this reason, we call the lines in X' *relevant*.

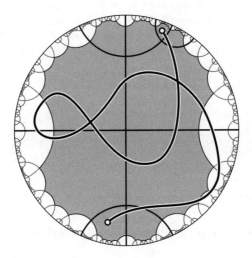

FIGURE 11. The relevant lines and relevant region of a lifted path.

Let $R(X)$ denote the complex of octagons reachable from the initial point of $\tilde{\pi}$ by crossing only (a subset of) lines in X; see Figure 11. A classical result of Dehn [**59**] implies that $R(X)$ is the union of at most $O(|X|)$ octagons. Colin de Verdière and Erickson describe an incremental algorithm to construct $R(X)$ in $O(x)$ time, where x is the length of the signed crossing sequence of π with respect to the cycles γ_i [**48**]. Equivalently, x is the total number of times the lifted path $\tilde{\pi}$ crosses the lines in X. Because the cycles in the tight octagonal decomposition may share edges, a single edge in the input path may cross $\Theta(g)$ cycles simultaneously; thus, we have an upper bound $x = O(kg)$.

The subset X' can be identified in $O(x)$ time by performing a breadth-first search in the dual 1-skeleton of $R(X)$. Alternatively, X' can be identified directly by reducing the signed crossing sequence of π with respect to the cycles γ_i. The crossing sequence can no longer be reduced by merely canceling bigons—see the example in Figure 11—but the regular hyperbolic tiling allows fast reduction in $O(x)$ time using techniques from small cancellation theory [**134, 140**]. Let x' denote the length of the reduced crossing sequence; again, because $x' \leq x$, we have $x = O(kg)$.

Finally, let $R(X')$ denote the complex of octagons reachable from the initial point of $\tilde{\pi}$ by crossing only (a subset of) lines in X'; again, this complex can be constructed in $O(x')$ time. After filling each octagon in $R(X')$ with the corresponding planar portion of the input graph G, Colin de Verdière and Erickson's algorithm computes the shortest path $\tilde{\pi}'$ between the endpoints of $\tilde{\pi}$ in the resulting planar graph [**105**]. The projection of $\tilde{\pi}'$ to the original graph G is the shortest path homotopic to π. Altogether, the query phase of their algorithm runs in $O(x + x'n) = O(gnk)$ time, where as usual k is the number of edges in the input path.

THEOREM 4.4 (Colin de Verdière and Erickson [**48**]). *Let Σ be an orientable combinatorial surface with genus $g \geq 2$ and no boundary, and let π be a path of k*

edges in Σ. The shortest path in Σ homotopic to π can be computed in $O(gn \log n + gnk)$ time.

For shortest homotopic paths on the torus, where no hyperbolic structure is possible, a similar algorithm runs $O(n \log n + nk^2)$ time. Instead of a tight octagonal decomposition, the algorithm cuts the surface into a disk with a system of loops, where each loop is as short as possible in its homotopy class. Shortest homotopic paths in any non-orientable surface Σ can be computed by searching the oriented double cover of Σ.

4.4. Shortest Homotopic Cycles. As promised, we now describe the necessary modifications to find shortest homotopic cycles. We cannot directly apply the previous algorithm, because any lift of a non-contractible cycle to the universal cover is an *infinite* path; see Figure 9.

Let γ be a non-contractible cycle in some combinatorial surface Σ. In the simplest nontrivial special case of this problem, Σ is a combinatorial *annulus* and γ is one of its boundary cycles; we call any cycle homotopic to γ a *generating* cycle. Using an argument similar to Lemma 4.2, Itai and Shiloach [**110**] observed that the shortest generating cycle crosses any shortest path between the two boundary cycles exactly once. Thus, one can compute the shortest generating cycle by cutting Σ along a shortest path σ between the two boundaries; duplicating every vertex and edge of σ; and then, for each vertex v of σ, computing the shortest path between the two copies of v in the resulting planar graph. See Figure 12.

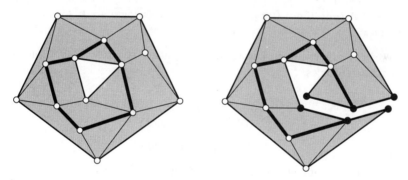

FIGURE 12. Finding the shortest nontrivial cycle in an annulus by cutting it along a shortest boundary-to-boundary path.

Itai and Shiloach applied Dijkstra's shortest-path algorithm at each vertex of σ, immediately obtaining a running time of $O(n^2 \log n)$ [**110**]. Reif [**160**] improved the running time of this algorithm to $O(n \log^2 n)$ using a divide-and-conquer strategy. Frederickson [**85**] further improved the running time to $O(n \log n)$ using a recursive separator decomposition [**132**] to speed up the shortest-path computations. The same improvement can be obtained using more recent algorithms for shortest paths [**105, 121**] and maximum flows [**19, 77**] in planar graphs. Most recently, Italiano *et al.* [**111**] improved the running time to $O(n \log \log n)$ using a more careful separator decomposition and other algorithmic tools for planar graphs [**82, 85, 142**].

Colin de Verdière and Erickson [**48**] describe a reduction from the more general shortest homotopic cycle problem to this special case. At a very high level, the algorithm identifies an infinite *periodic* relevant region R in the universal cover that contains a lift of the shortest homotopic cycle. The algorithm actually constructs one period R_0 of this infinite relevant region, identifies corresponding boundary paths of R_0 to obtain a combinatorial annulus A, and finally computes the shortest generating cycle of A. Yin et al. [192] sketch a similar algorithm to compute shortest homotopic cycles; however, their description omits several key details and offers no time analysis.

More concretely, suppose the surface Σ has genus 2 and no boundary. Let $\gamma\colon \mathbb{R}/\mathbb{Z} \to \Sigma$ denote the input cycle, and let $\tilde{\gamma}\colon \mathbb{R} \to \tilde{\Sigma}$ be one of its lifts to the universal cover. There is a translation $\tau\colon \tilde{\Sigma} \to \tilde{\Sigma}$ such that $\tau(\tilde{\gamma}(t)) = \tilde{\gamma}(t+1)$ for all t. Fix an arbitrary lift $\tilde{v}_0$ of some vertex v of γ, and for each positive integer i, let $\tilde{v}_i = \tau(\tilde{v}_{i+1})$. The set X' of all lines in the tight octagonal tiling of $\tilde{\Sigma}$ that separate $\tilde{v}_0$ from $\tilde{v}_4$ can be computed using the same algorithm as for shortest homotopic paths. Let ℓ_2 be any line that separates $\tilde{v}_0$ and $\tilde{v}_1$ from $\tilde{v}_2$ and $\tilde{v}_3$, and let $\ell_3 = \tau(\ell_2)$; both of these lines lie in the set X'. Let $R(X')$ be the complex of octagons reachable from $\tilde{v}_0$ by crossing only lines in X'. Finally, let R_0 be the portion of $R(X')$ that lies between ℓ_2 and ℓ_3; identifying the segments of ℓ_2 and ℓ_3 on the boundary of R_0 gives us the annulus A.

Combining Colin de Verdière and Erickson's reduction with the $O(n \log \log n)$-time algorithm of Italiano et al. for the annulus [111] gives us the following result.

THEOREM 4.5. *Let Σ be a combinatorial surface with genus g and b boundaries, and let γ be a cycle of k edges in Σ. The shortest cycle homotopic to γ can be computed in $O(n \log n + (g+b)nk \log \log nk)$ time if $b > 0$, in $O(gn \log n + gnk \log \log gnk)$ time if $b = 0$ and $g > 1$, and in $O(n \log n + nk^2 \log \log nk)$ time if $g = 1$ and $b = 0$.*

No efficient algorithm is known for computing shortest homotopic paths in *non-orientable* surfaces.

—— PART II. HOMOLOGY ——

5. General Remarks

We now turn from homotopy to homology. Unlike the homotopy problems we have considered so far, questions about homology are decidable for any finite regular cell complex. Not surprisingly, however, these problems can be solved more efficiently for combinatorial 2-manifolds than for general complexes, so we consider algorithms for surfaces separately.

Before we consider any algorithms, we must carefully define the functions we wish to optimize. Fix a simplicial complex Σ and a coefficient ring R. A *p-chain* is a formal linear combination of oriented p-simplices in Σ, which we identify with a vector $\boldsymbol{c} = (c_1, c_2, \ldots, c_{n_p}) \in R^{n_p}$. Fix a vector $\boldsymbol{w} = (w_1, w_2, \ldots, w_{n_p}) \in \mathbb{R}^{n_p}$ that assigns a non-negative weight to each p-cell in Σ. In the interest of developing *exact* optimization algorithms, we restrict our attention to two definitions of the "weight" of a p-chain:

- The *weighted L_0-norm* is the sum of the weights of all cells with non-zero coefficients:
$$\|\boldsymbol{c}\|_{0,\boldsymbol{w}} := \sum_{i:c_i \neq 0} w_i,$$

- The *weighted L_1-norm* is the weighted sum of the absolute values of the coefficients:
$$\|\mathbf{c}\|_{1,\mathbf{w}} := \sum_i w_i |c_i|.$$

The weighted L_0-norm is well-defined for any coefficient ring R; the weighted L_1-norm is well-defined only when R is a sub-ring of the reals. We call a p-chain *unitary* if every coefficient lies in the set $\{-1, 0, 1\}$; the weighted L_0- and L_1-norms of a chain are equal if and only if the chain is unitary.

Formally, a p-cycle is any p-chain that lies in the kernel of the boundary map $\partial_p \colon R^{n_p} \to R^{n_{p-1}}$. However, when we seek p-cycles or homology bases with minimum weighted L_0-norm, it is convenient to conflate any p-cycle $\mathbf{c}$ with the subset of p-cells with non-zero coefficients c_i. To avoid confusion over the multiple meanings of the word "cycle", we consistently refer to the elements of a pth homology class as p-*cycles*, and closed walks in the 1-skeleton of Σ as *loops*.

Finally, a pth *homology basis* is a minimum-cardinality set of p-cycles whose homology classes generate the pth homology group $H_p(\Sigma; R)$. We define the *weight* of a homology basis to be the sum of the weighted L_0-norms of its constituent p-cycles. Although the total L_1-norm of a homology basis is well-defined when $R \subseteq \mathbb{R}$, the corresponding optimization problem is uninteresting. If $R = \mathbb{Z}$, then every p-cycle in the L_1-minimal homology basis is unitary; thus, the L_1-minimal homology basis is also the L_0-minimal homology basis. On the other hand, if R contains numbers arbitrarily close to zero (for example, if $R = \mathbb{Q}$), there are homology bases whose weighted L_1-norm is arbitrarily close to zero.

6. Homology Bases

Let Σ be an arbitrary simplicial complex with weighted cells, and let p be a positive integer. Our goal in this section is to find a pth homology basis whose total weighted L_0-norm is as small as possible. Because we are optimizing the weighted L_0-norm, we need not distinguish between a p-cycle and the subset of p-cells with non-zero coefficients. In particular, any *first* homology basis for Σ is a minimal set of simple *loops* in the 1-skeleton of Σ; we seek to minimize the total length of these loops. See Figure 13.

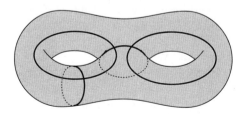

FIGURE 13. A first homology basis for a surface of genus 2.

6.1. Surfaces.
For combinatorial surfaces, any system of loops is also a first homology basis, for any coefficient ring; thus, Eppstein's tree-cotree algorithm constructs a homology basis for any combinatorial surface in $O(n + k) = O(gn)$ time. We can reduce the output size by replacing each loop $\ell_x(T, e)$ with the unique *simple* loop in the graph $T \cup \{e\}$, but the output size is still $\Theta(gn)$ in the worst case.

However, the minimum-weight first homology basis is not necessarily consistent with any tree-cotree decomposition.

Erickson and Whittlesey [80] describe an efficient algorithm to compute the optimal first homology basis with respect to any coefficient *field*, generalizing an earlier algorithm of Horton [108] to compute the shortest cycle basis of an edge-weighted graph. Their algorithm is simplified by the following observation of Dey et al. [66]: Every generator in the optimal *homology* basis is also a generator in the optimal *homotopy* basis for some basepoint. Thus, in $O(n^2 \log n + gn^2)$ time, we can compute a set of $O(gn)$ candidate loops that must include the optimal homology basis. Again, the $n^2 \log n$ term can be removed from the running time if $g = n^{1-\varepsilon}$ for any $\varepsilon > 0$. The homotopy basis algorithm automatically computes the length of each candidate loop.

We can also compute vectors of length $2g$ that encode the homology class of each candidate loop using an elementary form of Poincaré duality. Recall that we associate two *darts* $u \to v$ and $v \to u$ with each edge uv of the embedded graph G. We first construct a *dual* homology basis $\{\lambda_1^*, \lambda_2^*, \ldots, \lambda_{2g}^*\}$ using an arbitrary tree-cotree decomposition (T, L, C); each element λ_i^* of the dual homology basis is a loop in the dual graph G^*, obtained by adding an edge in L^* to the dual spanning tree C^*. We orient the loops λ_i^* arbitrarily. We then label each dart $u \to v$ in the primal graph with a vector $h(u \to v) \in R^{2g}$, called the *homology signature* of the edge, whose ith coordinate is $+1$ if the dart crosses λ_i^* from left to right, -1 if the dart crosses λ_i^* from right to left, and 0 otherwise. The homology class of any directed loop γ in G is then the sum of the homology signatures of its edges. By accumulating homology signatures along every root-to-leaf path in every shortest-path tree, we can compute the homology classes of all $O(gn)$ candidate loops in $O(gn^2)$ total time.

We are now faced with a standard *matroid optimization* problem: Given a collection of $O(gn)$ vectors, each with non-negative weight, find a subset of minimum total weight that generates the *vector space* R^{2g}. This problem can be solved by a greedy algorithm, similar to Kruskal's classical minimum-spanning-tree algorithm [124]. Starting with an empty basis, we consider the vectors one at a time in order of increasing weight; whenever we encounter a vector that is linearly independent of the vectors already in the basis, add it to the basis. Sorting the vectors takes $O(gn \log n)$ comparisons, and for each vector, we need $O(g^2)$ arithmetic operations to test linear independence.

THEOREM 6.1 (Erickson and Whittlesey [80]). *Let Σ be a combinatorial surface with complexity n and genus g. An optimal first homology basis for Σ, with respect to any coefficient field, can be computed in $O(n^2 \log n + gn^2 + g^3 n)$ time, or in $O(gn^2 + g^3 n)$ time if $g = O(n^{1-\varepsilon})$ for any $\varepsilon > 0$.*

The restriction to coefficient fields is unfortunately necessary. For coefficient rings without division, homology groups are not vector spaces, and thus a maximal linearly independent set of homology classes is not necessarily a basis. Gortler and Thurston [91] have shown that Erickson and Whittlesey's greedy algorithm can return a set of $2g$ loops that do *not* generate the first homology group, although they lie in independent $\mathbb{Z}$-homology classes, even when $g = 2$. It is natural to conjecture that computing an optimal $\mathbb{Z}$-homology basis is NP-hard, but no such result is known, even for more general complexes.

6.2. Complexes. If we do not care about optimization, we can compute a homology basis of any dimension, for any simplicial complex, over any coefficient ring R, using the classical Poincaré-Smith reduction algorithm [155, 173], which requires $O(n^3)$ arithmetic operations over the coefficient ring R. This is not the fastest algorithm known. Homology over any field can be computed by an algorithm of Bunch and Hopcroft [25], whose complexity is dominated by the time to multiply two $n \times n$ matrices; the fastest algorithm known for that problem requires only $O(n^{2.376})$ arithmetic operations [51]. On the other hand, these running times are misleading when $R = \mathbb{Z}$, as careless implementations can produce intermediate integers with exponentially many bits in the worst case [83]. The first polynomial-time algorithm for computing integer homology was described by Kannan and Bachem [114]; for a sample of more recent results, see Dumas et al. [68,69] and Eberly et al. [72].

Chen and Freedman [41] and Dey et al. [66] extend Erickson and Whittlesey's greedy strategy to compute optimal first homology bases with $\mathbb{Z}_2$ coefficients, for an arbitrary simplicial complex, in polynomial time. Instead, both algorithms modify the input complex by gluing disks to all loops in the evolving basis. A new loop γ is accepted into the evolving basis if and only if gluing a disk to γ decreases the rank β_1 of the first homology group. Chen and Freedman's algorithm [41] recomputes β_1 from scratch at each iteration, using an algorithm for sparse Gaussian elimination over finite fields [190]; their algorithm runs in $O(\beta_1 n^3 \log^2 n)$ time with high probability. Dey et al. [66] exploit persistent homology [73,74,195] to achieve a running time of $O(n^4)$. Both algorithms can be extended to compute optimal homology bases over any coefficient *field*; in this more general setting, the algorithm of Dey et al. [66] requires $O(n^4)$ arithmetic operations. Again, the restriction to coefficient fields is necessary; no polynomial-time algorithm or NP-hardness result is known for computing optimal $\mathbb{Z}$-homology bases.

Chen and Friedman further generalize their algorithm to compute minimum-cost bases for higher-dimensional homology groups, where the cost of a single generator is the *radius* of the smallest ball that contains it [41]. However, for the arguably more natural weighted L_0-norm, they also prove that computing optimal pth homology bases (over $\mathbb{Z}_2$) is NP-hard, for any $p \geq 2$ [40].

7. Homology Localization over $\mathbb{Z}_2$

We now turn to finding optimal representatives in a single given homology class. The complexity of this problem depends critically on the choice of coefficient ring. Somewhat surprisingly, in light of results for homology bases, optimization over finite fields turns out to be significantly harder than optimization either over the reals or (at least for manifolds) the integers.

We first consider one-dimensional homology over $\mathbb{Z}_2$. With this coefficient field, a 1-chain in any simplicial complex Σ is a *subgraph* of the 1-skeleton; a 1-cycle is a subgraph in which every vertex has even degree (henceforth, an *even subgraph*); a *1-boundary* is the boundary of the union of a subset of 2-cells; and two even subgraphs are $\mathbb{Z}_2$-homologous if and only if their symmetric difference is a boundary subgraph. The weight of an even subgraph is just the sum of the weights of its edges; this is equivalent to the weighted L_0-norm. Our goal in this section is to find a minimum-weight even subgraph $\mathbb{Z}_2$-homologous to a given even subgraph.

Unfortunately, any reasonable variant of this problem is NP-hard, even in combinatorial surfaces. An argument of Chambers et al. [**34**] can be modified to show that finding the minimum-weight *connected* even subgraph $\mathbb{Z}_2$-homologous to a given simple loop is NP-hard, by reduction from the Hamiltonian cycle problem in planar grid graphs [**109**]. A refinement of this argument by Cabello et al. [**28**] implies that finding the shortest *simple cycle* in a given $\mathbb{Z}_2$-homology class is NP-hard. Later Chambers et al. [**36**] proved that finding the optimal even subgraph $\mathbb{Z}_2$-homologous to a given simple loop is NP-hard, by reduction from the minimum cut problem in graphs with negative edges [**141**]. For more general complexes, Chen and Friedman [**40**] prove that even *approximating* the minimum-weight even subgraph in a given $\mathbb{Z}_2$-homology class by a constant factor is NP-hard, even when the rank of the first $\mathbb{Z}_2$-homology group is 1, by reduction from the nearest codeword problem [**9**].

However, for combinatorial surfaces with constant genus, it is possible to find minimal representatives in *every* $\mathbb{Z}_2$-homology class in $O(n \log n)$ time. Specifically, Erickson and Nayyeri [**79**] describe an algorithm to compute either the shortest loop or the shortest even subgraph in a given $\mathbb{Z}_2$-homology class in $2^{O(g)} n \log n$ time, simplifying and improving an earlier algorithm by Chambers et al. [**34**] that runs in $g^{O(g)} n \log n$ time.

Erickson and Nayyeri's algorithm first constructs the $\mathbb{Z}_2$-*homology cover* $\overline{\Sigma}$, which is the unique connected covering space of Σ whose group of deck transformations is $H_1(\Sigma; \mathbb{Z}_2) \cong (\mathbb{Z}_2)^{2g}$. They give two different but equivalent descriptions of the construction, one in terms of voltage graphs [**95**, Chapter 4], the other directly topological. We sketch the second formulation here. The construction is easier to visualize for a simple surface with boundary; see Figure 14.

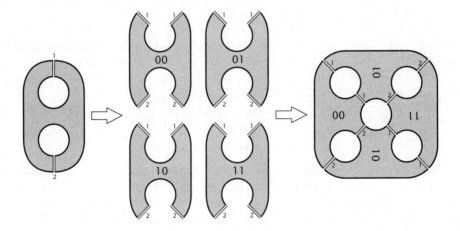

FIGURE 14. Constructing the $\mathbb{Z}_2$-homology cover of a pair of pants.

Let $\{\ell_1, \ell_2, \ldots, \ell_{2g}\}$ be any system of loops for Σ, such as the one constructed from a tree-cotree decomposition by Eppstein's algorithm; for surfaces with boundary, we use a system of arcs instead, as shown in Figure 14. The surface $D := \Sigma \setminus (\ell_1 \cup \cdots \cup \ell_{2g})$ is a topological disk; each loop ℓ_i appears on the boundary of D as two boundary segments ℓ_i^+ and ℓ_i^-. For each homology class $h \in (\mathbb{Z}_2)^{2g}$, we create a disjoint copy (D, h) of D; for each index i, let (ℓ_i^+, h) and (ℓ_i^-, h) denote the copies of ℓ_i^+ and ℓ_i^- in the disk (D, h). (In Figure 14, each copy (D, h) is labeled by a 2-bit string representing the homology class h.) For each index i, let b_i denote the $2g$-bit vector whose ith bit is equal 1 and whose other $2g - 1$ bits are all equal to 0. The $\mathbb{Z}_2$-homology cover $\overline{\Sigma}$ is constructed by gluing the copies of D together by identifying boundary paths (ℓ_i^+, h) and $(\ell_i^-, h \oplus b_i)$, for every index i and every homology class h, where $\oplus$ represents bitwise exclusive-or. (For example, in Figure 14, the copies D_{00} and D_{01} are glued together along a copy of ℓ_1.) The resulting combinatorial surface has $\overline{n} = 2^{2g}n$ vertices, each labeled by a pair (v, h) for some vertex v in Σ and some homology class h, and genus $\overline{g} = 2^{2g}(g - 1) + 1$. The entire construction takes $2^{O(g)}n$ time.

Because each connected component of an even subgraph has a closed Euler tour, we can reasonably regard any even subgraph as a collection of vertex-disjoint loops. We define the $\mathbb{Z}_2$-homology class of a loop as the $\mathbb{Z}_2$-homology class of its *carrier*: the subgraph of edges that the loop traverses an odd number of times. The weight of a loop is defined as the sum of the weights of its edges, *counted with appropriate multiplicity*; thus, if a loop traverses any edge more than once, its weight is larger than the weight of its carrier.

First consider the related problem of finding the shortest *loop* in a given $\mathbb{Z}_2$-homology class; we emphasize that we must consider loops that repeat edges, because the shortest loop in a given $\mathbb{Z}_2$-homology class need not be simple. Let ℓ be any loop in Σ, and let $[\ell]$ denote its $\mathbb{Z}_2$-homology class. The loop ℓ is the projection of a path in $\overline{\Sigma}$ from $(v, 0)$ to $(v, [\ell])$, where v is any vertex in ℓ. Thus, the shortest loop in homology class h is the projection of the shortest path in $\overline{\Sigma}$ from some vertex $(v, 0)$ to the corresponding vertex (v, h). This shortest path can be found in $O(n \cdot \overline{n} \log \overline{n}) = 2^{O(g)}n^2 \log n$ time by computing a shortest-path tree at every vertex $(v, 0)$. Erickson and Nayyeri [79] reduce the running time to $O(\overline{gn} \log \overline{n}) = 2^{O(g)}n \log n$ using more complex shortest-path data structures [26, 27, 121].

The minimum-weight even subgraph in any $\mathbb{Z}_2$-homology class has at most g connected components, each of which is (the carrier of) the shortest loop in its own $\mathbb{Z}_2$-homology class. To find the optimal even subgraph, Erickson and Nayyeri first compute the shortest loop in *every* $\mathbb{Z}_2$-homology class and then assemble the components using a simple dynamic programming algorithm. Specifically, let $C(h, k)$ denote the minimum total weight of any set of at most k loops whose homology classes sum to h. This function obeys the recurrence

$$C(h, k) = \min_{h'} \left(C(h', k - 1) + C(h \oplus h', 1) \right),$$

where h' ranges over all $\mathbb{Z}_2$-homology classes. The base cases are $C(0, k) = 0$ and $C(h, 1)$, which has already been computed for each h. The dynamic programming algorithm computes $C(h, g)$ in $2^{O(g)}$ additional time.

THEOREM 7.1 (Erickson and Nayyeri [79]). *Given a combinatorial surface Σ with complexity n and genus g, the minimum-weight even subgraph of Σ in any (in fact, every) $\mathbb{Z}_2$-homology class can be computed in $2^{O(g)}n \log n$ time.*

This algorithm can be used directly to compute minimum cuts in surface-embedded graphs. Fix a graph G, where every edge has a non-negative *capacity*, and two vertices s and t. An (s, t)-*cut* is a subset of edges of G that contains at least one edge in every path from s and t. Itai and Shiloach [110] proved that the minimum-capacity (s,t)-cut in an undirected planar graph G is dual to the minimum-cost cycle that separates faces s^* and t^* in the dual graph G^*. Thus, minimum cuts in undirected planar graphs can be computed by finding the shortest generating cycle in an annulus, as described in Section 4.4. Chambers et al. [36] generalized Itai and Shiloach's result to higher-genus surfaces, by proving that the minimum-capacity (s,t)-cut is dual to the minimum-weight even subgraph in G^* that is $\mathbb{Z}_2$-homologous with the boundary of s^* in the punctured surface $\Sigma \setminus (s^* \cup t^*)$. This result together with Theorem 7.1 immediately implies the following.

THEOREM 7.2 (Erickson and Nayyeri [79]). *Given an undirected graph G with non-negative edge capacities, embedded on a surface Σ with genus g, and two vertices s and t, a minimum (s,t)-cut in G can be computed in $2^{O(g)} n \log n$ time.*

Alternatively, recent results of Italiano et al. [111] also improve the running time of the algorithm of Chambers et al. [34] from $g^{O(g)} n \log n$ to $g^{O(g)} n \log \log n$. The resulting algorithm is faster than Erickson and Nayyeri's algorithm for graphs of constant genus, but it is also considerably more complex.

No similar algorithm is known for directed graphs. We consider the dual maximum-flow problem in Section 8.5.

Finally, Erickson and Nayyeri's algorithm can be generalized to p-manifold complexes of dimension any $p > 2$, using an arbitrary basis for first cohomology group $H^1(\Sigma; \mathbb{Z}_2) \cong H_{p-1}(\Sigma; \mathbb{Z}_2)$ (or the first relative cohomology group $H^1(\Sigma, \partial \Sigma; \mathbb{Z}_2) \cong H_{p-1}(\Sigma; \mathbb{Z}_2)$ if the manifold has boundary) in place of a system of loops or arcs. Each element of such a cohomology basis is a subgraph of the 1-skeleton of the complex. Thus, the cohomology basis can be used to construct the 1-skeleton of the $\mathbb{Z}_2$-homology cover using the same voltage-graph construction. The data structures used to accelerate shortest-path computations in surface graphs have no higher-dimensional analogue; otherwise, the remainder of the algorithm is unchanged. If we use the standard Poincaré-Smith reduction algorithm to compute the (relative) cohomology basis, the resulting algorithm requires $O(n^3) + 2^{O(\beta)} n^2 \log n$ real arithmetic operations, where β is the rank of the first $\mathbb{Z}_2$-homology group.

8. Homology Localization over $\mathbb{R}$ and $\mathbb{Z}$

We now switch to finding representatives in real and integer homology classes whose weighted L_1-norm is minimized. For homology over the *reals*, optimal homologous chains can be computed in polynomial time via linear programming. If the input complex satisfies certain conditions—in particular, if the input is an orientable $(p+1)$-manifold—the resulting linear programs actually have integral solutions and thus can be used without modification to find optimal representatives in *integer* homology classes. In general, however, finding optimal $\mathbb{Z}$-homologous chains is NP-hard.

8.1. Real homology via linear programming. Fix a simplicial complex Σ and a non-negative integer $p \geq 0$. To simplify notation, let m and n respectively denote the number of p-simplices and $(p+1)$-simplices in Σ. In real simplicial homology, a *p-chain* is a formal linear combination of oriented p-simplices in Σ, which we identify with a real vector $\boldsymbol{c} = (c_1, c_2, \ldots, c_m) \in \mathbb{R}^m$. Two p-chains are $\mathbb{R}$-homologous if their difference (as vectors) lies in the kernel of the boundary map $\partial_{p+1} \colon \mathbb{R}^n \to \mathbb{R}^m$. Given a p-chain $\boldsymbol{c}$ and a weight vector $\boldsymbol{w} \in \mathbb{R}^m$, our goal is to find a p-chain $\boldsymbol{x}$ with minimum weighted L_1-norm that is $\mathbb{R}$-homologous to $\boldsymbol{c}$.

This optimization problem can be solved in polynomial time by formulating it as a linear program as follows [62, 176]:

(LP)
$$\begin{aligned}\text{minimize} \quad & \sum_i (x_i^+ + x_i^-) \cdot w_i \\ \text{subject to} \quad & \bm{x}^+ - \bm{x}^- = \bm{c} + [\partial_{p+1}]\bm{y} \\ & \bm{x}^+ \geq 0 \\ & \bm{x}^- \geq 0\end{aligned}$$

Here, the variable $\bm{x} = \bm{x}^+ - \bm{x}^- \in \mathbb{R}^m$ represents the output p-chain, the variable $\bm{y} \in \mathbb{R}^n$ represents an unknown $(p+1)$-chain, and $[\partial_{p+1}]$ is the $m \times n$ boundary matrix. The output vector $\bm{x}$ is split into two *non-negative* vectors $\bm{x}^+$ and $\bm{x}^-$ so that the weighted L_1-norm can be represented as a linear objective function.

We emphasize that this linear programming formulation does *not* require the input p-chain $\bm{c}$ to have zero boundary. Thus, for example, we can use this formulation to find a minimal surface with prescribed boundary in (the 2-skeleton of) a triangulation of $\mathbb{R}^3$, by letting $\bm{c}$ be any 2-chain with the desired boundary [62, 71, 176].

8.2. Integer homology is hard. Althaus and Fink [8] proved that finding minimum $\mathbb{Z}$-homologous *unitary* 2-chains is NP-hard, by reduction from 3-dimensional matching [87, 115]. Dunfield and Hirani [71] removed the coefficient restriction, proving NP-hardness by reduction from 1-in-3-SAT [87, 163]; their proof simplifies an earlier proof by Agol et al. [3] that finding a minimum-area surface with a given boundary in a *piecewise-linear* 3-manifold is NP-hard.

The minimum-cost representative in a given *integer* homology class is the solution to an *integer* program, obtained by adding the constraints $\bm{x}^+, \bm{x}^- \in \mathbb{Z}^m$ and $\bm{y} \in \mathbb{Z}^n$ to the linear program (LP). Integer programming is well-known to be NP-hard in general [87, 115]. However, some interesting families of of integer programs can be solved in polynomial time, and these can be exploited to compute optimal $\mathbb{Z}$-homologous p-chains in polynomial time in certain families of spaces.

8.3. Total unimodularity. A matrix is *totally unimodular* if every square minor has determinant -1, 0, or 1. Cramer's rule implies that for any totally unimodular matrix A and any integer vector $\bm{b}$, every vertex of the polyhedron $\{\bm{x} \mid A\bm{x} = \bm{b}, \bm{x} \geq 0\}$ is integral [107]. Thus, any linear program with a totally unimodular constraint matrix A and an integral constraint vector $\bm{b}$ has an integral solution. In other words, a totally unimodular integer program can be solved in polynomial time by dropping the integrality constraint and solving the resulting linear program.

Call a simplicial complex *totally p-unimodular* if its $(p+1)$th boundary matrix $[\partial_{p+1}]$ is totally unimodular. For such complexes, the linear program (LP) automatically has integral solutions, and thus can be used to find optimal $\mathbb{Z}$-homologous chains in polynomial time [62].

Total unimodularity is of central importance in combinatorial optimization, but its first application was actually topological. In the same paper that first describes the standard reduction algorithm to compute homology [155], Poincaré observed (in modern terminology) that the pth homology group $H_p(\Sigma; \mathbb{Z})$ of any totally p-unimodular simplicial complex Σ is torsion-free. This observation is a straightforward consequence of the reduction algorithm; each diagonal element in the Smith normal form of any matrix is the greatest common divisor of all subdeterminants of a certain size [**173**].

Poincaré also described a simple condition involving cycles of elements in the boundary matrix that implies total unimodularity. Poincaré's condition is more easily explained in topological terms. Following Dey et al. [**62**], a *cycle complex* is a pure simplicial complex whose dual 1-skeleton is a cycle, and a *Möbius complex* is a non-orientable cycle complex. For example, a 2-dimensional Möbius complex is a triangulation of the Möbius band with all vertices on the boundary, and 1-dimensional Möbius complexes do not exist. Poincaré proved by induction that any simplicial complex Σ with no $(p+1)$-dimensional Möbius subcomplex is totally p-unimodular [**155**, Section 6]. (It follows immediately that the incidence matrix of any directed graph—that is, the 1st boundary matrix of any 1-dimensional simplicial complex—is totally unimodular; this theorem is often attributed to Heller and Tomkins [**104**].)

As observed by Dey et al. [**62**], Poincaré's theorem implies that all orientable $(p + 1)$-manifolds (possibly with boundary) are totally p-unimodular, as are all simplicial complexes embedded in $\mathbb{R}^{p+1}$. It follows that optimal $\mathbb{Z}$-homologous p-chains in such complexes can be computed in polynomial time. We discuss a more efficient algorithm for these special cases in the next section. Grady [**93**, **94**] sketches a slightly weaker condition than total p-unimodularity that still supports polynomial-time solutions.

Dey et al. [**62**] recently extended these results in several directions. First, they proved that a complex Σ is totally p-unimodular if and only if the relative homology group $H_p(L, L_0)$ is torsion-free for all pure subcomplexes $L_0 \subset L \subseteq \Sigma$ such that L_0 has dimension p and L has dimension $p+1$. Note that a $(p+1)$-dimensional cycle complex L is a Möbius complex if and only if $H_p(L, \partial L)$ has nontrivial torsion. They also proved that any 2-dimensional complex is totally 1-unimodular *if and only if* it has no Möbius subcomplex; this equivalence does not extend to higher dimensions.

Finally, Dey et al. [**62**] proved that optimal *unitary* $\mathbb{Z}$-homologous p-chains in totally p-unimodular complexes can be computed in polynomial time, by adding the constraints $\boldsymbol{x}^+ \leq 1$ and $\boldsymbol{x}^- \leq 1$ to (LP) and solving the resulting linear program. The solution $\boldsymbol{x}$ to this augmented linear program is not necessarily the homologous chain with minimum weighted L_0-norm [**62**, Remark 3.11].

8.4. Manifolds and circulations. Motivated by a problem in minimal surface construction, Sullivan [**176**] developed a polynomial-time algorithm for the special case where Σ is an orientable $(p+1)$-manifold, exploiting both linear programming duality and Poincaré duality. For example, when $p = 2$, Sullivan's algorithm finds minimum-weight homologous 2-chains (intuitively, discrete surfaces) in triangulated 3-manifolds. Essentially the same algorithm was rediscovered by Bueller et al. [**24**, **92**, **119**]; see also recent related results of Grady [**93**, **94**].

Let G denote the dual 1-skeleton of Σ; this graph has a vertex for every $(p+1)$-cell of Σ and an edge for every p-cell of Σ. Conveniently, G has n vertices and m edges. Recall that each edge e in G is represented by a symmetric pair of directed edges or *darts*; each dart is dual to one of the orientations of the p-cell whose dual edge is e. The dual of (LP) is another linear program, which has a dual variable for each dart:

(LP*) $$\begin{aligned}\text{maximize} \quad & \sum_{u\to v} \varphi_{u\to v}\, c_{u\to v} \\ \text{subject to} \quad & \sum_u \varphi_{u\to v} = \sum_u \varphi_{v\to u} && \text{for every vertex } v \\ & \varphi_{u\to v} \le w_{uv} && \text{for every dart } u\to v \\ & \varphi_{u\to v} \ge 0 && \text{for every dart } u\to v \end{aligned}$$

Here, w_{uv} the weight of the p-cell whose dual edge is uv, and $c_{u\to v}$ is the coefficient of the input chain c for the *oriented* p-cell whose dual dart is $u\to v$. In particular, we have $c_{u\to v} = -c_{v\to u}$ for every dart $u\to v$.

Up to a sign change, this dual linear program describes the standard *minimum-cost circulation* problem. Intuitively, the dual variable $\varphi_{u\to v}$ represents an amount of *flow* traversing edge uv from u to v; without loss of generality, we can assume that either $\varphi_{u\to v} = 0$ or $\varphi_{v\to u} = 0$ for every edge uv. The equality constraint states that the total flow into any vertex equals the total flow out of that vertex; a vector φ that satisfies this constraint is called a *circulation*. Restated in topological language, a circulation φ is a real 1-cycle in G, or equivalently, a real *1-cocycle* in the primal complex Σ. The weight of the corresponding p-cell in Σ is interpreted as the *capacity* of the dual edge uv; a circulation φ that satisfies the capacity and non-negativity constraints is said to be *feasible*. In the objective function, each coefficient of the input chain c is interpreted as the *cost* of sending one unit of flow in either direction across the corresponding dual edge.

Several specialized algorithms are known for the minimum-cost circulation problem that are faster than general-purpose linear programming algorithms [**4**, **166**]. Sullivan [**176**] described an algorithm (independently proposed by Röck [**162**] and by Bland and Jensen [**14**]) that runs in $O(mn^2 \log n)$ time if the input chain c is unitary, and a second algorithm that runs in $O(mn)$ time if in addition every p-cell has weight 1. The fastest algorithm known (in terms of m and n) for the general minimum-cost circulation problem, due to Orlin [**153**], runs in $O(m^2 \log n + mn \log^2 n)$ time. Each of these algorithms either returns or can be modified to return a solution x to the primal linear program (LP) along with the minimum-cost circulation φ. Moreover, if the input p-chain c is integral, the p-chain x output by these algorithms is also integral, which implies that x is also the optimal $\mathbb{Z}$-homologous p-chain.

THEOREM 8.1 (Sullivan [**176**] and Orlin [**153**]). *Let Σ be an orientable combinatorial $(p+1)$-manifold Σ with n $(p+1)$-cells and m p-cells, and let c be an integer p-chain in Σ. A p-chain $\mathbb{Z}$-homologous to c with minimum weighted L_1-norm can be computed in $O(m^2 \log n + mn \log^2 n)$ time.*

8.5. Back to surfaces. Finally, we we describe a recent algorithm of Chambers et al. [**37**] to compute optimal $\mathbb{R}$-homologous circulations (real 1-cycles) in combinatorial surfaces. Their algorithm is a generalization of algorithms for computing maximum flows in planar graphs, which have been an object of study for more than 50 years; see Weihe [**188**] or Borradaile and Klein [**15, 19**] for a detailed history. Unlike the algorithms described in the previous sections, this algorithm requires that the input chain is a 1-cycle; that is, its boundary *must* be empty.

Let G be a cellularly embedded graph on some orientable surface Σ. Given a circulation c in G, our goal is to compute another circulation ϕ with minimum weighted L_1-norm that is homologous with c. Like Sullivan's algorithm [**176**], the algorithm of Chambers *et al.* considers the LP-dual formulation as a *minimum-cost circulation* problem in the dual graph G^*, as described by (LP*). We emphasize that the primal and dual circulation problems are distinct. In the primal circulation problem, the edges of G have non-negative weights but infinite capacities; whereas, in the dual problem, the edges of G^* have both weights (costs) and non-negative capacities. Specifically, the cost of each dual dart in G^* is the coefficient of the input chain c for the corresponding primal dart in G, and the capacity of each dual dart is the weight of the corresponding primal dart. Moreover, in the primal problem, the solution ϕ is restricted to a particular homology class; whereas, the dual problem imposes no such restriction.

Homology between circulations in G^* can be characterized in terms of cycles in the primal graph G as follows. A *cocycle* λ^* in G^* is any subgraph dual to a directed cycle λ in G. For any flow φ, let $\varphi(\lambda^*) = \sum_{u \to v \in \lambda^*} \varphi_{u \to v}$ denote the total flow through the edges of λ^*. Chambers *et al.* [**37**] observe that two circulations φ and ψ are homologous if and only if $\phi(\lambda) = \psi(\lambda)$ for every cocycle λ.

Recall that a circulation φ is *feasible* if $\varphi_{u \to v} \leq w_{u \to v}$ for every dart $u \to v$ in G^*; call a homology class of circulations *feasible* if it contains a feasible circulation. Let $w(\lambda^*)$ denote sum of the capacities of the edges in any cocycle λ^*. Generalizing an observation of Venkatesan [**186**] for planar networks, Chambers *et al.* [**37**] prove that the homology class of a circulation φ is feasible if and only if $\varphi(\lambda^*) \leq w(\lambda^*)$ for every cocycle λ^*. Moreover, this condition can be checked by solving a single-source shortest path problem in the primal graph G, but with different, possibly negative, edge weights. Because the necessary edge weights may be negative, Dijkstra's algorithm cannot be used to solve this shortest-path problem; Chambers *et al.* describe a suitable shortest-path algorithm that runs in $O(g^2 n \log^2 n)$ time, generalizing an earlier algorithm for planar graphs by Klein *et al.* [**120, 146**]. Their algorithm returns either a feasible circulation homologous to φ or a cocycle that is over-saturated by φ.

To simplify notation, let $c(\varphi) := \sum_{u \to v} \varphi_{u \to v} c_{u \to v}$ denote the total cost of a circulation φ in G^*; this is the objective function in (LP*). The input circulation c can be expressed as a weighted sum of directed cycles. It follows that the dual cost function is *homology-invariant*; that is, $c(\varphi) = c(\psi)$ for any homologous circulations φ and ψ in G^*. Moreover, for each cocycle λ in G, the inequality $\varphi(\lambda^*) \leq w(\lambda^*)$ is a *linear* constraint on the homology class of φ. Thus, the set of all feasible homology classes is a convex polyhedron in $H_1(\Sigma, \mathbb{R}) \cong \mathbb{R}^{2g}$, and finding the feasible homology class of minimum cost is a $(2g)$-dimensional linear programming problem. More careful analysis reveals that this linear program is just a linear projection of the $O(n)$-dimensional min-cost circulation linear program (LP*) into the homology subspace $\mathbb{R}^{2g}$.

Unfortunately, this linear program appears to have $n^{O(g)}$ non-redundant constraints, so it cannot be solved directly, but it can be solved using implicit methods that apply the new shortest-path algorithm as a membership and separation oracle. If the edge capacities are integers less than C, the central-cut ellipsoid method [**96, 97**] solves the linear program in $O(g^8 n \log^2 n \log^2 C)$ time. Alternatively, multidimensional parametric search [**2, 44, 45, 152**], together with a parallel shortest-path algorithm of Cohen [**43**], gives us a combinatorial algorithm that runs in $g^{O(g)} n^{3/2}$ arithmetic operations, for arbitrary capacities. When g is constant, both time bounds are faster by roughly a factor of $\sqrt{n}$ than the fastest minimum-cost circulation algorithms for general sparse graphs [**67, 153**].

References

[1] Pankaj K. Agarwal, Sariel Har-Peled, Micha Sharir, and Kasturi R. Varadarajan, *Approximating shortest paths on a convex polytope in three dimensions*, J. ACM **44** (1997), no. 4, 567–584.

[2] Pankaj K. Agarwal, Micha Sharir, and Sivan Toledo, *An efficient multi-dimensional searching technique and its applications*, Tech. Rep. CS-1993-20, Dept. Comp. Sci., Duke Univ., August 1993.

[3] Ian Agol, Joel Hass, and William P. Thurston, *The computational complexity of knot genus and spanning area*, Trans. Amer. Math. Soc. **358** (2006), no. 9, 3821–3850.

[4] Ravindra K. Ahuja, Thomas L. Magnanti, and James Orlin, *Network flows: Theory, algorithms, and applications*, Prentice Hall, 1993.

[5] Lyudmil Aleksandrov, Anil Maheshwari, and Jörg-Rüdiger Sack, *Determining approximate shortest paths on weighted polyhedral surfaces*, J. ACM **52** (2005), no. 1, 25–53.

[6] James W. Alexander, II, *Normal forms for one- and two-sided surfaces*, Ann. Math. **16** (1914–1915), no. 1/4, 158–161.

[7] Alexandr D. Alexandrov, *Existence of a convex polyhedron and of a convex surface with a given metric*, Rec. Math. [Mat. Sbornik] N. S. **11(53)** (1942), no. 1–2, 15–65, In Russian, with English summary.

[8] Ernst Althaus and Christian Fink, *A polyhedral approach to surface reconstruction from planar contours*, Proc. 9th Int. Conf. Integer Prog. Combin. Optim., Lecture Notes Comput. Sci., vol. 2337, Springer-Verlag, 2006, pp. 258–272.

[9] Sanjeev Arora, Laszlo Babai, Jacques Stern, and Z Sweedyk, *The hardness of approximate optima in lattices, codes, and systems of linear equations*, J. Comput. Syst. Sci. **54** (1997), no. 2, 317–331.

[10] Dominique Attali, André Lieutier, and David Salinas, *Efficient data structure for representing and simplifying simplicial complexes in high dimensions*, Proc. 27th Ann. Symp. Comput. Geom., 2011, pp. 501–509.

[11] Reuven Bar-Yehuda and Bernard Chazelle, *Triangulating disjoint Jordan chains*, Int. J. Comput. Geom. Appl. **4** (1994), no. 4, 475–481.

[12] Bruce G. Baumgart, *Winged edge polyhedron representation*, Tech. Report CS-TR-72-320, Dept. Comput. Sci., Stanford Univ., 1972.

[13] Norman Biggs, *Spanning trees of dual graphs*, J. Comb. Theory **11** (1971), 127–131.

[14] Robert G. Bland and David L. Jensen, *On the computational behavior of a polynomial-time network flow algorithm*, Math. Program. **54** (1992), no. 1, 1–39.

[15] Glencora Borradaile, *Exploiting planarity for network flow and connectivity problems*, Ph.D. thesis, Brown University, May 2008.

[16] Glencora Borradaile, Erik D. Demaine, and Siamak Tazari, *Polynomial-time approximation schemes for subset-connectivity problems in bounded-genus graphs*, Proc. 26th Int. Symp. Theoretical Aspects Comput. Sci., Leibniz Int. Proc. Informatics, vol. 3, Schloss Dagstuhl–Leibniz-Zentrum für Informatik, 2009, pp. 171–182.

[17] Glencora Borradaile, Claire Kenyon-Mathieu, and Philip N. Klein, *A polynomial-time approximation scheme for Steiner tree in planar graphs*, Proc. 18th Ann. ACM-SIAM Symp. Discrete Algorithms, 2007, pp. 1285–1294.

[18] ———, *Steiner tree in planar graphs: An $O(n \log n)$ approximation scheme with singly-exponential dependence on epsilon*, Proc. 10th Workshop on Algorithms and Data Structures, 2007, pp. 275–286.

[19] Glencora Borradaile and Philip Klein, *An $O(n \log n)$ algorithm for maximum st-flow in a directed planar graph*, J. ACM **56** (2009), no. 2, 9:1–30.

[20] Glencora Borradaile, James R. Lee, and Anastasios Sidiropoulos, *Randomly removing g handles at once*, Proc. 25th Ann. Symp. Comput. Geom., 2009, pp. 371–376.

[21] Otakar Borůvka, *O jistém problému minimálním [About a certain minimal problem]*, Práce Moravské Přírodovědecké Společnosti v Brně III **3** (1926), 37–58, English translation in [**151**].

[22] Henry R. Brahana, *Systems of circuits on two-dimensional manifolds*, Ann. Math. **23** (1922), no. 2, 144–168.

[23] Martin R. Bridson, *Combings of semidirect products and 3-manifold groups*, Geom. Funct. Anal. **3** (1993), no. 3, 263–278.

[24] Chris Buehler, Steven J. Gortler, Michael F. Cohen, and Leonard McMillan, *Minimal surfaces for stereo*, Proc. 7th European Conf. Comput. Vision, vol. 3, 2002, pp. 885–899.

[25] James R. Bunch and John E. Hopcroft, *Triangular factorization and inversion by fast matrix multiplication*, Math. Comput. **28** (1974), no. 125, 231–236.

[26] Sergio Cabello and Erin W. Chambers, *Multiple source shortest paths in a genus g graph*, Proc. 18th Ann. ACM-SIAM Symp. Discrete Algorithms, 2007, pp. 89–97.

[27] Sergio Cabello, Erin W. Chambers, and Jeff Erickson, *Multiple-source shortest paths in embedded graphs*, Preprint, February 2012, ArXiv:1202.0314. Full version of [**26**].

[28] Sergio Cabello, Éric Colin de Verdière, and Francis Lazarus, *Finding cycles with topological properties in embedded graphs*, SIAM J. Discrete Math. **25** (2011), 1600–1614.

[29] Sergio Cabello, Matt DeVos, Jeff Erickson, and Bojan Mohar, *Finding one tight cycle*, ACM Trans. Algorithms **6** (2010), no. 4, article 61.

[30] David Canino, Leila De Floriani, and Kenneth Weiss, *IA*: An adjacency-based representation for non-manifold simplicial shapes in arbitrary dimensions*, Computers and Graphics **35** (2011), no. 3, 747–753, Proc. Shape Modeling International 2011.

[31] Erik Carlsson, Gunnar Carlsson, and Vin de Silva, *An algebraic topological method for feature identification*, Int. J. Comput. Geom. Appl. **16** (2006), no. 4, 291–314.

[32] Gunnar Carlsson, Tigran Ishkhanov, Vin de Silva, and Afra Zomorodian, *On the local behavior of spaces of natural images*, Int. J. Comput. Vision **76** (2008), no. 1, 1–12.

[33] Gunnar Carlsson, Afra Zomorodian, Anne Collins, and Leonidas J. Guibas, *Persistence barcodes for shapes*, Int. J. Shape Modeling **11** (2005), no. 2, 149–187.

[34] Erin W. Chambers, Éric Colin de Verdière, Jeff Erickson, Francis Lazarus, and Kim Whittlesey, *Splitting (complicated) surfaces is hard*, Comput. Geom. Theory Appl. **41** (2008), no. 1–2, 94–110.

[35] Erin W. Chambers, Jeff Erickson, and Amir Nayyeri, *Homology flows, cohomology cuts*, Proc. 42nd Ann. ACM Symp. Theory Comput., 2009, pp. 273–282.

[36] _____, *Minimum cuts and shortest homologous cycles*, Proc. 25th Ann. Symp. Comput. Geom., 2009, pp. 377–385.

[37] _____, *Homology flows, cohomology cuts*, SIAM J. Comput. (to appear), Full version of [**35**].

[38] Bernard Chazelle, *A theorem on polygon cutting with applications*, Proc. 23rd Ann. IEEE Symp. Found. Comput. Sci., 1982, pp. 339–349.

[39] _____, *Triangulating a simple polygon in linear time*, Discrete Comput. Geom. **6** (1991), 485–524.

[40] Chao Chen and Daniel Freedman, *Hardness results for homology localization*, Proc. 21st Ann. ACM-SIAM Symp. Discrete Algorithms, 2010, pp. 1594–1604.

[41] _____, *Measuring and computing natural generators for homology groups*, Comput. Geom. Theory Appl. **43** (2010), no. 2, 169–181.

[42] Jindong Chen and Yijie Han, *Shortest paths on a polyhedron, part I: Computing shortest paths*, Int. J. Comput. Geom. Appl. **6** (1996), no. 2, 127–144.

[43] Edith Cohen, *Efficient parallel shortest-paths in digraphs with a separator decomposition*, J. Algorithms **21** (1996), 331–357.

[44] Edith Cohen and Nimrod Megiddo, *Maximizing concave functions in fixed dimension*, Complexity in Numerical Optimization (Panos M. Pardalos, ed.), World Scientific, 1993, pp. 74–87.

[45] _____, *Strongly polynomial-time and NC algorithms for detecting cycles in periodic graphs*, J. Assoc. Comput. Mach. **40** (1993), no. 4, 791–830.

[46] Richard Cole and Alan Siegel, *River routing every which way, but loose*, Proc. 25th Ann. IEEE Symp. Found. Comput. Sci., 1984, pp. 65–73.

[47] Éric Colin de Verdière, *Shortest cut graph of a surface with prescribed vertex set*, Proc. 18th Ann. Europ. Symp. Algorithms, Lecture Notes Comput. Sci., vol. 6347, 2010, pp. 100–111.

[48] Éric Colin de Verdière and Jeff Erickson, *Tightening non-simple paths and cycles on surfaces*, SIAM J. Comput. **39** (2010), no. 8, 3784–3813.

[49] Éric Colin de Verdière and Francis Lazarus, *Optimal system of loops on an orientable surface*, Discrete Comput. Geom. **33** (2005), no. 3, 507–534.

[50] Anne Collins, Afra Zomorodian, Gunnar Carlsson, and Leonidas J. Guibas, *A barcode shape descriptor for curve point cloud data*, Comput. & Graphics **28** (2004), no. 6, 881–894.

[51] Don Coppersmith and Shmuel Winograd, *Matrix multiplication via arithmetic progressions*, J. Symb. Comput. **9** (1990), no. 3, 251–280.

[52] Thomas H. Cormen, Charles E. Leiseron, Ronald L. Rivest, and Clifford Stein, *Introduction to algorithms*, 3rd ed., MIT Press, 2009.

[53] Shervin Daneshpajouh, Mohammad Ali Abam, Lasse Deleuran, and Mohammad Ghodsi, *Computing strongly homotopic line simplification in the plane*, Proc. 27th Europ Workshop Comput. Geom., 2011, pp. 185–188.

[54] Marc de Berg, Marc van Kreveld, and Stefan Schirra, *Topologically correct subdivision simplification using the bandwidth criterion*, Cartography and Geographic Inform. Syst. **25** (1998), no. 4, 243–257.

[55] Mark de Berg, Otfried Cheong, Marc van Kreveld, and Mark Overmars, *Computational geometry: Algorithms and applications*, 3rd ed., Springer-Verlag, 2008.

[56] Leila de Floriani and Annie Hui, *Data structures for simplicial complexes: an analysis and a comparison*, Proc. 3rd Eurographics Symp. Geom. Processing, 2005, pp. 119–128.

[57] Vin de Silva and Robert Ghrist, *Homological sensor networks*, Notices Amer. Math. Soc. **54** (2007), no. 1, 10–17.

[58] Vin de Silva, Robert Ghrist, and Abubakr Muhammad, *Blind swarms for coverage in 2-D*, Proc. Robotics: Science and Systems, 2005, pp. 335–342.

[59] Max Dehn, *Transformation der Kurven auf zweiseitigen Flächen*, Math. Ann. **72** (1912), no. 3, 413–421.

[60] Erik D. Demaine, MohammadTaghi Hajiaghayi, and Bojan Mohar, *Approximation algorithms via contraction decomposition*, Proc. 18th Ann. ACM-SIAM Symp. Discrete Algorithms, 2007, pp. 278–287.

[61] Tamal K. Dey and Sumanta Guha, *Transforming curves on surfaces*, J. Comput. System Sci. **58** (1999), 297–325.

[62] Tamal K. Dey, Anil N. Hirani, and Bala Krishnamoorthy, *Optimal homologous cycles, total unimodularity, and linear programming*, Proc. 42nd Ann. ACM Symp. Theory Comput., 2010, pp. 221–230.

[63] Tamal K. Dey, Kuiyu Li, and Jian Sun, *On computing handle and tunnel loops*, IEEE Proc. Int. Conf. Cyberworlds, 2007, pp. 357–366.

[64] Tamal K. Dey, Kuiyu Li, Jian Sun, and David Cohen-Steiner, *Computing geometry-aware handle and tunnel loops in 3D models*, ACM Trans. Graphics **27** (2008), no. 3, 1–9, Proc. SIGGRAPH 2008.

[65] Tamal K. Dey and Haijo Schipper, *A new technique to compute polygonal schema for 2-manifolds with application to null-homotopy detection*, Discrete Comput. Geom. **14** (1995), no. 1, 93–110.

[66] Tamal K. Dey, Jian Sun, and Yusu Wang, *Approximating loops in a shortest homology basis from point data*, Proc. 26th Ann. Symp. Comput. Geom., 2010, pp. 166–175.

[67] Samuel I. Diatch and Daniel A. Spielman, *Faster lossy generalized flow via interior point algorithms*, Proc. 40th Ann. ACM Symp. Theory Comput., 2008, pp. 451–460.

[68] Jean-Guillaume Dumas, Frank Heckenbach, B. David Saunders, and Volkmar Welker, *Computing simplicial homology based on efficient Smith normal form algorithms*, Algebra, Geometry, and Software Systems (Michael Joswig and Nobuki Takayama, eds.), Springer-Verlag, 2003, pp. 177–206.

[69] Jean-Guillaume Dumas, B. David Saunders, and Gilles Villard, *On efficient sparse integer matrix Smith normal form computations*, J. Symb. Comput. **32** (2001), 71–99.

[70] Christian A. Duncan, Alon Efrat, Stephen G. Kobourov, and Carola Wenk, *Drawing with fat edges*, Int. J. Found. Comput. Sci. **17** (2006), no. 5, 1143–1164.

[71] Nathan M. Dunfield and Anil N. Hirani, *The least spanning area of a knot and the optimal bounding chain problem*, Proc. 27th Ann. Symp. Comput. Geom., 2011, pp. 135–144.

[72] Wayne Eberly, Mark Giesbrecht, and Gilled Villard, *On computing the determinant and Smith form of an integer matrix*, Proc. 41st IEEE Symp. Found. Comput. Sci., 2000, pp. 675–685.

[73] Herbert Edelsbrunner and John Harer, *Persistent homology—a survey*, Essays on Discrete and Computational Geometry: Twenty Years Later (Jacob E. Goodman, János Pach, and Richard Pollack, eds.), Contemporary Mathematics, no. 453, American Mathematical Society, 2008, pp. 257–282.

[74] Herbert Edelsbrunner, David Letscher, and Afra Zomorodian, *Topological persistence and simplification*, Discrete Comput. Geom. **28** (2002), 511–533.

[75] Alon Efrat, Stephen G. Kobourov, Michael Stepp, and Carola Wenk, *Growing fat graphs*, Proc. 18th Ann. Symp. Comput. Geom., 2002, pp. 277–278.

[76] David Eppstein, *Dynamic generators of topologically embedded graphs*, Proc. 14th Ann. ACM-SIAM Symp. Discrete Algorithms, 2003, pp. 599–608.

[77] Jeff Erickson, *Parametric shortest paths and maximum flows in planar graphs*, Proc. 21st Ann. ACM-SIAM Symp. Discrete Algorithms, 2010, pp. 794–804.

[78] Jeff Erickson and Sariel Har-Peled, *Optimally cutting a surface into a disk*, Discrete Comput. Geom. **31** (2004), 37–59.

[79] Jeff Erickson and Amir Nayyeri, *Minimum cuts and shortest non-separating cycles via homology covers*, Proc. 22nd Ann. ACM-SIAM Symp. Discrete Algorithms, 2011, pp. 1166–1176.

[80] Jeff Erickson and Kim Whittlesey, *Greedy optimal homotopy and homology generators*, Proc. 16th Ann. ACM-SIAM Symp. Discrete Algorithms, 2005, pp. 1038–1046.

[81] Regina Estowski and Joseph S. B. Mitchell, *Simplifying a polygonal subdivision while keeping it simple*, Proc. 17th Ann. Symp. Comput. Geom., 2001, pp. 40–49.

[82] Jittat Fakcharoenphol and Satish Rao, *Planar graphs, negative weight edges, shortest paths, and near linear time*, J. Comput. Syst. Sci. **72** (2006), no. 5, 868–889.

[83] Xin Gui Fang and Goerge Havas, *On the worst-case complexity of integer Gaussian elimination*, Proc. 1997 Int. Symp. Symb. Alg. Comput., 1997, pp. 28–31.

[84] Leila De Floriani, Annie Hui, Daniele Panozzo, and David Canino, *A dimension-independent data structure for simplicial complexes*, Proc. 19th International Meshing Roundtable (Suzanne Shontz, ed.), Springer-Verlag, 2010, pp. 403–420.

[85] Greg N. Frederickson, *Fast algorithms for shortest paths in planar graphs with applications*, SIAM J. Comput. **16** (1987), no. 6, 1004–1004.

[86] Shaodi Gao, Mark Jerrum, Michael Kaufmann, Kurt Mehlhorn, and Wolfgang Rülling, *On continuous homotopic one layer routing*, Proc. 4th Ann. Symp. Comput. Geom., 1988, pp. 392–402.

[87] Michael R. Garey and David S. Johnson, *Computers and intractability: A guide to the theory of NP-completeness*, W. H. Freeman, New York, NY, 1979.

[88] Shayan Oveis Gharan and Amin Saberi, *The asymmetric traveling salesman problem on graphs with bounded genus*, Proc. 22nd Ann. ACM-SIAM Symp. Discrete Algorithms, 2011, pp. 967–975.

[89] Robert W. Ghrist and Abubakr Muhammad, *Coverage and hole-detection in sensor networks via homology*, Proc 4th. Int. Symp. Inform. Proc. Sensor Networks, 2005, pp. 254–260.

[90] Peter Giblin, *Graphs, surfaces and homology*, 3rd ed., Cambridge Univ. Press, 2010.

[91] Steven Gortler and Dylan Thurston, *Personal communication*, 2005.

[92] Steven J. Gortler and Danil Kirsanov, *A discrete global minimization algorithm for continuous variational problems*, Comput. Sci. Tech. Rep. TR-14-04, Harvard Univ., 2004.

[93] Leo Grady, *Computing exact discrete minimal surfaces: Extending and solving the shortest path problem in 3D with applicaton to segmentation*, Proc. IEEE CS Conf. Comput. Vis. Pattern Recog., vol. 1, 2006, pp. 67–78.

[94] _____, *Minimal surfaces extend shortest path segmentation methods to 3D*, IEEE Trans. Pattern Anal. Mach. Intell. **32** (2010), no. 2, 321–334.

[95] Jonathan L. Gross and Thomas W. Tucker, *Topological graph theory*, Dover Publications, 2001.

[96] Martin Grötschel, László Lovász, and Alexander Schrijver, *The ellipsoid method and its consequences in combinatorial optimization*, Combinatorica **1** (1981), no. 2, 169–197.

[97] _____, *Geometric algorithms and combinatorial optimization*, 2nd ed., Algorithms and Combinatorics, no. 2, Springer-Verlag, 1993.

[98] Xianfeng Gu, Steven J. Gortler, and Hughes Hoppe, *Geometry images*, ACM Trans. Graphics **21** (2002), no. 3, 355–361.

[99] Xianfeng Gu and Shing-Tung Yau, *Global conformal surface parameterization*, Proc. Eurographics/ACM SIGGRAPH Symp. Geom. Process., 2003, pp. 127–137.
[100] Leonidas J. Guibas and Jorge Stolfi, *Primitives for the manipulation of general subdivisions and the computation of Voronoi diagrams*, ACM Trans. Graphics **4** (1985), no. 2, 75–123.
[101] Igor Guskov and Zoë Wood, *Topological noise removal*, Proc. Graphics Interface, 2001, pp. 19–26.
[102] Joel Hass and Peter Scott, *Intersections of curves on surfaces*, Israel J. Math. **51** (1985), 90–120.
[103] Allen Hatcher, *Algebraic topology*, Cambridge Univ. Press, 2002.
[104] Isidore Heller and Charles Brown Tomkins, *An extension of a theorem of Dantzig's*, Linear Inequalities and Related Systems (Harold W. Kuhn and Albert William Tucker, eds.), Annals of Mathematical Studies, no. 38, Princeton University Press, 1956, pp. 215–221.
[105] Monika R. Henzinger, Philip Klein, Satish Rao, and Sairam Subramanian, *Faster shortest-path algorithms for planar graphs*, J. Comput. Syst. Sci. **55** (1997), no. 1, 3–23.
[106] John Hershberger and Jack Snoeyink, *Computing minimum length paths of a given homotopy class*, Comput. Geom. Theory Appl. **4** (1994), 63–98.
[107] Alan J. Hoffman and Joseph B. Kruskal, *Integral boundary points of convex polyhedra*, Linear Inequalities and Related Systems (Harold W. Kuhn and Albert William Tucker, eds.), Annals of Mathematical Studies, no. 38, Princeton University Press, 1956, pp. 223–246.
[108] Joseph D. Horton, *A polynomial-time algorithm to find the shortest cycle basis of a graph*, SIAM J. Comput. **16** (1987), 358–366.
[109] Alon Itai, Christos H. Papadimitriou, and Jayme Luiz Szwarcfiter, *Hamilton paths in grid graphs*, SIAM J. Comput. **11** (1982), 676–686.
[110] Alon Itai and Yossi Shiloach, *Maximum flow in planar networks*, SIAM J. Comput. **8** (1979), 135–150.
[111] Giuseppe F. Italiano, Yahav Nussbaum, Piotr Sankowski, and Christian Wulff-Nilsen, *Improved algorithms for min cut and max flow in undirected planar graphs*, Proc. 43rd Ann. ACM Symp. Theory Comput., 2011, pp. 313–322.
[112] Donald B. Johnson and Shankar M. Venkatesan, *Partition of planar flow networks (preliminary version)*, Proc. 24th IEEE Symp. Found. Comput. Sci., 1983, pp. 259–264.
[113] Tomasz Kaczynski, Konstantin Mischaikow, and Marian Mrozek, *Computational homology*, Applied Mathematical Sciences, vol. 157, Springer-Verlag, 2004.
[114] Ravindran Kannan and Achim Bachem, *Polynomial algorithms for computing the Smith and Hermite normal forms of an integer matrix*, SIAM J. Comput. **8** (1979), no. 4, 499–507.
[115] Richard M. Karp, *Reducibility among combinatorial problems*, Complexity of Computer Computations (E. Miller and J. W. Thatcher, eds.), Plenum Press, New York, 1972, pp. 85–103.
[116] Ken-ichi Kawarabayashi and Bojan Mohar, *Graph and map isomorphism and all polyhedral embeddings in linear time*, Proc. 40th Ann. ACM Symp. Theory Comput., 2008, pp. 471–480.
[117] Ken-ichi Kawarabayashi and Bruce Reed, *Computing crossing number in linear time*, Proc. 39th Ann ACM Symp. Theory Comput., 2007, pp. 382–390.
[118] Ron Kimmel and James A. Sethian, *Computing geodesic paths on manifolds*, Proc. Nat. Acad. Sci. USA **95** (1998), 8431–8435.
[119] Danil Kirsanov, *Minimal discrete curves and surfaces*, Ph.D. thesis, Div. Engin. Appl. Sci., Harvard Univ., September 2004.
[120] Philip Klein, Shay Mozes, and Oren Weimann, *Shortest paths in directed planar graphs with negative lengths: A linear-space $O(n \log^2 n)$-time algorithm*, ACM Trans. Algorithms **6** (2010), no. 2, article 30.
[121] Philip N. Klein, *Multiple-source shortest paths in planar graphs*, Proc. 16th Ann. ACM-SIAM Symp. Discrete Algorithms, 2005, pp. 146–155.
[122] Jon Kleinberg and Éva Tardos, *Algorithm design*, Addison-Wesley, 2005.
[123] Bruce Kleiner and John Lott, *Notes on Perelman's papers*, Geom. Topol. **12** (2008), no. 5, 2587–2855.
[124] Joseph. B. Kruskal, *On the shortest spanning subtree of a graph and the traveling salesman problem*, Proc. Amer. Math. Soc. **7** (1956), no. 1, 48–50.

[125] Sergei K. Lando and Alexander K. Zvonkin, *Graphs on surfaces and their applications*, Low-Dimensional Topology, no. II, Springer-Verlag, 2004.

[126] Francis Lazarus, Michel Pocchiola, Gert Vegter, and Anne Verroust, *Computing a canonical polygonal schema of an orientable triangulated surface*, Proc. 17th Ann. Symp. Comput. Geom., 2001, pp. 80–89.

[127] Der-Tsai Lee and Franco P. Preparata, *Euclidean shortest paths in the presence of rectilinear barriers*, Networks **14** (1984), 393–410.

[128] James R. Lee and Anastasios Sidiropoulos, *Genus and the geometry of the cut graph*, Proc. 21st Ann. ACM-SIAM Symp. Discrete Algorithms, 2010, pp. 193–201.

[129] Charles E. Leiserson and F. Miller Maley, *Algorithms for routing and testing routability of planar VLSI layouts*, Proc. 17th Ann. ACM Symp. Theory Comput., 1985, pp. 69–78.

[130] Marc Levoy, Kari Pulli, Brian Curless, Szymon Rusinkiewicz, David Koller, Lucas Pereira, Matt Ginzton, Sean E. Anderson, James Davis, Jeremy Ginsberg, Jonathan Shade, and Duane Fulk, *The digital Michelangelo project: 3D scanning of large statues*, Proc. 27th Ann. Conf. Comput. Graph. (SIGGRAPH), 2000, pp. 131–144.

[131] Sóstenes Lins, *Graph-encoded maps*, J. Comb. Theory Ser. B **32** (1982), 171–181.

[132] Richard J. Lipton and Robert E. Tarjan, *A separator theorem for planar graphs*, SIAM J. Applied Math. **36** (1979), no. 2, 177–189.

[133] Yong-Jin Liu, Qian-Yi Zhou, and Shi-Min Hu, *Handling degenerate cases in exact geodesic computation on triangle meshes*, The Visual Computer **23** (2007), no. 9, 661–668.

[134] Roger C. Lyndon and Paul E. Schupp, *Combinatorial group theory*, Springer-Verlag, 1977.

[135] Martin Mareš, *Two linear time algorithms for MST on minor closed graph classes*, Archivum Mathematicum **40** (2004), no. 3, 315–320.

[136] Andrei Andreyevich Markov, *Impossibility of algorithms for recognizing some properties of associative systems*, Dokl. Akad. Nauk SSSR **77** (1951), 953–956, In Russian.

[137] _____, *The insolubility of the problem of homeomorphy*, Dokl. Akad. Nauk SSSR **121** (1958), 218–220.

[138] _____, *Unsolvability of certain problems in topology*, Dokl. Akad. Nauk SSSR **123** (1958), 978–980.

[139] Sergeĭ Vladomirivich Matveev, *Algorithmic topology and classification of 3-manifolds*, Algorithms and Computation in Mathematics, no. 9, Springer-Verlag, 2003.

[140] Jon McCammond and Daniel Wise, *Fans and ladders in combinatorial group theory*, Proc. London Math. Soc. **84** (2002), 499–644.

[141] S. Thomas McCormick, M. R. Rao, and Giovanni Rinaldi, *Easy and difficult objective functions for max cut*, Math. Program., Ser. B **94** (2003), 459–466.

[142] Gary L. Miller, *Finding small simple cycle separators for 2-connected planar graphs*, J. Comput. System Sci. **32** (1986), no. 3, 265–279.

[143] Joseph S. B. Mitchell, David M. Mount, and Christos H. Papadimitriou, *The discrete geodesic problem*, SIAM J. Comput. **16** (1987), 647–668.

[144] Bojan Mohar and Carsten Thomassen, *Graphs on surfaces*, Johns Hopkins Univ. Press, 2001.

[145] John Morgan and Gang Tian, *Ricci flow and the Poincaré conjecture*, Clay Mathematics Monographs, no. 3, American Mathematical Society, 2007.

[146] Shay Mozes and Christian Wulff-Nilsen, *Shortest paths in planar graphs with real lengths in $O(n \log^2 n / \log \log n)$ time*, Proc. 18th Ann. Europ. Symp. Algorithms, Lecture Notes Comput. Sci., no. 6347, Springer-Verlag, 2010, pp. 206–217.

[147] Marian Mrozek, *Čech type approach to computing homology of maps*, Discrete Comput. Geom. **44** (2010), 546–576.

[148] David E. Muller and Franco P. Preparata, *Finding the intersection of two convex polyhedra*, Theoret. Comput. Sci. **7** (1978), 217–236.

[149] Ketan Mulmuley, Umesh Vazirani, and Vijay Vazirani, *Matching is as easy as matrix inversion*, Combinatorica **7** (1987), 105–113.

[150] James R. Munkres, *Topology*, 2nd ed., Prentice-Hall, 2000.

[151] Jaroslav Nešetřil, Eva Milková, and Helena Nešetřilová, *Otakar Borůvka on minimum spanning tree problem: Translation of both the 1926 papers, comments, history*, Discrete Math. **233** (2001), no. 1–3, 3–36.

[152] Carolyn Haibt Norton, Serge A. Plotkin, and Éva Tardos, *Using separation algorithms in fixed dimension*, J. Algorithms **13** (1992), no. 1, 79–98.

[153] James B. Orlin, *A faster strongly polynomial minimum cost flow algorithm*, Oper. Res. **41** (1993), no. 2, 338–350.
[154] Steve Y. Oudot, Leonidas J. Guibas, Jie Gao, and Yue Wang, *Geodesic Delaunay triangulations in bounded planar domains*, ACM Trans. Algorithms **6** (2010), article 67.
[155] Henri Poincaré, *Second complément à l'Analysis Situs*, Proc. London Math. Soc. **32** (1900), 277–308, English translation in [**157**].
[156] _____, *Cinquième complement à l'analysis situs*, Rendiconti del Circulo Matematico di Palermo **18** (1904), 45—110, English translation in [**157**].
[157] _____, *Papers on topology: Analysis Situs and its five supplements*, History of Mathematics, vol. 37, American Mathematical Society, 2010, Translated from the French and with an introduction by John Stillwell.
[158] Konrad Polthier and Marchis Schmies, *Geodesic flow on polyhedral surfaces*, Data Visualization: Proc. Eurographics Worksh. Scientific Visualization, Springer Verlag, 1999, pp. 179–188.
[159] Jean-Philippe Préaux, *Congugacy problems in groups of orientable geometrizable 3-manifolds*, Topology **45** (2006), 171–208.
[160] John Reif, *Minimum s-t cut of a planar undirected network in $O(n \log^2 n)$ time*, SIAM J. Comput. **12** (1983), 71–81.
[161] R. Bruce Richter and Herbert Shank, *The cycle space of an embedded graph*, J. Graph Theory **8** (1984), 365–369.
[162] H. Röck, *Scaling techniques for minimal cost network flows*, Discrete Structures and Algorithms [Proc. 5th Workshop Graph-Theoretic Concepts Comput. Sci.] (U. Pape, ed.), Hanser, München, 1980, pp. 181–191.
[163] Thomas J. Schaefer, *The complexity of satisfiability problems*, Proc. 10th ACM Symp. Theory Comput., 1978, pp. 216–226.
[164] Yevgeny Schreiber, *An optimal-time algorithm for shortest paths on realistic polyhedra*, Discrete Comput. Geom. **43** (2010), no. 1, 21–53.
[165] Yevgeny Schreiber and Micha Sharir, *An optimal-time algorithm for shortest paths on a convex polytope in three dimensions*, Proc. 22nd Ann. Symp. Comput. Geom., 2006, pp. 30–39.
[166] Alexander Schrijver, *Combinatorial optimization: Polyhedra and efficiency*, Algorithms and Combinatorics, no. 24, Springer-Verlag, 2003.
[167] Raimund Seidel, *A simple and fast incremental randomized algorithm for computing trapezoidal decompositions and for triangulating polygons*, Comput. Geom. Theory Appl. **1** (1991), 51–64.
[168] Herbert Seifert and William Threlfall, *Lehrbook der Topologie*, Teubner, Leipzig, 1934, Reprinted by AMS Chelsea, 2003. English translation in [**169**].
[169] _____, *A textbook of topology*, Pure and Applied Mathematics, vol. 89, Academic Press, New York, 1980, Edited by Joan S. Birman and Julian Eisner. Translated from [**168**] by Michael A. Goldman.
[170] James A. Sethian and Alexander Vladimirsky, *Fast methods for the Eikonal and related Hamilton–Jacobi equations on unstructured meshes*, Proc. Nat. Acad. Sci. USA **97** (2000), no. 11, 5699–5703.
[171] Anastasios Sidiropoulos, *Optimal stochastic planarization*, Proc. 51st IEEE Symp. Found. Comput. Sci., 2010.
[172] Martin Škoviera, *Spanning subgraphs of embedded graphs*, Czech. Math. J. **42** (1992), no. 2, 235–239.
[173] Henry John Stephen Smith, *On systems of linear indeterminate equations and congruences*, Phil. Trans. Royal Soc. London **151** (1861), 293–326.
[174] Boris Springborn, Peter Schröder, and Ulrich Pinkall, *Conformal equivalence of triangle meshes*, ACM Trans. Graphics **27** (2008), no. 3, article 77, Proc. SIGGRAPH 2008.
[175] John Stillwell, *Classical topology and combinatorial group theory*, 2nd ed., Graduate Texts in Mathematics, no. 72, Springer-Verlag, 1993.
[176] John Matthew Sullivan, *A crystalline approximation theorem for hypersurfaces*, Ph.D. thesis, Princeton Univ., October 1990.
[177] Vitaly Surazhsky, Tatiana Surazhsky, Danil Kirsanov, Steven J. Gortler, and Hugues Hoppe, *Fast exact and approximate geodesics on meshes*, ACM Trans. Graph. **24** (2005), no. 3, 553–560.

[178] Alireza Tahbaz-Salehi and Ali Jadbabaie, *Distributed coverage verification in sensor networks without location information*, IEEE Trans. Automatic Control **55** (2010), no. 8, 1837–1849.

[179] Robert Endre Tarjan, *Data structures and network algorithms*, CBMS-NSF Regional Conference Series in Applied Mathematics, vol. 44, SIAM, 1983.

[180] Gabriel Taubin and Jarek Rossignak, *Geometric compression through topological surgery*, ACM Trans. Graphics **17** (1998), no. 2, 84–115.

[181] Carsten Thomassen, *Embeddings of graphs with no short noncontractible cycles*, J. Comb. Theory Ser. B **48** (1990), no. 2, 155–177.

[182] William P. Thurston, *The geometry and topology of 3-manifolds*, Princeton University lecture notes, 1980.

[183] Martin Tompa, *An optimal solution to a wire-routing problem*, J. Comput. Syst. Sci. **23** (1981), 127–150.

[184] Yiying Tong, Pierre Alliez, David Cohen-Steiner, and Mathieu Desbrun, *Designing quadrangulations with discrete harmonic forms*, Proc. Eurographics Symp. Geom. Proc., 2006, pp. 201–210.

[185] Gert Vegter and Chee-Keng Yap, *Computational complexity of combinatorial surfaces*, Proc. 6th Ann. Symp. Comput. Geom., 1990, pp. 102–111.

[186] Shankar M. Venkatesan, *Algorithms for network flows*, Ph.D. thesis, The Pennsylvania State University, 1983, Cited in [**112**].

[187] Karl Georg Christian von Staudt, *Geometrie der Lage*, Verlag von Bauer and Rapse (Julius Merz), Nürnberg, 1847.

[188] Karsten Weihe, *Maximum (s,t)-flows in planar networks in $O(|V|\log|V|)$-time*, J. Comput. Syst. Sci. **55** (1997), no. 3, 454–476.

[189] Kevin Weiler, *Edge-based data structures for solid modeling in curved-surface environments*, IEEE Comput. Graph. Appl. **5** (1985), no. 1, 21–40.

[190] Douglas H. Wiedemann, *Solving sparse linear equations over finite fields*, IEEE Trans. Inform. Theory **IT-32** (1986), no. 1, 54–62.

[191] Zoë Wood, Hughes Hoppe, Mathieu Desbrun, and Peter Schröder, *Removing excess topology from isosurfaces*, ACM Trans. Graphics **23** (2004), no. 2, 190–208.

[192] Xiaotian Yin, Miao Jin, and Xianfeng Gu, *Computing shortest cycles using universal covering space*, Vis. Comput. **23** (2007), no. 12, 999–1004.

[193] Qian-Yi Zhou, Tao Ju, and Shi-Min Hu, *Topology repair of solid models using skeletons*, IEEE Trans. Vis. Comput. Graph. **13** (2007), no. 4, 675–685.

[194] Afra Zomorodian, *The tidy set: A minimal simplicial set for computing homology of clique complexes*, Proc. 26th Ann. Symp. Comput. Geom., 2010, pp. 257–266.

[195] Afra Zomorodian and Gunnar Carlsson, *Computing persistent homology*, Discrete Comput. Geom. **33** (2005), no. 2, 249–274.

DEPARTMENT OF COMPUTER SCIENCE, UNIVERSITY OF ILLINOIS, URBANA-CHAMPAIGN
E-mail address: jeffe@cs.illinois.edu
URL: http://www.cs.illinois.edu/~jeffe

Index

$\mathfrak{F}(\Sigma)$, 174
$G_{\leftarrow s}$, 156
$\mathbb{S}^m$, 153
$\mathbb{S}^{n-2}$, 156, 157, 162
$\mathbb{Z}_2$-homology cover, 214
Δ_G, 152, 160
$\overline{\Delta}_G^*$, 166
$\overline{\Delta}_G$, 165
$\odot \models^n$, 157
$\simeq$, 156
$\not\models^n$, 157

action, 150, 159
acyclic, 33
acyclic-valued, 61
additive, 79
adjunction, 92
admissible order, 70
adv, 161
adversity, 161
Alexander-Spanier cohomology, 50
alpha and omega limit sets, 42
alpha complex, 9
analysis, 2
associated digraph, 52

backchain, 173, 188
backchaining, 148
barycentric subdivision, 155
base change, 100
Bessel transform, 117
Betti number, 17
boundary, 16
boundary homomorphism, 16, 33
boundary operator, 57

CAPD, 64
CAPD and CAPD::RedHom software projects, 64
CAPD::RedHom, 64
categorification, 81
category, 184
category of endomorphisms, 49
Čech cohomology, 92
Čech complex, 8
certainly attainable, 162
chain complex, 16
chain group, 16, 33
chain map, 85
chain selector, 61
circuit, 152, 160
classifying space, 184, 186, 188
clique, 10
clique complex, 10
closed, 6
co-index, 126
coboundary, 83

cochain complex, 83
cocycle condition, 112
cocycles, 83
coefficients, 16
coface, 6
cofaces, 83
cohomological Conley index, 48
cohomologous, 83
cohomology, 83
cohomology sheaf, 94
cohomology with compact supports, 83
collapse, 32
combinatorial enclosure, 53
combinatorial index pair, 54
combinatorial map, 52
combinatorial Morse decomposition, 70
combinatorial Morse sets, 70
combinatorial surfaces, 197
commutative diagram, 86
complete invariant, 15
complete strategy, 156, 157
components, 17
conformation space, 2
Conley index, 50
Conley, Charles, 48
Conley-Morse graph, 70
connecting map, 87
constructible, 79
constructible feature size, 139
constructible sheaf, 96
continuation classes, 71
continuation graph, 71
continuous, 6
contractible, 6, 157
contravariant, 89
convergent, 152, 160
convolution, 110
covering set, 162
crossing sequence, 204, 206
cubical complex, 12
cubical product, 56
cycle, 16, 82
cycle-preserving, 185

data, 2
Day, S., 67
decision tree, 180, 181
definable, 78
deformation retract, 47
degeneracy operator, 31
degenerate, 31
Delaunay complex, 9
deletion, 172
derived categories, 94
design, 149, 168
dimension, 6, 17, 52, 56

direct image, 91
discrete Morse theory, 65
discrete semidynamical system, 42
dl, 172
dual, 110
dual complex, 166
dualizing complex, 96

elementary cell, 60
elementary chain, 56
elementary contraction, 29
elementary cube, 52
elementary interval, 52
empty complex, 151
essential, 177
étale sheaf, 90
étale space, 90
Euclidean space, 6
Euler calculus, 75
Euler characteristic, 15, 77
Euler integral, 79, 97
Euler-Poincaré formula, 17
Euler-Poincaré index, 115, 126
evasive, 180, 182
exact, 86
execute, 152
executed, 150, 159
exit set, 47
extrinsic, 3

face, 6
face operator, 31
face poset, 174
filtration, 20
finite type, 27
flabby, 94
flag complex, 10
forward-chaining, 148, 186, 188
Fourier transform, 116
free, 17
free coface, 58
free face, 57
free pair, 29
Frongilo, R., 67
full, 52
full cubical set, 52
full shift, 43
fully controllable, 162
functor, 186, 187

Gaussian elimination, 18
general position, 9
generalized kernel, 49
generator, 42, 48
ghost solutions, 45
global sections, 90
goal-preserving, 185
good, 7
good cover, 93

grading, 82
graph, 148, 150, 160
greedy system of arcs, 205, 206
greedy system of loops, 201
group of q-chains, 56

Hénon map, 67
higher direct image, 94
homeomorphism, 6
Homeomorphism Problem, 14
homologous, 16, 82
homology, 15, 82
homology class, 82
homology group, 16, 57
homology localization, 213, 216
homology model, 63
homotopic, 156
homotopy, 6
homotopy basis, 199
homotopy Conley index, 48
homotopy equivalence, 6
homotopy type, 6
horseshoe map, 50

image, 47
incomplete, 15
indecomposable, 26, 27
index map, 50
index pair, 49
index quadruple, 49
induced topology, 6
injective resolutions, 92
interval extension, 46
intrinsic, 3
intrinsic volumes, 142
invariant part, 42, 48, 54
inverse image, 91
isolated invariant set, 48
isolating block, 47
isolating neighborhood, 48, 54
isospectral set, 116

join, 152
join semi-lattices, 185

k-cochain, 92
Kot-Schaffer growth-dispersal model, 68

landmarks, 11
Leray functor, 49
linear programming, 216
link, 111, 172
lk, 172
local Euler-Poincaré index, 97
local Morse data, 126
long exact sequence, 86
long exact sequence of the pair, 87
loopback actions, 156
loopback complex, 156

loopback complexes, 179
loopback graph, 156
Lorenz system, 45
Lorenz, Edward Norton, 44

manipulate, 147
matroid optimization, 212
maximal, 6, 10
maximum flows, 219
Mayer-Vietoris sequence, 87
Mayer-Vietoris sequence for sheaves, 99
microlocal Fourier transform, 115
minimal nonface, 168
minimal representation, 52
minimal strategy, 182
minimum cuts, 215
minimum-cost circulation, 219
Minkowski sum, 110
Möbius complex, 218
models, 147
Moore, R.E., 45
morphism, 91, 185
Morse, 125
Morse decomposition, 70
Morse index, 125
Morse sets, 70
motivic measure, 76
moves off, 160
moves off W subject to σ, 173
multidimensional persistence, 23
multifiltration, 20
multivalued map, 47

negatively invariant part, 54
neighborhood graph, 10
nerve, 7
non-degenerate, 31
nondeterministic, 150
nonevasive, 180, 182
NP-hard, 203, 212, 213, 214, 217
null space, 18

o-minimal structure, 78
one-critical, 20
open ball, 8
open cover, 7
open set, 6
optimal homology basis, 212
order complex, 174
orientation, 15
oriented simplex, 15

periodic, 42
persistence barcode, 22
persistent homology, 21
plan, 152
planner, 147, 186, 188
Poincaré map, 43
Poincaré section, 43

polyconvex, 78
poset, 174
poset of local strategies, 187
positively invariant part, 54
presheaf, 89
prestacks, 89
projection formula, 111
pullback, 91, 111
pure, 12
pushforward, 91

Quillen Fiber Lemma, 174
quiver, 26

Radon transform, 111
range space, 18
rank invariant, 24
reduction, 18
relative homology, 83
relevant region, 206, 208, 209
representable intervals, 46
representable numbers, 46
representation, 26, 52, 61
representation theory, 26
restricted Voronoi region, 9
return time map, 42
ring of coefficients, 17
Robbin, J., 50
robot, 147

Salamon, D., 50
scissors equivalence, 79
sd, 155
section, 90
sections, 89
selector, 53
semialgebraic, 78
semicontinuous, 47
(semi)dynamical system, 42
sensing relation, 101
sensitive dependence on initial conditions, 44
sensor supports, 101
shaving, 34
sheaf, 88, 89
sheafification, 91
shortest homotopic cycles, 209
shortest homotopic path, 204
simplex, 6
simplices, 6
simplicial complex, 6, 151
simplicial set, 30
skeleton, 7
Smale horseshoe map, 44
Smale, Stephen, 43
Smith diagonalization, 56
Smith normal form, 18
solution, 48, 53
source, 150, 159

source complex, 165, 187
sphere, 153, 156, 184
src, 160
stalk, 89
start region, 160
stationary, 42
stochastic, 159, 160
strategy, 152, 160
strategy complex, 152, 160
strong witness, 11
subanalytic, 78
subcomplex, 7
subgraph, 170
sublevel sets, 21
subshift of finite type, 67
subspace, 6
support, 57
system of arcs, 203
system of loops, 199

t-translation map, 42
tame, 78
target, 159
target supports, 101
targets, 101, 150
thinning, 34
tidy set, 34
tight octagonal decomposition, 206
topological data analysis, 5
topological invariant, 14
topological space, 6
topological type, 6
topology, 1, 6
torsion, 17
total unimodularity, 217
trace, 102
trajectory (orbit), 42
transition time, 161
tree-cotree decomposition, 201
Trevion, R., 67
trimming, 34
trivial invariant, 15
tunnels, 17

uncertainty, 148
underlying space, 7
underlying undirected graph, 26
unified semidynamical system, 69
upper semicontinuous, 47

Verdier duality, 92
Vietoris-Rips complex, 10
void complex, 151
voids, 17
Voronoi diagram, 9
Voronoi region, 9
VR complex, 10

Warmus, M., 45

Ważewski's Theorem, 47, 48
weak feature size, 139
weak preimage, 47
weak witness, 11
witness complex, 11
witness graph, 11
worst-case expected convergence times, 161

zero-skeleton, 151
zigzag, 26
zigzag persistence, 25

编辑手记

世界著名数学家 H. O. 波拉克(H. O. Pollak)曾指出：

数学的应用往往是从数学之外的一个不佳的定义开始的.这项工作是要尽可能好地理解所定义的内容,其工作程序是建立一个数学模型,这个模型将帮助我们搞清我们试图理解的内容.现在外部世界通常是如此的复杂,以至我们不能把它所有的相关特征都包括到数学模型中,也不能指望用那种包罗万象的模型做任何事情.我们将不得不简化事情,仅保留其重要成分.现在的危险是扔得太多了,而得到一个能够计算的数学公式.那么所得结果和原有情况的关系恐怕就很值得怀疑了.数学应用的最有趣的特征之一就是不断地追求数学的简单性,而事实上物理世界又是十分复杂的.

本书是一部与计算相关的英文专著,中文书名或可译为《应用与计算拓扑学进展》,它是 AMS(美国数学学会)计算拓扑短期课程(2011 年 1 月 4 日—5 日,路易斯安那州新奥尔良)的会议记录.

本书的主编为阿弗拉·佐莫罗迪安(Afra Zomorodian),美国人,美国达特茅斯学院计算机科学系的教授,曾是斯坦福大学 Bio-X 的博士后学者.其研究领域为计算拓扑与几何学、计算结构生物学、计算机图形学、可视化、程序设计方法学.

正如前言中所述：

本书是 AMS 计算拓扑短期课程的会议记录,该短期课程是为 2011 年 1 月 4 日至 5 日在新奥尔良举行的联合数学会议组织的.计算拓扑的出现是为了应对几何问题中的拓扑障碍,例如在计算机绘图和计算几何社区重建的表面上的额外孔和隧道.在很多科学领域中拓扑问题的出现是很自然的.在机器人学中,我们需要捕获机器人配置空间的连通性以进行规划。在传感器网络中，我们希望从局部感知中推断出全局信息。在动力学系统中，我们想通过计算来理解系统的定性性质。在数据分析中，我们在给定有限的噪声样本的情况下寻找底空间的稳健特征。

像大部分新兴领域一样，计算拓扑被分成了几个社区，每个社区在它们自己的图像中对计算拓扑进行了定义。通过这门课程，我想扩大该领域的定义，使其包括能够使用计算方法解决拓扑问题的任何领域。为了达到这个目的，我邀请了来自很多专业领域的演讲者，包括代数拓扑、动力系统、应用拓扑、机器人学和计算几何。我也想让该课程涵盖计算拓扑丰富的发展情况，包括从理论到算法设计与分析、快速软件的实施以及应用。

本书的结构反映了该课程的结构。课程的第一天进行拓扑数据分析，其中一位演讲者贡纳尔·卡尔松(Gunnar Garlsson)并没有为本书贡献一个章节，因为他正在编写一本有关本书主题的书。第一天的高潮是软件会议：亨利·亚当斯(Herry Adams)提供了一个关于JPlex的学习指南，它是一个拓扑数据分析的Java软件包。玛丽安·姆罗策克(Marian Mrozek)演示了RedHom，一个计算三次同调的C++库。该课程的第二天集中于拓扑在传感器网络、机器人学和几何中的应用方面。罗伯特·格力斯特(Robert Ghrist)贡献的章节是与他的两位同事合作完成的。

该短期课程以小组会议结束，演讲者和出席者讨论了计算拓扑的现状和未来。十年之前，发表有关该领域的出版物是困难的，因为其所需的数学知识对于计算机科学家来说是未知的，计算机应用程序的价值也没有得到数学家的重视。到现在为止，许多的会议和期刊已经把计算拓扑作为了一个子领域。随着特别研讨会和计划的成熟，计算拓扑将需要专门的会议和期刊，以便研究人员拥有集中的论坛来传播他们的研究成果。

感谢丹·罗克莫尔(Dan Rockmore)招募了该短期课程的提案，还要感谢他给出的许多有益的意见。感谢演讲者们出色的表现，以及他们去年贡献的章节。这本书经历了由演讲者和匿名审稿人参与的两个阶段的同行评审的过程。感谢所有审稿人及时且彻底的批评。最后，感谢美国数学学会的塞奇·盖尔范德(Sergei Gelfand)和克里斯汀·蒂维耶热(Christine Thivierge)指导了这个项目。

本杰明·曼(Benjamin Mann)在DARPA任职期间是应用和计算拓扑的研究者中精力最充沛的。我们将本书献给他，感谢他一直以来的支持。

编辑手记

本书的目录为:

前言
拓扑数据分析
拓扑动力学:通过三次同调的严格数值
欧拉微积分在信号和传感中的应用
带不确定性的拓扑离散计划
循环与基的组合优化
索引

关于计算拓扑学,在《古今数学趣话》中曾有一段这样描述:

在 1976 年的一次国际数学会议上,美国普林斯顿大学的哈罗德·库恩(Harold Kuhn)教授宣读了一篇奇特的论文. 大家知道,n 次方程在复数范围内有 n 个根,但除 $n=2$ 及少数特例外,要找出 n 个复数根是非常困难的. 你想解一个复数系数的 n 次方程 $z^n + a_1 z^{n-1} + \cdots + a_n = 0$ 吗? 那么, 请你看看库恩先生的表演吧:他准备了一个培养皿和一个立体大篱笆,篱笆越往上越密. 然后把你要解的方程的信息"告诉"培养皿. 皿内吐出几个新芽,芽变成藤,飞快地攀上篱笆,一层一层往上穿. 最后,每根藤恰好指向方程的一个根,于是方程的 n 个根就被找出来了(图1).

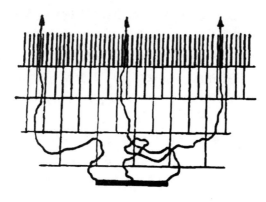

图 1

与会者无不目瞪口呆,惊奇万分. 植物竟会解方程,而且是很难解的方程! 这是在变魔术吗? 不是的! 任何高明的魔术师都只能变出他事先"知道"(暗中准备好了)的东西,绝不可能变出他事先不知道的东西. 这也不是神话,恰恰相反,这是科学. 这个奇迹是怎样创造出来的呢? 原来库恩先生运用了现代数学中一个极为重要的定理,这就是拓扑学中著名的"不动点定理".

我们先从一个有趣的问题谈起.某学生进城,早晨六点从家里出发,下午六点到达.第二天沿原路返回,早晨六点离城,下午六点到家.他和老师谈起了上述经过,老师告诉他:"你知道吗? 途中有一个地点,你昨天进城和今天回来经过那个地方时,所用的时间完全相同."学生说:"没有这么巧的事吧? 我在路上走得时快时慢,有时还停下休息、吃东西,两次经过某地的时间怎么可能完全相同呢?"老师说:"不是可能,而是肯定有这一点,虽然我不知道它到底在哪里."究竟谁是正确的呢?

看起来,学生理由充足,振振有词;而老师既然"肯定"有这一点,又"不知道"这点在哪里,似乎自相矛盾.但其实老师是正确的.道理很简单,设想进城和回家发生在同一天——学生离家往城走,而学生的"替身"则同时离城往家走(途中经过情况与学生回家时完全相同),那么两人必定在路上某地相遇,进城和回家经过这个相遇点的时间不是完全相同的吗? 所以老师是正确的.

这个有趣的问题给著名的"拓扑不动点定理"提供了一个极为生动简明的例证.

高中学生都知道:A,B 两个集合,如果按某种对应关系,使 A 中的任何元素在 B 中仅有唯一的元素和它对应,这样的对应关系称为从 A 到 B 的单值对应,也叫作映射或变换.如果 A 中的某元素通过某种变换后仍变为自己,我们就说这个元素是该变换下的一个不动点.

函数关系 $g(x)$ 就是从一个集合 A(定义域)到另一个集合 B(值域)的一个变换.如果 $g(x_1)=x_1$,即 x_1 通过变换 g 后仍变为自己,x_1 就是函数 $g(x)$ 的一个不动点.所以要找函数的不动点,只需找出满足关系 $g(x)=x$ 的 x 值就可以.如 $g(x)=x^2-2$,由 $x^2-2=x$ 得 $x^2-x-2=0$,即 $(x+1)(x-2)=0$,求出 $x_1=-1,x_2=2$.所以 $g(x)=x^2-2$ 有两个不动点:-1 和 2.它们满足 $g(-1)=-1,g(2)=2$.

变换不一定限于数的对应关系,向量、矩阵、行列式、几何图形甚至时间和空间都可以作为元素进行变换.那么,各种变换下不动点是否存在? 有何特点和用途? 这是现代数学各领域内极有价值而且引人入胜的课题.

拓扑学是几何的一个分支.现代几何学门类繁多,按什么原则来分门别类呢? 原先,人们着重于它们的基础结构——公理系统,将其分成欧几里得(Euclid)几何、罗巴切夫斯基(Lobachevskiĭ)几何……另一位德国数学家克莱因(Klein)从几何的研究对象出发,给了几何学一个新的定义:几何学是研究图形在各种变换群下不变性质的一门科学.例如:研究刚体运动变换群下不变性质的几何通常称为"初等几何"(包括欧几里得几何、罗巴切夫斯基几何和黎曼几何);研究仿射变换群下不变性质的几何称为"仿射几何";还有"射影几何""保角几何""连续几何"等.

连续几何又称"拓扑学"(topology),是研究一对一连续变换(拓扑变换)下不变性质的几何. 21 世纪以来,拓扑学发展很快,引起了许多数学家的兴趣,我国著名数学家吴文俊院士、江泽涵教授等人在拓扑学研究方面做出了许多重大贡献.

什么是拓扑变换呢?一个橡皮筋做的圆,可以把它拉成大圆,也可以压缩成小圆,还可以拉成椭圆、三角形、矩形或任意形状的封闭曲线(但是不能拉断,也不能扭成 ∞ 字形,因为断开或两点贴在一起将改变曲线上各点之间一一对应的连续变换关系),这就是拓扑变换.在拓扑变换下,三角形、矩形、椭圆以及一切自身不相交的简单封闭曲线均可看作"相同"(称为"拓扑等价"或"同胚").这里,简单封闭曲线是拓扑变换下的不变性质,而长度、夹角、面积、体积在这里都没有意义,也就不是拓扑学研究的内容了.

再看看球面,在拓扑变换下,如民间艺人吹糖葫芦,可以吹成各种形状,但是绝不可能吹成救生圈一样的圆环,因为环面和球面不是拓扑等价的,它们有不同的拓扑性质.比如球面上的任何一条闭曲线都可以将球面分成互不相连的两部分,而环面则无此性质.在图 2 中,沿 C 或 D 切开,环面都不能被分为两部.

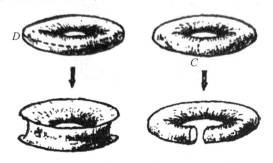

图 2

1912 年,荷兰数学家布劳威尔(Brouwer)证明:任意一个把 n 维球体映入自己的连续映象(即拓扑变换)至少有一个不动点.这就是著名的拓扑不动点定理.

我们知道,直线是一维空间,平面是二维空间,普通空间是三维空间,而四维、五维以至 n 维空间就很抽象了(抽象是数学的一大特点,因为越抽象就越有普遍性).因此,一维球体就是直线上的线段,二维球体是平面上的圆形区域,三维球体则是通常所说的球……

前面谈到的学生的进城路线便可看成一维球体(拓扑学里的线不

分曲直).如果进城经过点 A 与回家经过点 B 的时间相同,我们就说点 A 与点 B 对应,这种对应关系显然是把此线段变换成了自己——即是把一维球体映入自己的连续映象.按照布劳威尔定理,这个变换至少有一个不动点,就是学生与"替身"相遇的地方.

布劳威尔定理的严格证明是很抽象、很艰深的.我们已经对一维球体举出了有趣的例证,下面再举一个二维球体的有趣例证:

取一个浅纸盒与一张纸,这张纸恰好能盖住盒的底面.此时,纸上每个点正好与它下面盒底上的点一一对应.把纸拿起来随便揉成一个小纸球,再把小纸球扔进盒里(图3).不动点定理说:不管小纸球是怎样揉的,也不管它落在盒底的什么地方,揉成小纸球的纸上至少有这样的一个点,它恰好处在盒底原来与它对应的点的正上方.就是说,此点经过拓扑变换后仍变为自己,它是一个"不动点".

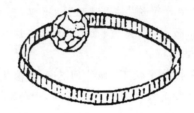

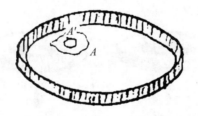

图 3

我们可以给出浅显的证明:小纸球在盒底的水平投影为区域 A,显然,原纸片上和区域 A 内某点相对应的点一定在区域 A 的上方.纸球上原先盖在盒底区域 A 上的那一部分纸片,被随意卷成纸球的一小部分 A'_1,设 A'_1 在盒底的水平投影为区域 A_1(图3),显然有 $A_1 < A$.再考虑原先盖在盒底区域 A_1 上的小纸片,它则被随意卷成纸球中更小的部分 A'_2,设 A'_2 在盒底水平投影为区域 A_2,于是又有 $A_2 < A_1$.这样反复做下去(理论上可作无穷多次),得到一连串区域 $A > A_1 > A_2 > \cdots > A_n > \cdots$.这些区域必然一个比一个小(否则,如果有 $A_i = A_{i+1}$,即小纸球上有一片与盒底上的一片重合,定理显然得证),最后必然缩小到"一点",这一点就是定理所断言存在的"不动点".

这个证明方法我们不妨称为"缩小包围圈",即把点所在的范围越缩越小,最后此点便无可逃遁了.

不动点定理问世以来,引起了科学家的极大兴趣,它有着广泛而奇妙的应用. 在数学中,很多问题的求解可以作为相应的不动点来处理. 例如求方程 $f(x)=0$ 的根,就可以化为求变换 $g(x)=f(x)+x$ 的不动点 x_0,因为它既满足 $g(x_0)=x_0$,又适合 $g(x_0)=f(x_0)+x_0$,于是就有 $f(x_0)=0$,所以 x_0 是方程的根. 我们知道,计算方程 $z^n+a_1z^{n-1}+\cdots+a_n=0$ 在复数范围内的几个根一般相当困难. 1967 年,赫伯特·斯卡弗提出了一种不动点方法,作为一种有效的计算方法,取得了一系列成果. 前面提到的库恩教授的"神奇"植物,便是根据这种方法设计的模型. 我们看看这个模型是怎样设计出来的呢? 这个立体大篱笆是把一系列复数平面 $C_{-1},C_0,C_1,C_2,\cdots$ 像盖楼房那样一层层地排好(图 4),在每个平面上划线,全部剖分成三角形,称为三角剖分(图 4). 每上一层,三角形的边缩小一半(面积缩小四分之三). 相邻两层间用竖的或斜的"钢筋"把层中的空间全部分成四面体,留下这些"钢筋"架,便成了越往上越密的大篱笆. 当你把给定方程的信息传达后,培养皿内的 n 根藤按指定规则绕篱笆一层层往上穿. 因为越往上三角形面积越小(无限减小到接近于零),于是每根藤恰好指向方程的一个复数根. 这正是用"缩小包围圈"的方法寻找不动点的又一个例子.

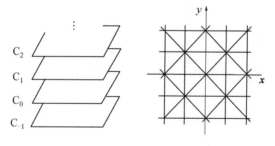

图 4

不动点定理的有趣应用还多着呢!根据这个定理可以断言:在任一时刻,地球上至少有一个地点没有风. 用它还可以证明:如果一个球面完全被毛发覆盖,则无论如何也不能把毛发梳平,但有趣的是,却可以把覆盖在整个圆环面上的毛发梳平.

这个计算不动点的拓扑方法最早是由中山大学的王则柯教授从美国访学回国后介绍到国内的. 后来王则柯先生从一位拓扑学专家转行成为了一位活跃的经济学家.

我们数学工作室从成立至今已有 20 余年. 以后什么样不敢说,起码现在还生存的不错. 借用王则柯先生研究的西方经济学之父亚当·斯密(Adam Smith)的一句话解释其原因.

亚当·斯密说,我们在交易的过程中从来不会说我需要什么,而是说我能够给别人什么,所以英文里面有一句俗语叫"If it pays, it stays",如果你对别人有贡献,你就能够活得久.

我们要持续给社会提供价值,社会就会让我们活得久,活得好!

刘培杰
2025.1.20
于哈工大

刘培杰数学工作室
已出版(即将出版)图书目录——原版影印

书　名	出版时间	定　价	编号
数学物理大百科全书.第1卷(英文)	2016—01	418.00	508
数学物理大百科全书.第2卷(英文)	2016—01	408.00	509
数学物理大百科全书.第3卷(英文)	2016—01	396.00	510
数学物理大百科全书.第4卷(英文)	2016—01	408.00	511
数学物理大百科全书.第5卷(英文)	2016—01	368.00	512
zeta函数,q-zeta函数,相伴级数与积分(英文)	2015—08	88.00	513
微分形式:理论与练习(英文)	2015—08	58.00	514
离散与微分包含的逼近和优化(英文)	2015—08	58.00	515
艾伦·图灵:他的工作与影响(英文)	2016—01	98.00	560
测度理论概率导论,第2版(英文)	2016—01	88.00	561
带有潜在故障恢复系统的半马尔柯夫模型控制(英文)	2016—01	98.00	562
数学分析原理(英文)	2016—01	88.00	563
随机偏微分方程的有效动力学(英文)	2016—01	88.00	564
图的谱半径(英文)	2016—01	58.00	565
量子机器学习中数据挖掘的量子计算方法(英文)	2016—01	98.00	566
量子物理的非常规方法(英文)	2016—01	118.00	567
运输过程的统一非局部理论:广义波尔兹曼物理动力学,第2版(英文)	2016—01	198.00	568
量子力学与经典力学之间的联系在原子、分子及电动力学系统建模中的应用(英文)	2016—01	58.00	569
算术域(英文)	2018—01	158.00	821
高等数学竞赛:1962—1991年的米洛克斯·史怀哲竞赛(英文)	2018—01	128.00	822
用数学奥林匹克精神解决数论问题(英文)	2018—01	108.00	823
代数几何(德文)	2018—04	68.00	824
丢番图逼近论(英文)	2018—01	78.00	825
代数几何学基础教程(英文)	2018—01	98.00	826
解析数论入门课程(英文)	2018—01	78.00	827
数论中的丢番图问题(英文)	2018—01	78.00	829
数论(梦幻之旅):第五届中日数论研讨会演讲集(英文)	2018—01	68.00	830
数论新应用(英文)	2018—01	68.00	831
数论(英文)	2018—01	78.00	832

刘培杰数学工作室
已出版(即将出版)图书目录——原版影印

书　名	出版时间	定　价	编号
湍流十讲(英文)	2018—04	108.00	886
无穷维李代数:第3版(英文)	2018—04	98.00	887
等值、不变量和对称性(英文)	2018—04	78.00	888
解析数论(英文)	2018—09	78.00	889
《数学原理》的演化:伯特兰·罗素撰写第二版时的手稿与笔记(英文)	2018—04	108.00	890
哈密尔顿数学论文集(第4卷):几何学、分析学、天文学、概率和有限差分等(英文)	2019—05	108.00	891
偏微分方程全局吸引子的特性(英文)	2018—09	108.00	979
整函数与下调和函数(英文)	2018—09	118.00	980
幂等分析(英文)	2018—09	118.00	981
李群,离散子群与不变量理论(英文)	2018—09	108.00	982
动力系统与统计力学(英文)	2018—09	118.00	983
表示论与动力系统(英文)	2018—09	118.00	984
分析学练习.第1部分(英文)	2021—01	88.00	1247
分析学练习.第2部分,非线性分析(英文)	2021—01	88.00	1248
初级统计学:循序渐进的方法:第10版(英文)	2019—05	68.00	1067
工程师与科学家微分方程用书:第4版(英文)	2019—07	58.00	1068
大学代数与三角学(英文)	2019—06	78.00	1069
培养数学能力的途径(英文)	2019—07	38.00	1070
工程师与科学家统计学:第4版(英文)	2019—06	58.00	1071
贸易与经济中的应用统计学:第6版(英文)	2019—06	58.00	1072
傅立叶级数和边值问题:第8版(英文)	2019—05	48.00	1073
通往天文学的途径:第5版(英文)	2019—05	58.00	1074
拉马努金笔记.第1卷(英文)	2019—06	165.00	1078
拉马努金笔记.第2卷(英文)	2019—06	165.00	1079
拉马努金笔记.第3卷(英文)	2019—06	165.00	1080
拉马努金笔记.第4卷(英文)	2019—06	165.00	1081
拉马努金笔记.第5卷(英文)	2019—06	165.00	1082
拉马努金遗失笔记.第1卷(英文)	2019—06	109.00	1083
拉马努金遗失笔记.第2卷(英文)	2019—06	109.00	1084
拉马努金遗失笔记.第3卷(英文)	2019—06	109.00	1085
拉马努金遗失笔记.第4卷(英文)	2019—06	109.00	1086
数论:1976年纽约洛克菲勒大学数论会议记录(英文)	2020—06	68.00	1145
数论:卡本代尔1979:1979年在南伊利诺伊卡本代尔大学举行的数论会议记录(英文)	2020—06	78.00	1146
数论:诺德韦克豪特1983:1983年在诺德韦克豪特举行的Journees Arithmetiques数论大会会议记录(英文)	2020—06	68.00	1147
数论:1985—1988年在纽约城市大学研究生院和大学中心举办的研讨会(英文)	2020—06	68.00	1148

刘培杰数学工作室
已出版（即将出版）图书目录——原版影印

书　名	出版时间	定　价	编号
数论:1987年在乌尔姆举行的Journees Arithmetiques数论大会会议记录(英文)	2020—06	68.00	1149
数论:马德拉斯1987:1987年在马德拉斯安娜大学举行的国际拉马努金百年纪念大会会议记录(英文)	2020—06	68.00	1150
解析数论:1988年在东京举行的日法研讨会会议记录(英文)	2020—06	68.00	1151
解析数论:2002年在意大利切特拉罗举行的C.I.M.E.暑期班演讲集(英文)	2020—06	68.00	1152
量子世界中的蝴蝶:最迷人的量子分形故事(英文)	2020—06	118.00	1157
走进量子力学(英文)	2020—06	118.00	1158
计算物理学概论(英文)	2020—06	48.00	1159
物质,空间和时间的理论:量子理论(英文)	2020—10	48.00	1160
物质,空间和时间的理论:经典理论(英文)	2020—10	48.00	1161
量子场理论:解释世界的神秘背景(英文)	2020—07	38.00	1162
计算物理学概论(英文)	2020—06	48.00	1163
行星状星云(英文)	2020—10	38.00	1164
基本宇宙学:从亚里士多德的宇宙到大爆炸(英文)	2020—08	58.00	1165
数学磁流体力学(英文)	2020—07	58.00	1166
计算科学:第1卷,计算的科学(日文)	2020—07	88.00	1167
计算科学:第2卷,计算与宇宙(日文)	2020—07	88.00	1168
计算科学:第3卷,计算与物质(日文)	2020—07	88.00	1169
计算科学:第4卷,计算与生命(日文)	2020—07	88.00	1170
计算科学:第5卷,计算与地球环境(日文)	2020—07	88.00	1171
计算科学:第6卷,计算与社会(日文)	2020—07	88.00	1172
计算科学:别卷,超级计算机(日文)	2020—07	88.00	1173
多复变函数论(日文)	2022—06	78.00	1518
复变函数入门(日文)	2022—06	78.00	1523
代数与数论:综合方法(英文)	2020—10	78.00	1185
复分析:现代函数理论第一课(英文)	2020—07	58.00	1186
斐波那契数列和卡特兰数:导论(英文)	2020—10	68.00	1187
组合推理:计数艺术介绍(英文)	2020—07	88.00	1188
二次互反律的傅里叶分析证明(英文)	2020—07	48.00	1189
旋瓦兹分布的希尔伯特变换与应用(英文)	2020—07	58.00	1190
泛函分析:巴拿赫空间理论入门(英文)	2020—07	48.00	1191
卡塔兰数入门(英文)	2019—05	68.00	1060
测度与积分(英文)	2019—04	68.00	1059
组合学手册.第一卷(英文)	2020—06	128.00	1153
-代数、局部紧群和巴拿赫-代数丛的表示.第一卷,群和代数的基本表示理论(英文)	2020—05	148.00	1154
电磁理论(英文)	2020—08	48.00	1193
连续介质力学中的非线性问题(英文)	2020—09	78.00	1195
多变量数学入门(英文)	2021—05	68.00	1317
偏微分方程入门(英文)	2021—05	88.00	1318
若尔当典范性:理论与实践(英文)	2021—07	68.00	1366
伽罗瓦理论.第4版(英文)	2021—08	88.00	1408
R统计学概论	2023—03	88.00	1614
基于不确定静态和动态问题解的仿射算术(英文)	2023—03	38.00	1618

刘培杰数学工作室
已出版(即将出版)图书目录——原版影印

书　名	出版时间	定　价	编号
典型群,错排与素数(英文)	2020—11	58.00	1204
李代数的表示:通过gln进行介绍(英文)	2020—10	38.00	1205
实分析演讲集(英文)	2020—10	38.00	1206
现代分析及其应用的课程(英文)	2020—10	58.00	1207
运动中的抛射物数学(英文)	2020—10	38.00	1208
2—纽结与它们的群(英文)	2020—10	38.00	1209
概率,策略和选择:博弈与选举中的数学(英文)	2020—11	58.00	1210
分析学引论(英文)	2020—11	58.00	1211
量子群:通往流代数的路径(英文)	2020—11	38.00	1212
集合论入门(英文)	2020—11	48.00	1213
酉反射群(英文)	2020—11	58.00	1214
探索数学:吸引人的证明方式(英文)	2020—11	58.00	1215
微分拓扑短期课程(英文)	2020—10	48.00	1216
抽象凸分析(英文)	2020—11	68.00	1222
费马大定理笔记(英文)	2021—03	48.00	1223
高斯与雅可比和(英文)	2021—03	78.00	1224
π与算术几何平均:关于解析数论和计算复杂性的研究(英文)	2021—01	58.00	1225
复分析入门(英文)	2021—03	48.00	1226
爱德华·卢卡斯与素性测定(英文)	2021—03	78.00	1227
通往凸分析及其应用的简单路径(英文)	2021—01	68.00	1229
微分几何的各个方面.第一卷(英文)	2021—01	58.00	1230
微分几何的各个方面.第二卷(英文)	2020—12	58.00	1231
微分几何的各个方面.第三卷(英文)	2020—12	58.00	1232
沃克流形几何学(英文)	2020—11	58.00	1233
彷射和韦尔几何应用(英文)	2020—12	58.00	1234
双曲几何学的旋转向量空间方法(英文)	2021—02	58.00	1235
积分:分析学的关键(英文)	2020—12	48.00	1236
为有天分的新生准备的分析学基础教材(英文)	2020—11	48.00	1237
数学不等式.第一卷.对称多项式不等式(英文)	2021—03	108.00	1273
数学不等式.第二卷.对称有理不等式与对称无理不等式(英文)	2021—03	108.00	1274
数学不等式.第三卷.循环不等式与非循环不等式(英文)	2021—03	108.00	1275
数学不等式.第四卷.Jensen不等式的扩展与加细(英文)	2021—03	108.00	1276
数学不等式.第五卷.创建不等式与解不等式的其他方法(英文)	2021—04	108.00	1277

刘培杰数学工作室
已出版(即将出版)图书目录——原版影印

书　名	出版时间	定　价	编号
冯·诺依曼代数中的谱位移函数:半有限冯·诺依曼代数中的谱位移函数与谱流(英文)	2021—06	98.00	1308
链接结构:关于嵌入完全图的直线中链接单形的组合结构(英文)	2021—05	58.00	1309
代数几何方法.第1卷(英文)	2021—06	68.00	1310
代数几何方法.第2卷(英文)	2021—06	68.00	1311
代数几何方法.第3卷(英文)	2021—06	58.00	1312
代数、生物信息和机器人技术的算法问题.第四卷,独立恒等式系统(俄文)	2020—08	118.00	1199
代数、生物信息和机器人技术的算法问题.第五卷,相对覆盖性和独立可拆分恒等式系统(俄文)	2020—08	118.00	1200
代数、生物信息和机器人技术的算法问题.第六卷,恒等式和准恒等式的相等 问题、可推导性和可实现性(俄文)	2020—08	128.00	1201
分数阶微积分的应用:非局部动态过程,分数阶导热系数(俄文)	2021—01	68.00	1241
泛函分析问题与练习:第2版(俄文)	2021—01	98.00	1242
集合论、数学逻辑和算法论问题:第5版(俄文)	2021—01	98.00	1243
微分几何和拓扑短期课程(俄文)	2021—01	98.00	1244
素数规律(俄文)	2021—01	88.00	1245
无穷边值问题解的递减:无界域中的拟线性椭圆和抛物方程(俄文)	2021—01	48.00	1246
微分几何讲义(俄文)	2020—12	98.00	1253
二次型和矩阵(俄文)	2021—01	98.00	1255
积分和级数.第2卷,特殊函数(俄文)	2021—01	168.00	1258
积分和级数.第3卷,特殊函数补充:第2版(俄文)	2021—01	178.00	1264
几何图上的微分方程(俄文)	2021—01	138.00	1259
数论教程:第2版(俄文)	2021—01	98.00	1260
非阿基米德分析及其应用(俄文)	2021—03	98.00	1261
古典群和量子群的压缩(俄文)	2021—03	98.00	1263
数学分析习题集.第3卷,多元函数:第3版(俄文)	2021—03	98.00	1266
数学习题:乌拉尔国立大学数学力学系大学生奥林匹克(俄文)	2021—03	98.00	1267
柯西定理和微分方程的特解(俄文)	2021—03	98.00	1268
组合极值问题及其应用:第3版(俄文)	2021—03	98.00	1269
数学词典(俄文)	2021—01	98.00	1271
确定性混沌分析模型(俄文)	2021—06	168.00	1307
精选初等数学习题和定理.立体几何.第3版(俄文)	2021—03	68.00	1316
微分几何习题:第3版(俄文)	2021—05	98.00	1336
精选初等数学习题和定理.平面几何.第4版(俄文)	2021—05	68.00	1335
曲面理论在欧氏空间 E_n 中的直接表示(俄文)	2022—01	68.00	1444
维纳-霍普夫离散算子和托普利兹算子:某些可数赋范空间中的诺特性和可逆性(俄文)	2022—03	108.00	1496
Maple 中的数论:数论中的计算机计算(俄文)	2022—03	88.00	1497
贝尔曼和克努特问题及其概括:加法运算的复杂性(俄文)	2022—03	138.00	1498

刘培杰数学工作室
已出版(即将出版)图书目录——原版影印

书 名	出版时间	定 价	编号
复分析:共形映射(俄文)	2022—07	48.00	1542
微积分代数样条和多项式及其在数值方法中的应用(俄文)	2022—08	128.00	1543
蒙特卡罗方法中的随机过程和场模型:算法和应用(俄文)	2022—08	88.00	1544
线性椭圆型方程组:论二阶椭圆型方程的迪利克雷问题(俄文)	2022—08	98.00	1561
动态系统解的增长特性:估值、稳定性、应用(俄文)	2022—08	118.00	1565
群的自由积分解:建立和应用(俄文)	2022—08	78.00	1570
混合方程和偏差自变数方程问题:解的存在和唯一性(俄文)	2023—01	78.00	1582
拟度量空间分析:存在和逼近定理(俄文)	2023—01	108.00	1583
二维和三维流形上函数的拓扑分类(俄文)	2023—03	68.00	1584
齐次马尔科夫过程建模的矩函数方法:此类方法能够用于不同目上的的复杂系统研究、设计和完善(俄文)	2023—03	68.00	1594
周期函数的近似方法和特性:特殊课程(俄文)	2023—04	158.00	1622
扩散方程解的矩函数:变分法(俄文)	2023—03	58.00	1623
多赋范空间和广义函数:理论及应用(俄文)	2023—03	98.00	1632
分析中的多值映射:部分应用(俄文)	2023—06	98.00	1634
数学物理问题(俄文)	2023—03	78.00	1636
函数的幂级数与三角级数分解(俄文)	2024—01	58.00	1695
星体理论的数学基础:原子三元组(俄文)	2024—01	98.00	1696
素数规律:专著(俄文)	2024—01	118.00	1697
狭义相对论与广义相对论:时空与引力导论(英文)	2021—07	88.00	1319
束流物理学和粒子加速器的实践介绍:第2版(英文)	2021—07	88.00	1320
凝聚态物理中的拓扑和微分几何简介(英文)	2021—05	88.00	1321
混沌映射:动力学、分形学和快速涨落(英文)	2021—05	128.00	1322
广义相对论:黑洞、引力波和宇宙学介绍(英文)	2021—06	68.00	1323
现代分析电磁均质化(英文)	2021—06	68.00	1324
为科学家提供的基本流体动力学(英文)	2021—06	88.00	1325
视觉天文学:理解夜空的指南(英文)	2021—06	68.00	1326
物理学中的计算方法(英文)	2021—06	68.00	1327
单星的结构与演化:导论(英文)	2021—06	108.00	1328
超越居里:1903年至1963年物理界四位女性及其著名发现(英文)	2021—06	68.00	1329
范德瓦尔斯流体热力学的进展(英文)	2021—06	68.00	1330
先进的托卡马克稳定性理论(英文)	2021—06	88.00	1331
经典场论导论:基本相互作用的过程(英文)	2021—07	88.00	1332
光致电离量子动力学方法原理(英文)	2021—07	108.00	1333
经典域论和应力:能量张量(英文)	2021—05	88.00	1334
非线性太赫兹光谱的概念与应用(英文)	2021—06	68.00	1337
电磁学中的无穷空间并矢格林函数(英文)	2021—06	88.00	1338
物理科学基础数学.第1卷,齐次边值问题、傅里叶方法和特殊函数(英文)	2021—07	108.00	1339
离散量子力学(英文)	2021—07	68.00	1340
核磁共振的物理学和数学(英文)	2021—07	108.00	1341
分子水平的静电学(英文)	2021—08	68.00	1342
非线性波:理论、计算机模拟、实验(英文)	2021—06	108.00	1343
石墨烯光学:经典问题的电解解决方案(英文)	2021—06	68.00	1344
超材料多元宇宙(英文)	2021—07	68.00	1345
银河系外的天体物理学(英文)	2021—07	68.00	1346
原子物理学(英文)	2021—07	68.00	1347
将光打结:将拓扑学应用于光学(英文)	2021—07	68.00	1348
电磁学:问题与解法(英文)	2021—07	88.00	1364
海浪的原理:介绍量子力学的技巧与应用(英文)	2021—07	108.00	1365

刘培杰数学工作室
已出版(即将出版)图书目录——原版影印

书 名	出版时间	定 价	编号
多孔介质中的流体:输运与相变(英文)	2021—07	68.00	1372
洛伦兹群的物理学(英文)	2021—08	68.00	1373
物理导论的数学方法和解决方法手册(英文)	2021—08	68.00	1374
非线性波数学物理学入门(英文)	2021—08	88.00	1376
波:基本原理和动力学(英文)	2021—07	68.00	1377
光电子量子计量学.第1卷,基础(英文)	2021—07	88.00	1383
光电子量子计量学.第2卷,应用与进展(英文)	2021—07	68.00	1384
复杂流的格子玻尔兹曼建模的工程应用(英文)	2021—08	68.00	1393
电偶极矩挑战(英文)	2021—08	108.00	1394
电动力学:问题与解法(英文)	2021—09	68.00	1395
自由电子激光的经典理论(英文)	2021—08	68.00	1397
曼哈顿计划——核武器物理学简介(英文)	2021—09	68.00	1401
粒子物理学(英文)	2021—09	68.00	1402
引力场中的量子信息(英文)	2021—09	128.00	1403
器件物理学的基本经典力学(英文)	2021—09	68.00	1404
等离子体物理及其空间应用导论.第1卷,基本原理和初步过程(英文)	2021—09	68.00	1405
磁约束聚变等离子体物理:理想MHD理论(英文)	2023—03	68.00	1613
相对论量子场论.第1卷,典范形式体系(英文)	2023—03	38.00	1615
相对论量子场论.第2卷,路径积分形式(英文)	2023—06	38.00	1616
相对论量子场论.第3卷,量子场论的应用(英文)	2023—06	38.00	1617
涌现的物理学(英文)	2023—05	58.00	1619
量子化旋涡:一本拓扑激发手册(英文)	2023—04	68.00	1620
非线性动力学:实践的介绍性调查(英文)	2023—05	68.00	1621
静电加速器:一个多功能工具(英文)	2023—06	58.00	1625
相对论多体理论与统计力学(英文)	2023—06	58.00	1626
经典力学.第1卷,工具与向量(英文)	2023—04	38.00	1627
经典力学.第2卷,运动学和匀加速运动(英文)	2023—04	58.00	1628
经典力学.第3卷,牛顿定律和匀速圆周运动(英文)	2023—04	58.00	1629
经典力学.第4卷,万有引力定律(英文)	2023—04	38.00	1630
经典力学.第5卷,守恒定律与旋转运动(英文)	2023—04	38.00	1631
对称问题:纳维—斯托克斯问题(英文)	2023—04	38.00	1638
摄影的物理和艺术.第1卷,几何与光的本质(英文)	2023—04	78.00	1639
摄影的物理和艺术.第2卷,能量与色彩(英文)	2023—04	78.00	1640
摄影的物理和艺术.第3卷,探测器与数码的意义(英文)	2023—04	78.00	1641
拓扑与超弦理论焦点问题(英文)	2021—07	58.00	1349
应用数学:理论、方法与实践(英文)	2021—07	78.00	1350
非线性特征值问题:牛顿型方法与非线性瑞利函数(英文)	2021—07	58.00	1351
广义膨胀和齐性:利用齐性构造齐次系统的李雅普诺夫函数和控制律(英文)	2021—06	48.00	1352
解析数论焦点问题(英文)	2021—07	58.00	1353
随机微分方程:动态系统方法(英文)	2021—07	58.00	1354
经典力学与微分几何(英文)	2021—07	58.00	1355
负定相交形式流形上的瞬子模空间几何(英文)	2021—07	68.00	1356
广义卡塔兰轨道分析:广义卡塔兰轨道计算数字的方法(英文)	2021—07	48.00	1367
洛伦兹方法的变分:二维与三维洛伦兹方法(英文)	2021—08	38.00	1378
几何、分析和数论精编(英文)	2021—08	68.00	1380
从一个新角度看数论:通过遗传方法引入现实的概念(英文)	2021—07	58.00	1387
动力系统:短期课程(英文)	2021—08	68.00	1382
几何路径:理论与实践(英文)	2021—08	48.00	1385

刘培杰数学工作室
已出版(即将出版)图书目录——原版影印

书 名	出版时间	定 价	编号
论天体力学中某些问题的不可积性(英文)	2021—07	88.00	1396
广义斐波那契数列及其性质(英文)	2021—08	38.00	1386
对称函数和麦克唐纳多项式:余代数结构与 Kawanaka 恒等式(英文)	2021—09	38.00	1400
杰弗里·英格拉姆·泰勒科学论文集:第1卷.固体力学(英文)	2021—05	78.00	1360
杰弗里·英格拉姆·泰勒科学论文集:第2卷.气象学、海洋学和湍流(英文)	2021—05	68.00	1361
杰弗里·英格拉姆·泰勒科学论文集:第3卷.空气动力学以及落弹数和爆炸的力学(英文)	2021—05	68.00	1362
杰弗里·英格拉姆·泰勒科学论文集:第4卷.有关流体力学(英文)	2021—05	58.00	1363
非局域泛函演化方程:积分与分数阶(英文)	2021—08	48.00	1390
理论工作者的高等微分几何:纤维丛、射流流形和拉格朗日理论(英文)	2021—08	68.00	1391
半线性退化椭圆微分方程:局部定理与整体定理(英文)	2021—07	48.00	1392
非交换几何、规范理论和重整化:一般简介与非交换量子场论的重整化(英文)	2021—09	78.00	1406
数论论文集:拉普拉斯变换和带有数论系数的幂级数(俄文)	2021—09	48.00	1407
挠理论专题:相对极大值,单射与扩充模(英文)	2021—09	88.00	1410
强正则图与欧几里得若尔当代数:非通常关系中的启示(英文)	2021—10	48.00	1411
拉格朗日几何和哈密顿几何:力学的应用(英文)	2021—10	48.00	1412
时滞微分方程与差分方程的振动理论:二阶与三阶(英文)	2021—10	98.00	1417
卷积结构与几何函数理论:用以研究特定几何函数理论方向的分数阶微积分算子与卷积结构(英文)	2021—10	48.00	1418
经典数学物理的历史发展(英文)	2021—10	78.00	1419
扩展线性丢番图问题(英文)	2021—10	38.00	1420
一类混沌动力系统的分歧分析与控制:分歧分析与控制(英文)	2021—11	38.00	1421
伽利略空间和伪伽利略空间中一些特殊曲线的几何性质(英文)	2022—01	68.00	1422
一阶偏微分方程:哈密尔顿—雅可比理论(英文)	2021—11	48.00	1424
各向异性黎曼多面体的反问题:分段光滑的各向异性黎曼多面体反边界谱问题:唯一性(英文)	2021—11	38.00	1425
项目反应理论手册.第一卷,模型(英文)	2021—11	138.00	1431
项目反应理论手册.第二卷,统计工具(英文)	2021—11	118.00	1432
项目反应理论手册.第三卷,应用(英文)	2021—11	138.00	1433
二次无理数:经典数论入门(英文)	2022—05	138.00	1434

刘培杰数学工作室
已出版(即将出版)图书目录——原版影印

书　　名	出版时间	定　价	编号
数,形与对称性:数论,几何和群论导论(英文)	2022—05	128.00	1435
有限域手册(英文)	2021—11	178.00	1436
计算数论(英文)	2021—11	148.00	1437
拟群与其表示简介(英文)	2021—11	88.00	1438
数论与密码学导论:第二版(英文)	2022—01	148.00	1423
几何分析中的柯西变换与黎兹变换:解析调和容量和李普希兹调和容量、变化和振荡以及一致可求长性(英文)	2021—12	38.00	1465
近似不动点定理及其应用(英文)	2022—05	28.00	1466
局部域的相关内容解析:对局部域的扩展及其伽罗瓦群的研究(英文)	2022—01	38.00	1467
反问题的二进制恢复方法(英文)	2022—03	28.00	1468
对几何函数中某些类的各个方面的研究:复变量理论(英文)	2022—01	38.00	1469
覆盖、对应和非交换几何(英文)	2022—01	28.00	1470
最优控制理论中的随机线性调节器问题:随机最优线性调节器问题(英文)	2022—01	38.00	1473
正交分解法:涡流流体动力学应用的正交分解法(英文)	2022—01	38.00	1475
芬斯勒几何的某些问题(英文)	2022—03	38.00	1476
受限三体问题(英文)	2022—05	38.00	1477
利用马利亚万微积分进行 Greeks 的计算:连续过程、跳跃过程中的马利亚万微积分和金融领域中的 Greeks(英文)	2022—05	48.00	1478
经典分析和泛函分析的应用:分析学的应用(英文)	2022—03	38.00	1479
特殊芬斯勒空间的探究(英文)	2022—03	48.00	1480
某些图形的施泰纳距离的细谷多项式:细谷多项式与图的维纳指数(英文)	2022—05	38.00	1481
图论问题的遗传算法:在新鲜与模糊的环境中(英文)	2022—05	48.00	1482
多项式映射的渐近簇(英文)	2022—05	38.00	1483
一维系统中的混沌:符号动力学,映射序列,一致收敛和沙可夫斯基定理(英文)	2022—05	38.00	1509
多维边界层流动与传热分析:粘性流体流动的数学建模与分析(英文)	2022—05	38.00	1510
演绎理论物理学的原理:一种基于量子力学波函数的逐次置信估计的一般理论的提议(英文)	2022—05	38.00	1511
R^2 和 R^3 中的仿射弹性曲线:概念和方法(英文)	2022—08	38.00	1512
算术数列中除数函数的分布:基本内容、调查、方法、第二矩、新结果(英文)	2022—05	28.00	1513
抛物型狄拉克算子和薛定谔方程:不定常薛定谔方程的抛物型狄拉克算子及其应用(英文)	2022—07	28.00	1514
黎曼-希尔伯特问题与量子场论:可积重正化、戴森-施温格方程(英文)	2022—08	38.00	1515
代数结构和几何结构的形变理论(英文)	2022—08	48.00	1516
概率结构和模糊结构上的不动点:概率结构和直觉模糊度量空间的不动点定理(英文)	2022—08	38.00	1517

刘培杰数学工作室
已出版(即将出版)图书目录——原版影印

书 名	出版时间	定 价	编号
反若尔当对:简单反若尔当对的自同构(英文)	2022—07	28.00	1533
对某些黎曼—芬斯勒空间变换的研究:芬斯勒几何中的某些变换(英文)	2022—07	38.00	1534
内诣零流形映射的尼尔森数的阿诺索夫关系(英文)	2023—01	38.00	1535
与广义积分变换有关的分数次演算:对分数次演算的研究(英文)	2023—01	48.00	1536
强子的芬斯勒几何和吕拉几何(宇宙学方面):强子结构的芬斯勒几何和吕拉几何(拓扑缺陷)(英文)	2022—08	38.00	1537
一种基于混沌的非线性最优化问题:作业调度问题(英文)	2023—03	38.00	1538
广义概率论发展前景:关于趣味数学与置信函数实际应用的一些原创观点(英文)	2023—03	48.00	1539
纽结与物理学:第二版(英文)	2022—09	118.00	1547
正交多项式和 q—级数的前沿(英文)	2022—09	98.00	1548
算子理论问题集(英文)	2022—09	108.00	1549
抽象代数:群、环与域的应用导论:第二版(英文)	2023—01	98.00	1550
菲尔兹奖得主演讲集:第三版(英文)	2023—01	138.00	1551
多元实函数教程(英文)	2022—09	118.00	1552
球面空间形式群的几何学:第二版(英文)	2022—09	98.00	1566
对称群的表示论(英文)	2023—01	98.00	1585
纽结理论:第二版(英文)	2023—01	88.00	1586
拟群理论的基础与应用(英文)	2023—01	88.00	1587
组合学:第二版(英文)	2023—01	98.00	1588
加性组合学:研究问题手册(英文)	2023—01	68.00	1589
扭曲、平铺与镶嵌:几何折纸中的数学方法(英文)	2023—01	98.00	1590
离散与计算几何手册:第三版(英文)	2023—01	248.00	1591
离散与组合数学手册:第二版(英文)	2023—01	248.00	1592
分析学教程.第1卷,一元实变量函数的微积分分析学介绍(英文)	2023—01	118.00	1595
分析学教程.第2卷,多元函数的微分和积分,向量微积分(英文)	2023—01	118.00	1596
分析学教程.第3卷,测度与积分理论,复变量的复值函数(英文)	2023—01	118.00	1597
分析学教程.第4卷,傅里叶分析,常微分方程,变分法(英文)	2023—01	118.00	1598

刘培杰数学工作室
已出版（即将出版）图书目录——原版影印

书　　名	出版时间	定　价	编号
共形映射及其应用手册(英文)	2024—01	158.00	1674
广义三角函数与双曲函数(英文)	2024—01	78.00	1675
振动与波：概论：第二版(英文)	2024—01	88.00	1676
几何约束系统原理手册(英文)	2024—01	120.00	1677
微分方程与包含的拓扑方法(英文)	2024—01	98.00	1678
数学分析中的前沿话题(英文)	2024—01	198.00	1679
流体力学建模：不稳定性与湍流(英文)	2024—03	88.00	1680
动力系统：理论与应用(英文)	2024—03	108.00	1711
空间统计学理论：概述(英文)	2024—03	68.00	1712
梅林变换手册(英文)	2024—03	128.00	1713
非线性系统及其绝妙的数学结构.第1卷(英文)	2024—03	88.00	1714
非线性系统及其绝妙的数学结构.第2卷(英文)	2024—03	108.00	1715
Chip-firing 中的数学(英文)	2024—04	88.00	1716
阿贝尔群的可确定性：问题、研究、概述(俄文)	2024—05	716.00(全7册)	1727
素数规律：专著(俄文)	2024—05	716.00(全7册)	1728
函数的幂级数与三角级数分解(俄文)	2024—05	716.00(全7册)	1729
星体理论的数学基础：原子三元组(俄文)	2024—05	716.00(全7册)	1730
技术问题中的数学物理微分方程(俄文)	2024—05	716.00(全7册)	1731
概率论边界问题：随机过程边界穿越问题(俄文)	2024—05	716.00(全7册)	1732
代数和幂等配置的正交分解：不可交换组合(俄文)	2024—05	716.00(全7册)	1733
数学物理精选专题讲座：李理论的进一步应用(英文)	2024—10	252.00(全4册)	1775
工程师和科学家应用数学概论：第二版(英文)	2024—10	252.00(全4册)	1775
高等微积分快速入门(英文)	2024—10	252.00(全4册)	1775
微分几何的各个方面.第四卷(英文)	2024—10	252.00(全4册)	1775
具有连续变量的量子信息形式主义概论(英文)	2024—10	378.00(全6册)	1776
拓扑绝缘体(英文)	2024—10	378.00(全6册)	1776
论全息度量原则：从大学物理到黑洞热力学(英文)	2024—10	378.00(全6册)	1776
量化测量：无所不在的数字(英文)	2024—10	378.00(全6册)	1776
21世纪的彗星：体验下一颗伟大彗星的个人指南(英文)	2024—10	378.00(全6册)	1776
激光及其在玻色—爱因斯坦凝聚态观测中的应用(英文)	2024—10	378.00(全6册)	1776
随机矩阵理论的最新进展(英文)	2025—02	78.00	1797
计算代数几何的应用(英文)	2025—02	78.00	1798
纽结与物理学的交界(英文)	即将出版		1799
公钥密码学(英文)	即将出版		1800
量子计算：一个对21世纪和千禧年的宏大的数学挑战(英文)	即将出版		1801
信息流的数学基础(英文)	即将出版		1802

刘培杰数学工作室
已出版(即将出版)图书目录——原版影印

书　名	出版时间	定　价	编号
偏微分方程的最新研究进展:威尼斯1996(英文)	即将出版		1803
拉东变换、反问题及断层成像(英文)	即将出版		1804
应用与计算拓扑学进展(英文)	2025—02	98.00	1805
复动力系统:芒德布罗集与朱利亚集背后的数学(英文)	2025—02	98.00	1806
双曲问题:理论、数值数据及应用(全2册)(英文)	即将出版		1807

联系地址:哈尔滨市南岗区复华四道街10号　哈尔滨工业大学出版社刘培杰数学工作室
邮　编:150006
联系电话:0451—86281378　　13904613167
E-mail:lpj1378@163.com